AF594955

Civil and Military Airworthiness

Civil and Military Airworthiness

Recent Developments and Challenges (Volume II)

Editor

Kyriakos I. Kourousis

MDPI • Basel • Beijing • Wuhan • Barcelona • Belgrade • Manchester • Tokyo • Cluj • Tianjin

Editor
Kyriakos I. Kourousis
School of Engineering,
University of Limerick
Ireland

Editorial Office
MDPI
St. Alban-Anlage 66
4052 Basel, Switzerland

This is a reprint of articles from the Special Issue published online in the open access journal *Aerospace* (ISSN 2226-4310) (available at: https://www.mdpi.com/journal/aerospace/special_issues/airworthiness_ii).

For citation purposes, cite each article independently as indicated on the article page online and as indicated below:

LastName, A.A.; LastName, B.B.; LastName, C.C. Article Title. *Journal Name* **Year**, *Volume Number*, Page Range.

ISBN 978-3-0365-0744-6 (Hbk)
ISBN 978-3-0365-0745-3 (PDF)

Cover image courtesy of Kyriakos Kourousis.

Contents

About the Editor

Kyriakos I. Kourousis is a tenured Senior Lecturer (Associate Professor) at the School of Engineering of the University of Limerick in Ireland. He also serves as the Director of the undergraduate and postgraduate airworthiness programmes offered by the University of Limerick. Dr. Kourousis holds a B.Sc. in Aeronautical Engineering from the Hellenic Air Force Academy, and an M.Sc. and Ph.D. in solid mechanics and metal plasticity, both from the National Technical University of Athens. Dr. Kourousis has more than 20 years of professional and research experience in the fields of metal plasticity, metal additive manufacturing and airworthiness, both as an Aeronautical Engineer Officer and a university faculty member in Ireland and Australia. His professional background includes 12 years in the Hellenic Air Force as a maintenance officer, quality control and assurance manager and airworthiness manager for the Mirage 2000 fleet. Dr. Kourousis has led various research projects funded by civil and defence companies and state organisations in defence airworthiness, aircraft fatigue, metal plasticity, aviation training and other areas. To date, he has authored more than 100 scholarly journal papers, conference papers and technical and engineering reports in his fields of research and professional expertise. He is a Chartered Engineer, registered with the United Kingdom Engineering Council, and a Fellow of the Royal Aeronautical Society. Dr. Kourousis is also the chair of the RAeS Airworthiness & Maintenance Specialist Group.

Editorial

Special Issue: Civil and Military Airworthiness: Recent Developments and Challenges (Volume II)

Kyriakos I. Kourousis [1,2]

[1] School of Engineering, University of Limerick, V94 T9PX Limerick, Ireland; kyriakos.kourousis@ul.ie
[2] School of Engineering, RMIT University, Melbourne, VIC 3001, Australia

Received: 5 February 2021; Accepted: 8 February 2021; Published: 8 February 2021

Effective safety management has always been a key objective for the broader airworthiness sector. This Special Issue is focused on safety themes with implications on airworthiness management. It offers a diverse set of analyses on aircraft maintenance accidents [1–4], empirical and systematic investigations on important continuing airworthiness matters [5–7] and research studies on methodologies for risk and safety assessment in continuing and initial airworthiness [8–10]. Overall, this collection of papers is a valuable addition to the published literature, and I am confident that the readers of Aerospace will find that useful.

Funding: I have not received external funding.

Acknowledgments: I wish to thank all authors for their contributions.

Conflicts of Interest: I declare no conflict of interest.

References

1. Insley, J.; Turkoglu, C. A Contemporary Analysis of Aircraft Maintenance-Related Accidents and Serious Incidents. *Aerospace* **2020**, *7*, 81. [CrossRef]
2. Khan, F.N.; Ayiei, A.; Murray, J.; Baxter, G.; Wild, G. A Preliminary Investigation of Maintenance Contributions to Commercial Air Transport Accidents. *Aerospace* **2020**, *7*, 129. [CrossRef]
3. Habib, K.A.; Turkoglu, C. Analysis of Aircraft Maintenance Related Accidents and Serious Incidents in Nigeria. *Aerospace* **2020**, *7*, 178. [CrossRef]
4. Clare, J.; Kourousis, K.I. Learning from Incidents: A Qualitative Study in the Continuing Airworthiness Sector. *Aerospace* **2021**, *8*, 27. [CrossRef]
5. Naweed, A.; Kourousis, K.I. Winging It: Key Issues and Perceptions around Regulation and Practice of Aircraft Maintenance in Australian General Aviation. *Aerospace* **2020**, *7*, 84. [CrossRef]
6. Obadimu, S.O.; Karanikas, N.; Kourousis, K.I. Development of the Minimum Equipment List: Current Practice and the Need for Standardisation. *Aerospace* **2020**, *7*, 7. [CrossRef]
7. Kourousis, K.I. Airlift Maintenance and Sustainment: The Indirect Costs. *Aerospace* **2020**, *7*, 130. [CrossRef]
8. Aust, J.; Pons, D. A Systematic Methodology for Developing Bowtie in Risk Assessment: Application to Borescope Inspection. *Aerospace* **2020**, *7*, 86. [CrossRef]
9. Thomas, J.; Davis, A.; Samuel, M.P. Integration-In-Totality: The 7th System Safety Principle Based on Systems Thinking in Aerospace Safety. *Aerospace* **2020**, *7*, 149. [CrossRef]
10. Clare, J.; Kourousis, K.I. Analysis of Continuing Airworthiness Occurrences under the Prism of a Learning Framework. *Aerospace* **2021**, *8*, 41. [CrossRef]

Article

A Contemporary Analysis of Aircraft Maintenance-Related Accidents and Serious Incidents

Jennifer Insley [1] and Cengiz Turkoglu [2,*]

[1] Lockheed Martin UK Ampthill, Bedford MK45 2HD, UK; jennifer.j.insley@gmail.com
[2] Cranfield Safety and Accident Investigation Centre, Martell House, Cranfield, Bedfordshire MK43 0TR, UK
* Correspondence: cengiz.turkoglu@cranfield.ac.uk; Tel.: +44-1234-754019

Received: 24 March 2020; Accepted: 31 May 2020; Published: 17 June 2020

Abstract: Aircraft maintenance has been identified as a key point of concern within many high-risk areas of aviation; still being a casual/contributory factor in a number of accidents and serious incidents in commercial air transport industry. The purpose of this study is to review and analyse the aircraft maintenance-related accidents and serious incidents which occurred between 2003 and 2017, to provide a better understanding of the causal and contributory factors. To achieve this, a dataset of maintenance-related accidents and serious incidents was compiled and then qualitatively analysed by thematic analysis method. Coding these events by using NVivo software enabled the development of a taxonomy, MxFACS. The coded output was then evaluated by subject matter experts, and an inter-rater concordance value determined to demonstrate the rigour of the research process. Subsequently, the events were evaluated in terms of their relationship to known accident categories such as loss of control, runway excursions. The most frequent maintenance event consequences were found to be runway excursions and air turnbacks, with the second level categories being related to failures in engine and landing gear systems. The greatest maintenance factor issues were 'inadequate maintenance procedures' and 'inspections not identifying defects'. In terms of fatalities, 'collision events' were the most prominent consequence, 'engine-related events' were the most significant event, and 'inadequate maintenance procedures' were the most concerning maintenance factor. The study's findings may be used in conjunction with existing risk analysis methodologies and enable the stakeholders to develop generic or customised bowties. This may identify the existing barriers in the system as well as weaknesses which will enable the development of mitigation strategies on both organisational and industry-wide levels.

Keywords: flight safety; aviation accidents; airworthiness; aircraft maintenance; MxFACS

---

1. Introduction

Managing safety risks in the Commercial Air Transport System in Europe is achieved by a 5-step 'safety risk management process' (as shown in Figure 1), which requires collaborative efforts by European Aviation Safety Agency (EASA), National Aviation Authorities of EU member states and—most importantly—all the other stakeholders in the industry.

In order to address the industry-wide risks, EASA annually publishes the following two key documents:

1. Annual Safety Review.
2. European Plan for Aviation Safety (four year rolling plan) which now also includes Rulemaking and Safety Promotion Programme.

Annual safety reviews include multiple, domain specific 'safety risk portfolios', which are developed based on the analysis of accidents, serious incidents, and other reportable occurrence data.

This analysis is further reviewed and assessed by the domain focused 'collaborative analysis groups' (CAGs), which include representatives from the industry. CAGs function is to contribute to the first and second steps in the 'safety risk management process', which ultimately aims for the development of the 'European Plan for Aviation Safety' [1] (pp. 15, 16).

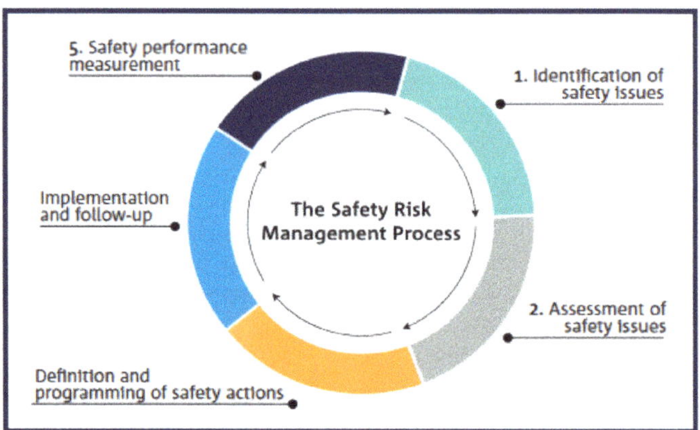

Figure 1. European Safety Risk Management Process (Annual Safety Review 2019 [1] (p. 21)).

1.1. Study Rationale

The most recent in-depth study into the nature of aircraft maintenance error, by the UK CAA [2], was published in 2015, yet only made use of data up to 2011. This shows that there is an underlying need to provide an up-to-date analysis of maintenance error types in order to understand the trends as well as emerging issues. Additionally, a majority of the most recent available studies are only from the 1990s, warranting the scope of this study to look at maintenance accidents and serious incidents from the early millennium onwards.

There is also suggestion that the most popular aircraft maintenance taxonomies presently in use can be difficult to apply to retrospective analyses. Therefore, exploration of a new taxonomy to further aid the process may assist in the categorisation of aircraft maintenance-related occurrences. Further to this, it may be of benefit to review how the discerned maintenance-specific issues correlate with key risks identified for CAT as a whole, as discussed by the UK CAA [2,3] and EASA [4].

Problem statement: since 2015, the European Aviation Safety Agency consistently identified aircraft maintenance as one of the safety issues and included it in the safety risk portfolios for "Commercial Air Transport (CAT)–Large Aeroplanes" [4–7]. However, there has been no further analysis conducted to enable the stakeholders to develop any mitigation strategies to address risks associated with aircraft maintenance and continuing airworthiness. Therefore, further analysis of accidents, serious incidents, and occurrence data was essential to better understand the causal and contributory factors in this area.

While the analysis of occurrence data extracted from European Central Repository was subject to another study [8], this paper focuses on the accidents and serious incidents where maintenance actions or continuing airworthiness processes played a causal or contributory role. The ICAO Annex 13 definitions of 'accident', and 'serious incident' apply, which can be found in Appendix B. Also, the details of the nature of flights are shown in Table 1 below and the aircraft types can be found in Appendix A.

Table 1. Breakdown of Events by Nature of the Flights (the full list of categories for the 'nature of flights' used by the Aviation Safety Network can be accessed @ https://aviation-safety.net/about/ASN-standards.doc).

Nature of Flights (Analysed by the Researchers)	Number of Events
Domestic Scheduled Passenger	31
Cargo	23
International Scheduled Passenger	17
Passenger	12
Scheduled Passenger	12
Domestic Non-Scheduled Passenger	10
Executive	2
Ferry/Positioning	2
International Non-Scheduled Passenger	2
Non-Scheduled Passenger	1

1.2. Aim and Objectives

The aim of this research is to explore the nature of aircraft maintenance-related accidents and serious incidents between 2003 and 2017, in order to better understand this safety-critical function.

In order to achieve this aim, the following objectives have been developed:

- Identify the maintenance-related accidents and serious incidents for CAT category aeroplanes, between 2003–2017;
- Qualitatively analyse the accident/serious incident data using thematic analysis;
- Develop and validate a taxonomy which stems from this qualitative analysis; and
- Propose next steps for how such data analysis output may be used to aid in identifying high-risk areas and mitigation strategies.

Achieving the above aim and objectives will enable the development of some specific safety issues related to aircraft maintenance with clear focus. They can be further scrutinised by industry experts in CAG and then included in the Safety Risk Portfolio for Commercial Air Transport. Subsequently, mitigation strategies can be developed and included in the next *'European Plan for Aviation Safety'*.

1.3. Scope

The accident rate in aviation is very low. The EU member states' accident statistics for each domain can be seen in Table 2 below. Global statistics also show similar trends [9]. It is clear that accident rate in commercial air transport is much lower than non-commercial operations. The analysis of all accidents and serious incidents (including non-commercial and military events) related to aircraft maintenance could have produced a larger data set and potentially revealed interesting and beneficial results; however, there are considerable differences between how large commercial aeroplanes and small airplanes involved in non-commercial operations or military aircraft are maintained.

Table 2. Cross-Domain Comparison of EASA MS Aircraft Fatal Accidents and Fatalities, 2008-2018 (Annual Safety Review 2019 [1] (p. 27)).

Aircraft Domain	Fatal Accidents 2018	Fatal Accidents 2008–2017 Mean	Fatalities 2018	Fatalities 2008–2017 Mean	Fatalities 2008–2017 Median
CAT Airlines	0	0.8	0	66.1	4.0
NCC Business	1	0.4	1	0.9	0
Specialised Operations	6	6.8	7	13.8	13.0
Non-commercial Operations	49	47.1	95	86.0	82.0

Considering the problem statement described in the introduction, and the aim and objectives defined above, the scope of this study was determined to be limited to the global commercial air transport accidents and serious incidents related to aircraft maintenance and continuing airworthiness.

2. Methodology

Firstly, based on the problem statement, a conscious decision was made to analyse secondary data related to accidents and serious incidents related to aircraft maintenance.

Secondly, an 'Interpretivist' philosophical approach formed the basis of this research.

> " ... . interpretative and constructionist research does not only focus on the content of empirical data, but also on how the content is produced through language practices. Furthermore, research done from these philosophical positions does not predefine dependent and independent variables, but focuses on the full complexity of human sense making as the situations emerge. It is also assumed that there are many possible interpretations of the same data, all of which are potentially meaningful." Eriksson and Kovalainen [10] (p. 21)

Thirdly, an inductive approach was used during the design of this study. According to Eriksson and Kovalainen [10] (p. 24], "the research process develops, starting from empirical materials, not from theoretical propositions".

Considering these chosen philosophical positions and methods in relation to research, the data was not analysed by using existing taxonomies such as ICAO ADREP, ECCAIRS, HFACS, etc., but coded by using thematic analysis. In order to apply rigour to the research process, primary data was also collected from subject matter experts (SME's) to receive feedback not only about the outcome of the analysis but also about the methods used and the taxonomy developed. The overall research design and key steps followed, can be seen in Figure 2 below.

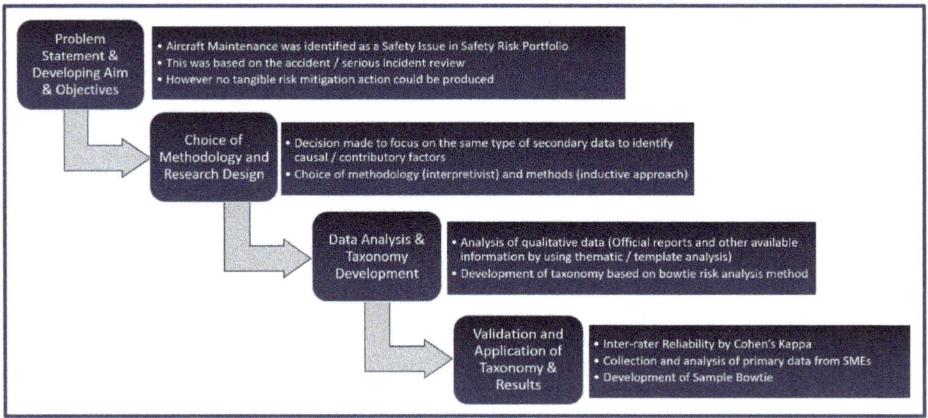

Figure 2. Research design and key steps.

2.1. Secondary Data: Accident and Serious Incidents

In order to discern the maintenance-related accidents and serious incidents, two sources were consulted: Aviation Safety Network's (ASN) Accident Database supported by Flight Safety Foundation; and SKYbrary's Accidents and Incidents database. Generating the data set for the analysis involved the review of all accidents and serious incidents to identify the aircraft maintenance-related events for CAT category aeroplanes occurred between 2003 and 2017. These events, once identified, were then compiled within a singular dataset, which can be provided on request.

The ASN database contains data on worldwide accidents and hijackings involving airliners (of 12 or more passengers), military transport aircraft and corporate jets since 1919. In order to refine this data to appropriately match the scope of the study, it was first filtered for the date range of relevance and then further filtered for accidents only relating to CAT aeroplanes. The remaining data was then reviewed to ensure that only maintenance-related accidents were contained within the dataset.

The second source, SKYbrary's Accidents and Incidents Database, was also refined for the time period of interest, and then filtered for airworthiness events related to maintenance. Where accidents were identified in both the ASN and SKYbrary databases, the relevant information was merged within the dataset.

The official investigation reports for these events were then sourced and consulted to ensure the validity of the data provided for each of the events within the dataset. Some events listed on the ASN database did not have traceable official investigation reports. The majority of these events were omitted from the dataset as it was not possible to assure their validity. However, a small selection of these events were allowed into the dataset, when there was significant indication within the narrative of maintenance-related contributory factors.

2.2. Primary Data Collection: Subject-Matter Experts (SME)

In order to scrutinise the results of the data analysis, SME feedback was sought. The method for this data collection was an online questionnaire, which was delivered through Qualtrics survey software.

The questionnaire had four open-ended questions and was distributed to five participants, who are from International Federation of Airworthiness and had extensive experience in design, production, operation, and maintenance domains including regulatory oversight and safety data analysis. The topics covered by the questions were:

1. Experience of and opinion on existing taxonomies;
2. Feedback on the study's methodology and taxonomy coding;
3. How they would approach classifying risk from coded events; and
4. Suggesting use for the research findings in helping regulators to plan better mitigation action or increased/targeted oversight.

The questionnaire was accompanied by a PowerPoint presentation which detailed the aim and objectives of the study alongside the project methodology. The participants were also provided with an Excel spreadsheet which gave an overview of the accidents/serious incident data, as coded by the developed taxonomy at the time.

2.3. Data Analysis

The decision to develop and validate a taxonomy suitable to code the events dataset was made due to other commonly utilised taxonomies being identified as lacking applicability for retrospective analyses. Consequently, initial qualitative analysis was required in order to create a basis for the development of this taxonomy.

Thematic analysis, which Braun and Clarke [11] explain, is a method for identifying, analysing and reporting patterns (or themes) within data, was chosen as the primary qualitative analysis method for this study. A specific type of thematic analysis, known as template analysis, was utilised for the purpose of taxonomy creation and development.

Brooks et al. [12] describe template analysis as:

> *"a form of thematic analysis which emphasises the use of hierarchical coding but balances a relatively high degree of structure in the process of analysing textual data with the flexibility to adapt it to the needs of a particular study."*

The coded themes which emerge from the data during this analysis are known as the "template" [13] and it is this template which forms the impetus for the taxonomy creation. The template analysis structure tends to be hierarchical with sub-themes emerging within themes [13], ideal for the development of a taxonomy.

NVivo 12 Plus qualitative analysis software was selected as the main tool for the template analysis. The events contained within the dataset were uploaded into the software as individual "cases", where each event was analysed for key themes, known as "nodes". This inductive thematic analysis is described by Braun and Clarke [11] as a *"process of coding data whereby no attempt is made to fit it into a pre-existing coding frame, or the researcher's analytic preconceptions"*.

During the generation of the baseline themes, the language of other taxonomies was consciously not utilised, in order to encourage the development of a new taxonomy which would classify the event categories purely from the narrative of the official accident/serious incident reports, and without interpretations or assumptions.

Once the baseline themes were identified within NVivo, as the template, it was then possible to begin creation of the taxonomy, named 'MxFACS' (Maintenance Factors Analysis and Classification System). This process initially involved separation of the node coding into a three-level hierarchy:

Level 1—Event outcome;
Level 2—System/component failure causing the accident/serious incident; and
Level 3—The maintenance contributing factor(s) which led to the system/component failure and the ultimate accident/serious incident

The selection of this hierarchical structure was developed from the Bowtie Model (illustrated in Figure 3) to complement risk analysis and assessment processes. While using this model, inevitably there is considerable level of subjectivity and interpretation involved. For example, Level 1 Event Outcomes can be considered as 'Top Events' or alternatively if Level 2 'System/Component Failures' are considered as top events, then the Level 1 'Event Outcomes' will be considered at 'Consequences' on the right-hand side of the bowtie. Subsequently, Level 3 factors can be considered as weaknesses in the barriers or escalation factors. In this paper, it is not our aim to define all of the specific components of bowtie model with a rigid approach but ultimately for each event that a bowtie analysis can be conducted, the 'hazard' would be the work(s) undertaken by the maintenance personnel which resulted in the accident or serious incident. We aim to make the strong link between the maintenance factors (high risk areas within maintenance environment) and the key risk areas identified in 'Safety Risk Portfolios' published by EASA. Without presenting this information to industry representatives contributing to the CAT-CAG, it is extremely challenging to influence the decision making for taking risk mitigation actions and include them in the European Plan for Aviation Safety.

Once distinguished into three levels, each event was then re-evaluated and coded in accordance with the template. This process allowed for evolution of the taxonomy as more appropriate themes became apparent throughout the coding. Once the MxFACS taxonomy had been fully established, each event was assessed and classified in accordance with MxFACS in order to allow for further analysis of the output, and associated high risk areas, to be undertaken.

Once the taxonomy and resultant output had been finalised, it was then possible to utilise this data to produce a sample Bowtie model. This was achieved using BowTieXP software and allowed for a demonstration of the applicability of MxFACS to existing risk assessment methodologies.

Following the collation of SME responses, their data was entered into NVivo. This allowed for the identification of the themes within each question answer in order to reflect upon the study's methodological framework and to provide guidance on future utilisation of the coded data.

Whilst inferential statistical analysis of other variables were considered (for example: country or operator type), it was concluded that the combination of the number of events within the dataset over an extended period of time would not allow for statistically significant results to subsequent from such analyses.

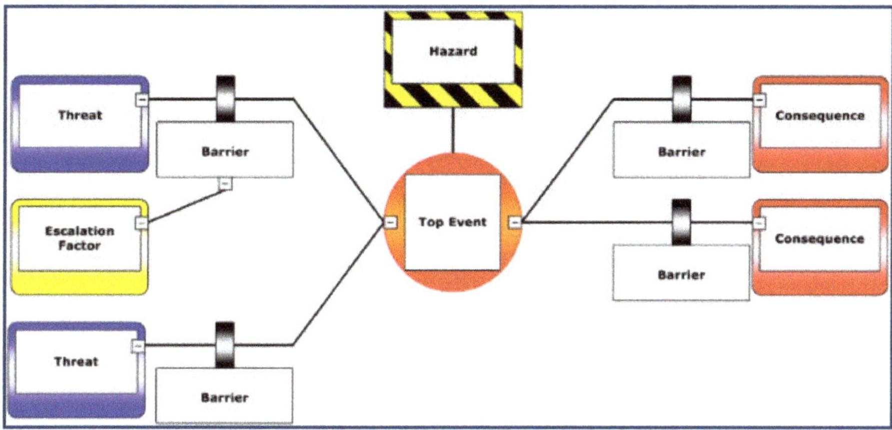

Figure 3. Bowtie risk assessment model [14].

2.4. Assessing the Rigour of the Research Process

Whilst a study's validity and reliability are key concerns for any piece of research, Liamputtong and Ezzy [15] argue that such terms are problematic in their application to qualitative research, suggesting that they are more suited to quantitative methods. The term 'rigour' is therefore used as a more appropriate, conceptualised measure of the underlying themes addressed by 'validity' and 'reliability' [15].

Inter-rater reliability, or inter-rater concordance, is a tool used within qualitative analysis to assess the level of agreement amongst two or more 'raters' [16]. However, as highlighted by Liamputtong and Ezzy [15], it does not guarantee the reliability or validity of interpretations but is a useful tool in assessing the rigour of qualitative research.

Cohen's Kappa [17] is a popular statistic for inter-rater concordance; it shows the proportion of agreement, corrected for chance. Equation (1) demonstrates how Cohen's Kappa is derived, while Equations (2) and (3) detail how the components of the formula are determined.

$$\kappa = \frac{P_o - P_e}{1 - P_e} \tag{1}$$

where κ = Cohen's Kappa; P_o = joint probability of agreement; and P_e = chance agreement.

$$P_o = \frac{\sum_{i=1}^{n} R}{n} \tag{2}$$

where P_o = joint probability of agreement; R = rater agreements; and n = total number of ratings.

$$P_e = \frac{\sum_{i=1}^{n} \left(\frac{c_i \times r_i}{n} \right)}{n} \tag{3}$$

where P_e = chance agreement; c = column marginal; r = row marginal; and n = total number of ratings.

In order to assess the rigour of this study, a SME coded a sample of 10 events using the MxFACS taxonomy. The SME's responses were then compared with the researcher's so that Cohen's Kappa could ultimately be determined. IBM SPSS statistics software was used to aid in the determination of a Cohen's Kappa value for the taxonomy as a whole.

2.5. Ethical Considerations

Whilst ethically low-risk, the study does contain survey elements, so it is therefore essential that the confidentiality of participants is maintained throughout. In complement of this, it was necessary to acquire informed consent prior to conducting the questionnaire. Research ethical approval was sought and granted by the university (Reference: CURES/6042/2018).

3. Results and Discussion

The results of the study, alongside a critical discussion of their implications, are presented herein.

3.1. Taxonomy Inter-Rater Concordance

Altman [18] categorises levels of agreement for Cohen's Kappa as shown in Table 3.

Table 3. Levels of agreement, adapted from Altman [18].

Value of Kappa	Strength of Agreement
<0.20	Poor
0.21–0.40	Fair
0.41–0.60	Moderate
0.61–0.80	Good
0.81–1.00	Very good

The Kappa value for this study's inter-rater concordance, derived using IBM SPSS Statistical Software, is given in Table 4. The derived agreement statistics are provided in Table 5.

Table 4. Determined inter-rater concordance.

	κ	Level of Agreement
Researcher and SME	0.90	Very good

Table 5. Derived agreement statistics across all levels.

Researcher and SME	κ	P_o	P_e
Overall	0.90	0.90	0.03
Level 1	0.80	0.80	0.002
Level 2	0.70	0.70	0.001
Level 3	1	1	0.0001

This shows a good to very good strength of agreement between the researcher and SME when utilising the MxFACS taxonomy and consequently indicates a high level of rigour for this qualitative research.

3.2. Distribution of Serious Incidents and Accidents for 2003–2017

From the data collection process, 112 accidents and serious incidents were identified as maintenance-related for 2003–2017. The global distribution for these events can be seen in Figure 4a,b.

The results shown in Figure 4a vary quite significantly from the maintenance threat levels given by IATA [19]. IATA [20], do however highlight that air accident investigations require greater cooperation on global standards; of around 1000 accidents over the last ten years, accident reports are only available for approximately 300. They state that of those reports, many contain insufficient information or lack rigorous analysis [20]. Consequently, this may provide some justification for the disparity.

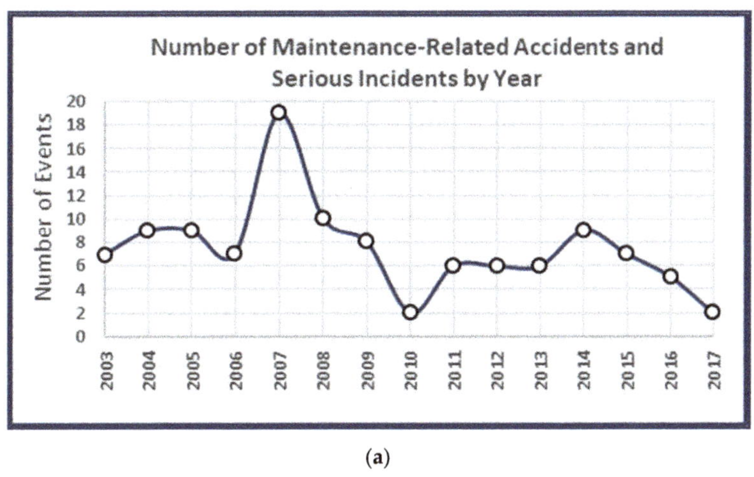

(**a**)

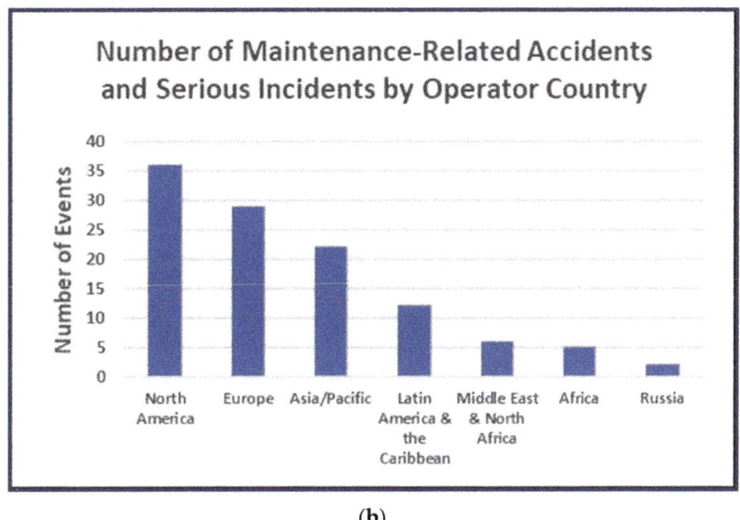

(**b**)

Figure 4. (**a**) Number of serious incidents and accidents identified for 2003–2017. (**b**) Number of serious incidents and accidents (2003–2017) by operator location.

A comparison of the number of EASA Member State events identified for the same time period as EASA's 2014–2017 Annual Safety Reviews (ASR) is shown in Table 6.

Table 6. Comparison of study data with EASA ASRs.

Time Period	ASR Date	n (Study)	n (ASR)	Difference
2012–2016	2017	13	14	−1
2011–2015	2016	13	8	+5
2009–2013	2014	23	3	+20

A plausible explanation for this disagreement may be due to interpretation of "CAT" used within the studies. EASA's ASR focus on CAT aeroplane airline operations for aircraft greater than 5700 Kg maximum take-off weight (MTOW), which they describe as covering "the bulk of the commercial air transport activity" [7]. Comparatively, this study analysed events for any flight which could be acknowledged as CAT under Commission Regulation (EU) no. 965/2012, regardless of number of passengers, aircraft type, or MTOW.

3.3. Level 1—Event Outcomes

112 top-level event outcomes were coded with the MxFACS taxonomy. Figure 5 illustrates the top-level event outcomes. Once the initial outcome category for each event was identified, they were coded in further depth to better detail the nature of the event.

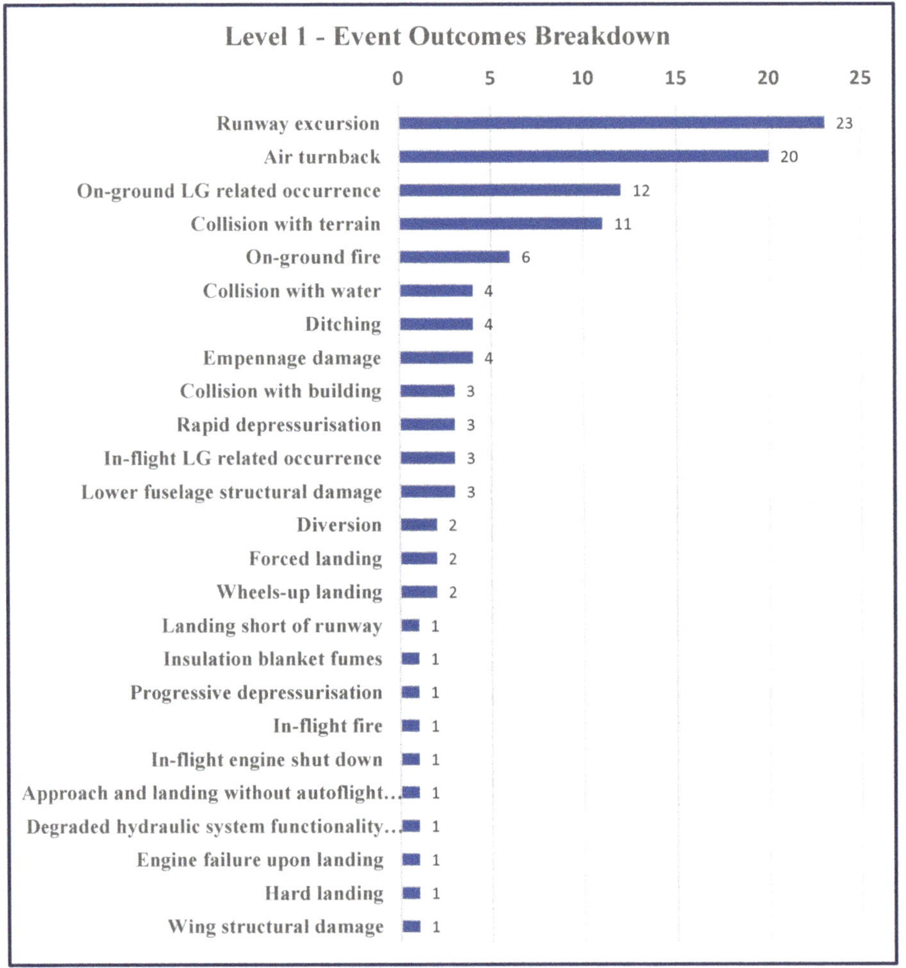

Figure 5. Level 1 top level event outcomes.

Trends within this data are in complement of the UK CAA's Significant Seven [21], and the Key Risk Areas (KRA) identified by EASA [4]. The KRA's present the outcome of accidents in the Safety Risk Portfolios published by EASA. Of the applicable top safety risks the UK CAA [21] present, three can be seen as of relevance from this study: runway overrun or excursion; airborne and post-crash fire; and loss of control (discussed further in Section 3.4). Whilst possible for maintenance actions to result in a runway incursion or ground collision, none of these types of events were identified within the study's time period.

Further exploring these Level 1 results alongside the five maintenance KRAs, runway excursion can be seen as the highest-ranking event outcomes in Figure 5. Terrain collision is fifth overall, and aircraft environment demonstrates some relevance within the study also.

3.4. *Level 2—System/Component Failure*

When coding the dataset at Level 2, 112 events were coded at the system/component level, as demonstrated by Figure 6.

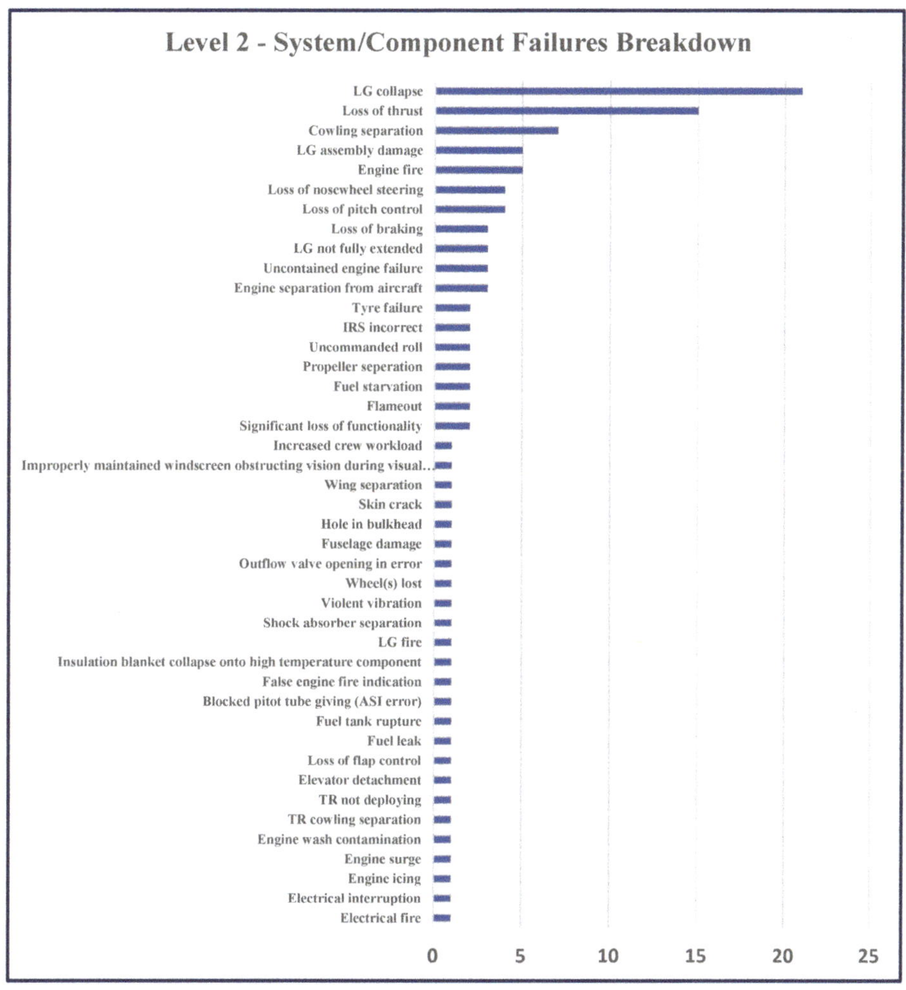

Figure 6. Level 2 top level system/component failures.

Inevitably, due to differing taxonomies, it is not possible to directly compare the study's findings with previous research. However, when looking at similarities between the results of the UK CAA [2,3], considerable homogeneity exists. In both UK CAA [2,3] studies, powerplant, landing gear, and flight controls are respectively ranked as the second to fourth most populous areas for maintenance error.

These three systems are ranked in the same order, but as the first to third most coded, within the top-level coding for MxFACS Level 2 (the 'engine' category can be used interchangeably with 'powerplant').

Given that the CAA studies were of low-level occurrences identified in MORs, and this comparison evidences the trend continuing to permeate through to the higher-level serious incidents and accidents, these three systems can be said to be of continuing significance from 1996 through to present day. Without further research into the targeted actions of maintenance organisations, it is not possible to postulate whether or not this trend continuation is due to inattention to this area of aircraft maintenance. It does however highlight key areas for regulators to target in their risk identification and assessment processes and when proposing oversight measures.

A plausible explanation for equipment and furnishings not attracting a higher frequency of coding within the findings of this study is the severity of the events analysed. In the analysis of low-level, low-severity MORs, it is understandable that frequent occurrences involving the equipment and furnishings listed in ATA 25 will arise. It would, however, be rare for such events to propagate into a serious enough outcome for a serious incident, or even accident, to transpire. Thus, one can expect to see a significantly lower number of occurrences related to ATA 25 within this study.

After identifying the initial event type for each occurrence, more detailed coding of the nature of these events was undertaken. The loss of control events affiliated with 'flight controls' can be seen as the third-highest ranking event type. This area is identified as of high risk by the UK CAA [2] and EASA [4], so this prevalence within the dataset therefore indicates a high frequency of event for a high-risk event.

3.5. Level 3—Maintenance Factors

Coding at the third level of the MxFACS taxonomy revealed 221 maintenance factors which were identified as contributory to the events in the dataset. The high-level maintenance factors are shown in Figure 7.

As with previous studies [2,3,22] cited in [23–25] omission errors, particularly incorrect installations, remain prevalent. However, the number of commission errors (procedures undertaken but not to an appropriate level) are also of note. Particularly inadequate maintenance procedures, which are not only the largest commission error but also the leading maintenance factor overall. Inspections, incorrect procedures, and operator oversight are further areas of significance within the dataset.

An initially surprising result is the low presence of human factors (HF) analysis within the published accident/serious incident official reports. The data was coded based only on the information available; the researcher did not make any assumptions about the HF nature of maintenance factors. Whilst many of the reports detailed HF information for how the flight crew responded to the situation created as a result of the maintenance factors, little or—in many cases—no attention was given to the human performance issues from the maintenance perspective. This raises the question as to why in-depth HF analysis related to maintenance personnel is not conducted during the investigations. This is one of the most significant findings of this study. Just like the flight crew, maintenance personnel performance relies on a wide range of factors from a personal and organisational perspective, to an industry-wide sociotechnical system level.

There is no doubt the investigators are constrained by many factors and finding out facts about the issues impacting on performance of the individuals involved during an investigation can be very challenging and—in some cases—impossible but as Burban [26] suggests there is a reluctance amongst some investigation bodies to fully embrace HF and address potentially important HF issues in detail in their investigations.

Many of the more-detailed maintenance factors have interdependencies where the combination of two or more of these factors led to the event. Identifying these interdependencies allows for an understanding of the barrier failings which lead to an event. One example of this is the 30 July 2008 event where incorrect maintenance procedures lead to a Boeing 777 fuel supply hose O-ring being damaged, thus causing an engine fire from the associated fuel leak. Following the aircraft maintenance manual (AMM) procedure should normally have prevented this from taking place, however the AMM did not contain the appropriate information for this procedure and thus contributed to the event [27].

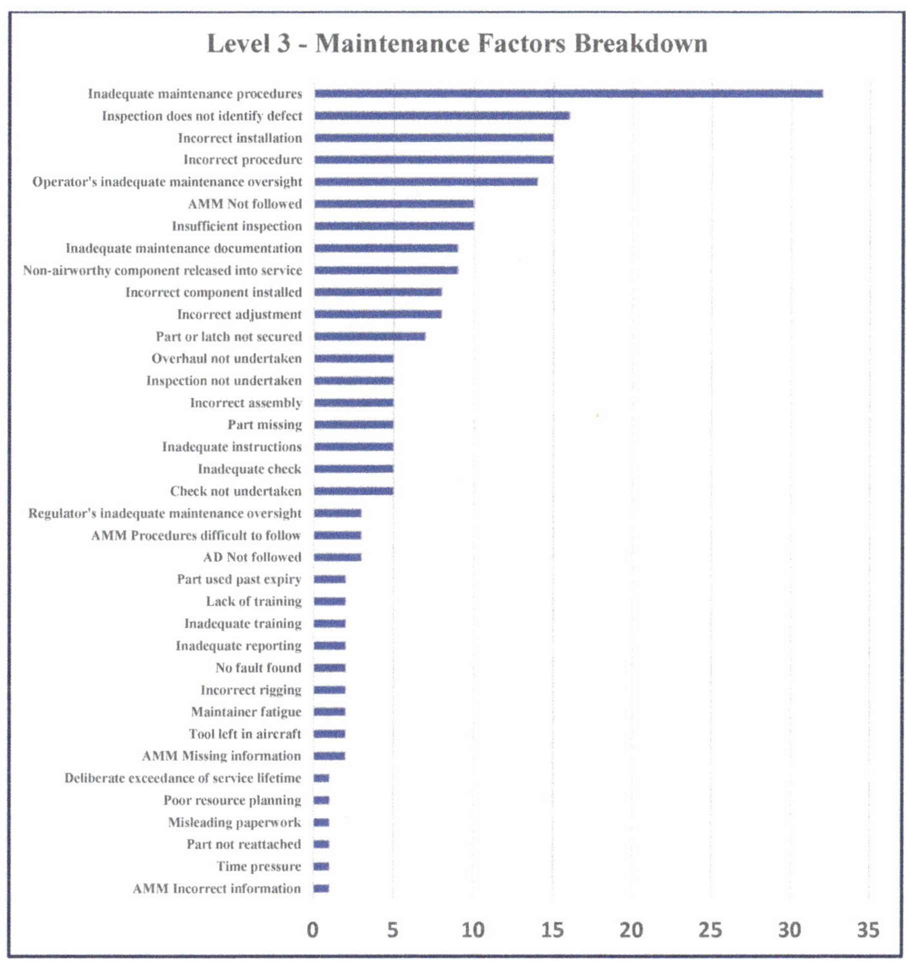

Figure 7. Level 3 top level maintenance factors.

3.6. Sample Bowtie Application

A sample event, Engine Fire, was selected and analysed by developing a Bowtie shown in Figure 8 in order to demonstrate the applicability of MxFACS output to existing risk methodologies, currently utilised by regulators. This allows users of the MxFACS data to better understand what further analysis may be required or can be done as well as the interdependencies between maintenance factors. Consequently, this aids the risk analysis and assessment processes through qualitative analysis of the control barriers which are in place.

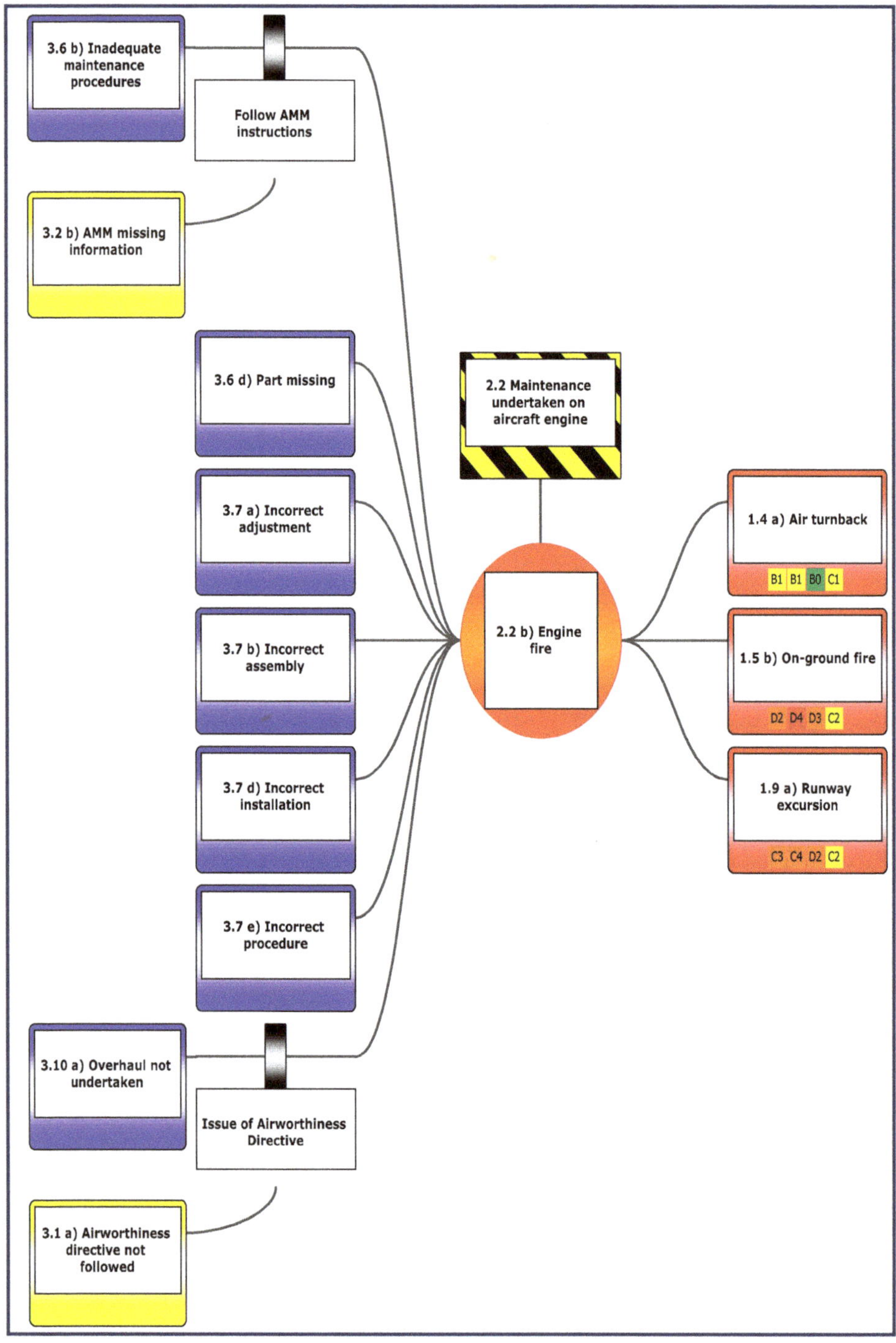

Figure 8. Sample bowtie analysis.

It should be noted that the sample shown in Figure 8 shows only the threats, consequences, and escalating factors derived directly from the MxFACS qualitative analysis process. Such bowties would require analysts to further evaluate the barriers which failed in order for the top event to occur, as part of their risk analysis processes.

By developing and maintaining such Bowties, it would be possible for users to continually monitor the effectiveness of barriers as part of their risk management process, as well as aiding in identifying the particularly high-risk areas which require further attention. This process is of course reliant upon an appropriate level of detail being captured following an event so as to be able to accurately depict the barriers which were in place and failed. Otherwise, the process becomes an exercise of interpretation as opposed to use of factual knowledge.

3.7. SME Survey

A multitude of information was collected within the responses to the SME questionnaire. The most noteworthy elements from the survey were extracted to be further discussed.

3.7.1. Challenges with Taxonomies

A particularly insightful description of the challenges faced when using taxonomies to code aircraft maintenance-related events was provided by this SME:

> *"All taxonomies suffer the problem that categorisation can condition results ... [Classifying] events after the event often requires a lot of imagination. There's an inverse relationship. Rare fatal accidents provide much detail whereas frequent occurrences can be one line in a log book."*

The challenge of attaining detailed information from low level occurrences was certainly faced in the initial data collection process for this study. It was behind the reasoning to scope the research to focus on the lower quantity, but higher quality, serious incidents and accidents as opposed to high volume, minimal detail, low-level occurrences. That does not mean to say that these low-level occurrences should be ignored, to the contrary, but rather that further action is required to ensure the associated reporting processes capture adequate, actionable information.

Regarding the statement on categorisation, it can be argued that using peer review to determine inter-rater concordance could certainly aid in reducing the subjectivity of taxonomies. It was also highlighted by another SME that this use of peer review to assess the inter-rater concordance of the taxonomy categorisation, as was undertaken for the purpose of this study, matched the methodology of the UK CAA [3] which had a peer review to try and validate the initial categorisation."

One SME argued that the importance in learning from occurrences does not lie in coding the existing data for interpretation, but rather in comparing theory with reality: *"Taxonomy is unimportant. What is important is comparing practice with prediction."*

These SME opinions suggest that the utility of pre-existing aircraft maintenance taxonomies may perhaps be a source of some contention.

3.7.2. Feedback on the Study's Methodology

Feedback on the study's methodological framework was largely positive, as this extract demonstrates: "The study has done a great job in collecting the data and classifying it into useful information."

Another SMEs detected the traits of Bowtie within the taxonomy, referencing the "threats" and "escalating factors" as: "causal factors (remove them and the accident is avoided) and circumstantial factors (increased the probability of the event)."

Whilst not directly referring to bowtie, this statement does reflect the thinking that there are different types of factors which can contribute to accidents, serious incidents and occurrences. The decision to name MxFACS Level 3 "maintenance factors" was done so for precisely this reason, so it is therefore encouraging to hear a SME mirror this sentiment.

3.7.3. Assigning Risk and Identifying High Risk Areas from Coded Event Data

One SME proposed an assessment of the effectiveness of remaining safety barriers for non-accident level occurrences as a means of assigning risk, acknowledging existing EASA methodology:

> *"The risk should be based on the effectiveness of the remaining barriers left before it ended up as a credible accident. See ARMS methodology." [28]*

This statement brings about the consideration of the use of MxFACS output in conjunction with the ARMS Event Risk Classification (ERC). The UK CAA [2] highlight that Bowtie is often used within the ARMS methodology regarding the ERC barrier effectiveness assessment. This shows a possible pathway for the integration of MxFACS with the ARMS ERC methodology as both have strong applicability to bowtie.

Another SME proposed the use of an expert panel to effectively assess and allocate risk:

> *"Use a team of experts. [It is] hard work to get consensus but expert challenge is a good way of getting a realistic classification. Use a problem statement to get everyone on the same page."*

The use of MxFACS output, alongside the associated Bowties, could aid this approach by providing the experts with an outline of the risks and barriers involved in the events to be analysed.

3.7.4. Using the Findings of the Study

It was suggested by one SME that the key to better targeted action may be to address near misses rather than accidents: *"Focus on the near miss events rather than the accidents to try and determine how close to an accident we are."*

In contrast, another SME, who highlighted a regulatory resource shortage as being challenging to acting on occurrences, suggested a different approach:

> *"[Regulators] all struggle with limited resources-being driven by events provides only the basics. Continuous improvement means being proactive. Uncover the trends, pick off 5 'low hanging' fruit and work them through. A Pareto analysis was used by US CAST to good effect ... Target priorities on the higher risk items that are easiest fixed first."*

This complements the thinking that prevalence of high frequency occurrences does not necessarily indicate a propensity for accident propagation from their associated risks, particularly if sufficient barriers are in place. By instead focusing on identifying high risk areas within maintenance, regardless of the frequency of actual catastrophic outcome, it could be argued that the resultant targeted prevention strategies may be more effective.

3.8. Identifying High Risk Areas

Maurino [29] proposes that the findings of investigations should encourage error tolerance and error recovery, as opposed to error suppression. By identifying the high-risk areas in aircraft maintenance, it is possible to understand what factors shaped particular human errors.

Upon reviewing the fatalities and damage count for the 112 analysed events, it was found that 16 flights had a fatal outcome and 77 lead to aircraft damage. EASA [4] identify damage to be of a medium level risk within their key risk areas. As 69% of the events identified within the dataset resulted in some level of damage, it can be said that the likelihood of event for this KRA was substantial over the past 15 years.

In order to better relate the frequency of events data from Figures 5–7 to risk, the events within the dataset were evaluated at each level of the taxonomy for number of fatal accidents and instances of aircraft damage. The fatal accident figures were then plotted alongside number of events and number of fatal events (represented as the size of the bubble), to replicate the same risk visualisation approach used by EASA [4] in Figure 9.

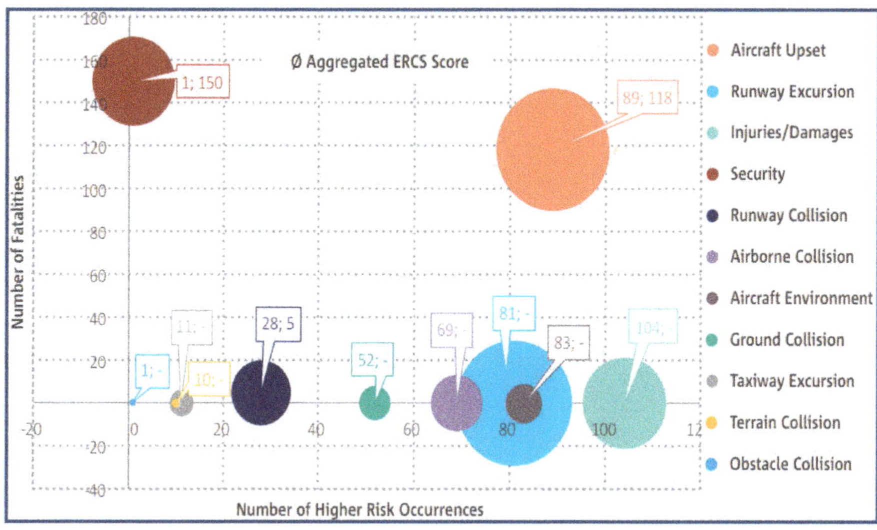

Figure 9. Key risk areas for CAT aeroplanes by fatalities 2013–2017, adapted from EASA (2018).

Figure 10 shows this chart for the Level 1 results, with more detail about the relationship between fatalities, the number of fatal events (represented by the size of the bubble), and the total number of Level 1 outcomes.

Three of the four event outcomes identified as having fatal outcomes are congruent with three of the maintenance KRAs listed in EASA ASR [4]. One particular point of significance is the positioning of collision: this area has a large proportion of fatalities, damage and frequency; it could be a key area of focus for further risk analysis processes. The coded MxFACS data may be used in conjunction with analysis methodologies such as bowtie to examine the particular barrier failings which lead to these kinds of accidents. Further information about aircraft damage can be found in Table 7.

Table 7. Level 1 fatal accident and aircraft damage relationships.

Event Outcomes	Fatal Accident		Aircraft Damage	
	n (Fatal Accidents)	% of Total Fatalities	n (Aircraft Damage)	% of Total Damage
Runway-related events	1	6.3%	19	24.7%
Collision	11	68.8%	17	22.1%
Diversion or air turnback			10	13.0%
Landing-related events	3	18.8%	10	13.0%
Fire	1	6.3%	6	7.8%
LG-related events			6	7.8%
Structural damage			6	7.8%
Depressurisation			2	2.6%

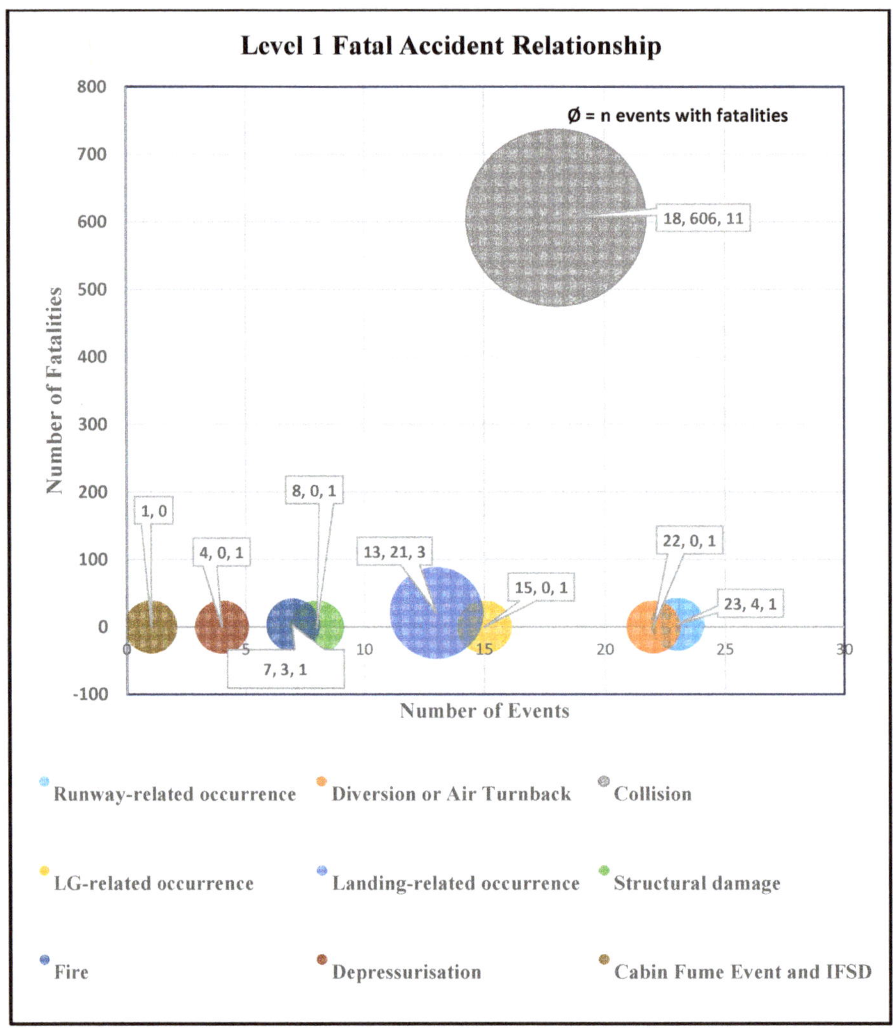

Figure 10. Level 1 fatal accident relationship.

The ranked orders for greatest number of fatal accidents and aircraft damage for Level 2 of the dataset are given in Table 8.

Figure 11 shows the relationship between fatalities, number of fatal events (represented by the size of the bubble) and the total number of Level 2 events.

Engine-related events can be seen to have the largest propensity for both fatalities and aircraft damage. As shown in Figure 11, these events were also the most frequently occurring across the dataset. This would suggest that maintenance related to aircraft powerplants should be placed high on the agenda for regulators when proposing better-targeted action and oversight, as well as being a key focus for maintenance organisations. Similar comment can be made in relation to landing gear and flight controls, which also rank highly across all three areas.

Table 8. Level 2 fatal accident and aircraft damage relationships.

System/Component Failures	Fatal Accident		Aircraft Damage	
	n (Fatal Accidents)	% of Total Fatalities	n (Aircraft Damage)	% of Total Damage
Engine	8	50.0%	29	37.7%
Landing gear	1	6.3%	29	37.7%
Flight controls	2	12.5%	5	6.5%
Steering			3	3.9%
Electrical power	2	12.5%	2	2.6%
Fuel			2	2.6%
Instrumentation and indication	1	6.3%	2	2.6%
Structure	1	6.3%	2	2.6%
Windscreen			1	1.3%
Workload	1	6.3%	1	1.3%

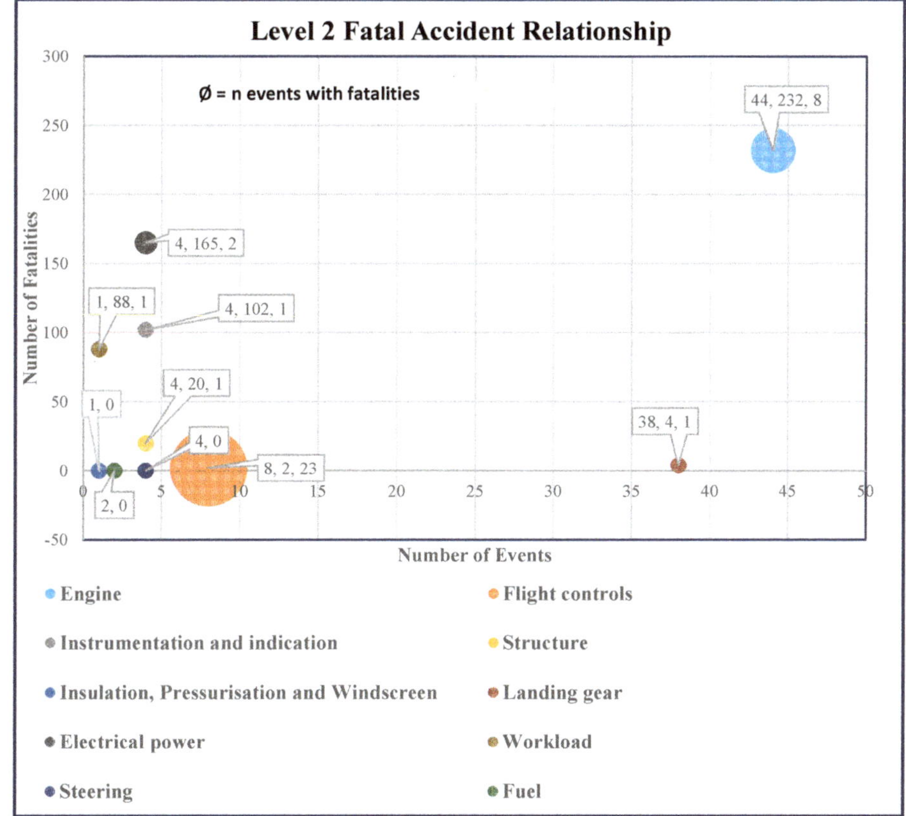

Figure 11. Level 2 fatal accident relationship.

It is more difficult to directly compare the Level 3 events with the fatality and damage figures as many of the events have multiple maintenance factor categorisations assigned to them. Therefore, the events at this level were analysed as a percentage of the total number of instances where a maintenance factor was attributed to a fatal accident or aircraft damage. The maintenance factors related to fatal accidents and aircraft damage are shown in Figure 12, with a full breakdown given in Table 9.

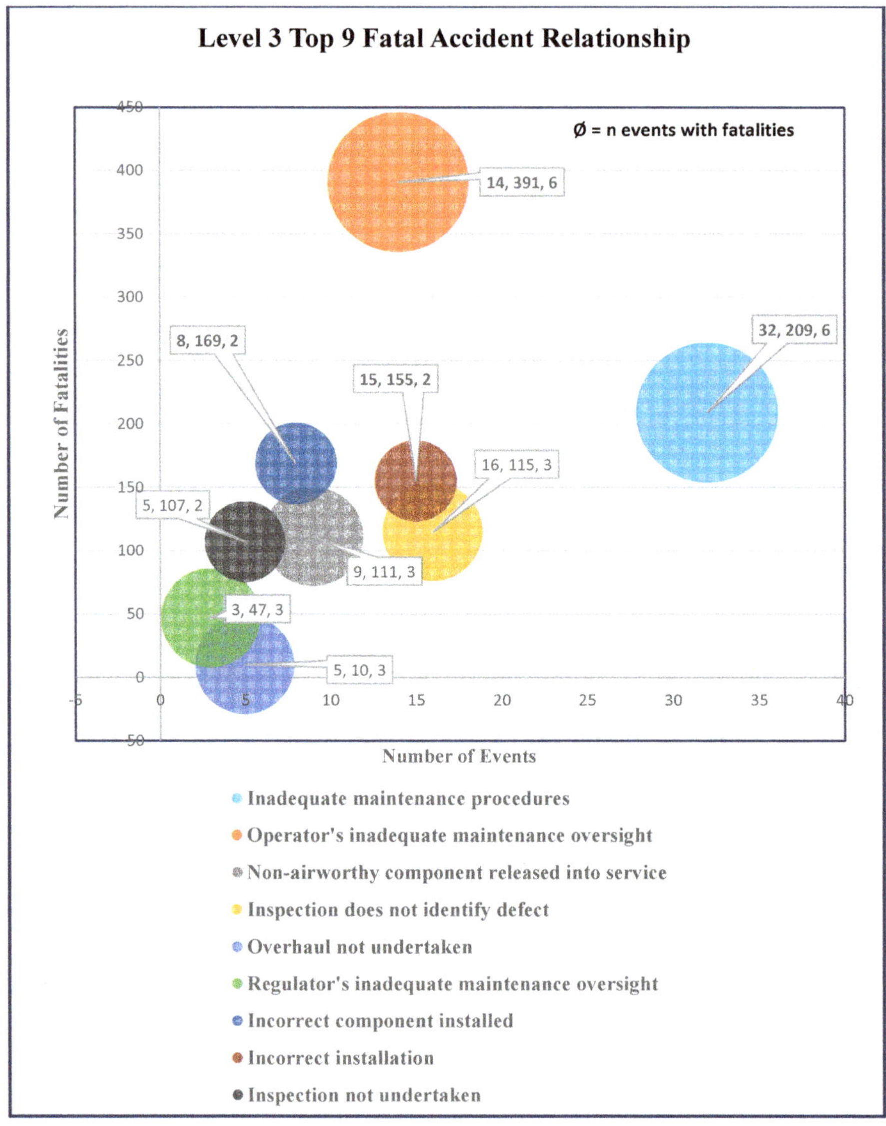

Figure 12. Level 3 fatal accident relationship.

Table 9. Breakdown of maintenance factors related to fatalities and aircraft damage.

Maintenance Factor	% Related to Fatalities	% Related to Damage
Inadequate maintenance procedures	14.6%	14.4%
Inspection does not identify defect	7.3%	9.2%
Incorrect procedure	2.4%	7.8%
Incorrect installation	4.9%	7.2%
Operator's inadequate maintenance oversight	14.6%	6.5%
Incorrect component installed	4.9%	4.6%
Insufficient inspection	2.4%	4.6%
AMM Not followed	2.4%	3.9%
Non-airworthy component released into service	7.3%	3.9%
Inadequate maintenance documentation	2.4%	3.3%
Overhaul not undertaken	7.3%	3.3%
Check not undertaken	2.4%	2.6%
Inadequate instructions		2.6%
Part or latch not secured	2.4%	2.6%
Incorrect adjustment		2.6%
Incorrect assembly	2.4%	2.6%
Part missing		2.0%
Inspection not undertaken	4.9%	2.0%
Regulator's inadequate maintenance oversight	7.3%	2.0%
Not followed		1.3%
AMM procedures difficult to follow		1.3%
Incorrect rigging	2.4%	1.3%
No fault found		1.3%
Inadequate reporting	2.4%	1.3%
Part used past expiry		1.3%
AMM incorrect information	2.4%	0.7%
AMM missing information		0.7%
Inadequate check	2.4%	0.7%
Tool left in aircraft		0.7%
Inadequate training		0.7%
Deliberate exceedance of service lifetime		0.7%

Inadequate maintenance procedures and operator oversight can be seen as the two predominating areas within the top nine maintenance factors for fatalities. This may suggest that organisational safety management requires particular attention and would perhaps warrant further risk analysis to identify the interdependencies which interact with these maintenance factors.

4. Conclusions

In order to conclude this study, it is of relevance to first evaluate the challenges and limitations before drawing together a series of final concluding remarks.

4.1. Challenges and Limitations

A number of challenges and limitations were encountered within the study. These were in relation to data collection and analysis, SME survey, and the MxFACS taxonomy as a whole.

4.1.1. Using the Selected Data Sources

One limitation of the data collection for this study was the sourcing of credible aircraft maintenance events. The data was collected from English language databases, and relied on English translations of investigation reports being available where the investigating body was not English-speaking. This means that the dataset, although very much international, may not be 100% globally representative.

The broad nature of the databases used to collect the data was a further challenge to the data collection process. Manual extraction of maintenance-related accidents and serious incidents was the only means of identifying these events, amongst a vast quantity of events which were irrelevant to the study.

4.1.2. Data Analysis

It was challenging to analyse the data in a way that would provide meaningful and insightful output as the focus of the study was somewhat narrow, and there is little existing literature for means of comparison. By evaluating the most recent and relevant studies [2–5] it was subsequently possible to identify key areas to initially focus upon.

4.1.3. Taxonomy Development and Inter-Rater Reliability

As mentioned in Section 2.4, only one subject matter expert coded a sample of 10 events out of 121 using the MxFACS taxonomy. Despite the Cohen's Kappa was calculated and revealed a certain level of confidence, this is still one of the limitations of this study. Nevertheless, subsequent to this study, MxFACS taxonomy was used during another research project analysing the accidents and serious incidents only in Nigeria and it was found to be beneficial to analyse all events in the dataset.

4.1.4. SME Survey

Whilst the SME survey provided a number of insightful and helpful comments, it proved difficult to extract the desired level of information through open-ended questions than would have been available through more-structured interviews.

It was initially the intention to conduct semi-structured interviews with these SMEs, however the logistics of such interviews proved to be challenging to arrange, particularly due to differing time zones. Consequently, a questionnaire was decided as a more efficient means of collecting this data.

4.1.5. MxFACS

The MxFACS taxonomy aids in identifying three out of four of the basic elements of error identified by Reason [30]; the action, the outcome, and the context. However, it is not always possible to determine from investigative reports the intention of those who produced the errors.

As such, interview (perhaps in the MEDA format) with the maintenance personnel involved with particular events, shortly after the event, would complement a more representative coding of the accident or serious incident. This is not without its challenges, particularly for lower level occurrences where investigation may be undertaken sometime after the fact, therefore making it difficult for the personnel to accurately recall the intentions which lead to the error.

4.2. Study Conclusions

The study provides a modern-day and maintenance-specific viewpoint on CAT accidents and serious incidents. The development of the MxFACS taxonomy brings about a contemporary approach for exploring the nature of these maintenance events, by looking at a combination of maintenance-specific causations, system/component failures, and event outcomes. Such taxonomy output is demonstrated as being applicable to existing risk analysis processes and methods and could be used to complement existing taxonomy and methodologies.

The results of the study show that the three most frequent maintenance events for 2003–2017 are runway excursions, air turnbacks, and on-ground landing gear-related events. The most common system/component failures were related to engine, landing gear, and flight controls. At the causation level, the largest maintenance factor issues were inadequate maintenance procedures, inspections not identifying defects, incorrect installation, and incorrect procedures.

By combining the frequency of event data with number of fatalities, it was possible to begin to create a picture of the higher risk areas within maintenance for this time period. Collision events were the most prominent consequence, engine-related events were the largest event type, and inadequate maintenance procedures were the most concerning maintenance factor.

The study's findings could be used to aid in a Safety-II approach to understanding where barrier weaknesses lie within maintenance safety management systems, particularly when integrated with bowtie analysis. Such an approach may allow regulators and maintenance organisations to develop means of ensuring as much as possible goes right within the maintenance system, as opposed to looking solely at just what went wrong.

5. Recommendations

In acknowledgement of SME feedback, ARMS and ERC methodologies may be explored for means of better identifying the high-risk areas within the MxFACS output. By combining the MxFACS output and fatality data with bowtie models to understand the effectiveness of the barriers in place, it would then be possible to develop an ERC score and consequently substantiate the higher risk areas for this data. This would then allow for a maintenance-specific depiction of key risk areas akin to the work of EASA in Figure 9 [4].

Further to this, it would be advisable to create a number of bowtie models for the high-risk areas and continually maintain and update these models as time progresses. This would allow for continual monitoring of the barriers in place for higher risk areas. Additionally, the MxFACS taxonomy and database should also be continually maintained and updated to ensure relevance for new accidents and serious incidents as they evolve.

Author Contributions: Conceptualization, J.I. and C.T.; methodology, J.I. and C.T.; formal analysis, J.I.; data curation, J.I.; writing—original draft preparation, J.I.; writing—review and editing, J.I. and C.T.; supervision, C.T. All authors have read and agreed to the published version of the manuscript.

Funding: This research received no external funding.

Acknowledgments: We would like to thank all of the SMEs for their contribution by their advice and opinions; the International Federation of Airworthiness (IFA) for supporting this study and CGE Risk Management Solutions for providing access to BowTieXP. Finally, we would like to recognise the financial support Jenni received from the Royal Aeronautical Society through their Centennial Scholarship for studying the MSc course and producing this thesis as part of her research project.

Conflicts of Interest: The authors declare no conflict of interest.

Appendix A

Table A1. Breakdown of events by aircraft type.

Aircraft Type	ICAO Wake Turbulence Category (WTC)	Number of Occurrences
Boeing 737 Classic	M	7
Boeing 747	H	6
Airbus A320	M	4
Airbus A330	H	4
Boeing 737 NG	M	4
Swearingen SA226-TC Metro II	L/M	4
Bae 146/Avro RJ	M	3
Beech 1900D	M	3
Beech 200 Super King Air	L/M	3
Boeing 777	H	3
Boeing/MD-83	M	3

Table A1. *Cont.*

Aircraft Type	ICAO Wake Turbulence Category (WTC)	Number of Occurrences
Cessna 208B Grand Caravan	L	3
DHC-8-402 Q400	M	3
Airbus A300	H	2
Airbus A319	M	2
Beech 100 King Air	L	2
BN-2A Islander	L	2
Boeing 757	M	2
Canadair CRJ-200LR	M	2
Cessna F406 Caravan II	L	2
DHC-3 Otter	L	2
DHC-8-301	M	2
Embraer EMB-110 Bandeirante	M	2
Fokker F-27 Friendship	M	2
Learjet 35	M	2
Swearingen SA227-AC Metro III	L/M	2
Airbus A321	M	1
Airbus A340	H	1
Airbus A380	H	1
ATR 42	M	1
ATR 72	M	1
Beech 99A	L	1
Boeing 707	H	1
Boeing 737 Original	M	1
Boeing 767	H	1
Boeing/MD-10	H	1
Boeing/MD-11F	H	1
Bombardier CRJ-100	M	1
Bombardier CRJ-200	M	1
Bombardier DHC8-300	M	1
Canadair CRJ-100ER	M	1
CASA/Nurtanio NC-212 Aviocar	M	1
Cessna 208B Super Cargomaster	L	1
Cessna 402 Businessliner	L	1
Cessna 560XL Citation Excel	M	1
Convair CV-340-70	M	1
DC-8-71F	H	1
DC-9-81 (MD-81)	M	1
DC-9-82 (MD-82)	M	1
DHC-8-202	M	1
Dornier Do 28D-2 Skyservant	L	1
Douglas C-54G (DC-4)	M	1
Douglas Super R4D-8 (DC-3S)	M	1
Fokker 70	M	1
Grumman G-73T Turbo Mallard	L	1
IAI 1125 Astra SPX	M	1
Ilyushin Il-76TD	H	1
Learjet 60	M	1
Lockheed L-100-30 Hercules	M	1
Saab 2000	M	1
Tupolev Tu-154B-2	M	1
Xian MA60	M	1

Please note the following paragraph is extracted from ICAO Doc. 8643 'Aircraft Type Designators' to provide clarification about the WTC categories.

"Wake Turbulence Category (WTC)

The wake turbulence category (WTC) indicator will follow the aircraft type designator and is provided on the basis of the maximum certificated take-off mass, as follows:

- H (Heavy) aircraft types of 136,000 kg (300,000 lb) or more;
- M (Medium) aircraft types less than 136,000 kg (300,000 lb) and more than 7000 kg (15,500 lb)
- L (Light) aircraft types of 7000 kg (15,500 lb) or less.

Note. Where variants of an aircraft type fall into different wake turbulence categories, both categories are listed (e.g., L/M or M/H). In these cases, it is the responsibility of the pilot or operator to enter the appropriate single character wake turbulence category indicator in Item 9 of the ICAO model flight plan form."

Appendix B

Definitions of 'Accident', 'Serious Incident', and 'Incident'

For the purpose of this study, the following definitions which were extracted from ICAO Annex 13 apply.

"Accident. An occurrence associated with the operation of an aircraft which, in the case of a manned aircraft, takes place between the time any person boards the aircraft with the intention of flight until such time as all such persons have disembarked, or in the case of an unmanned aircraft, takes place between the time the aircraft is ready to move with the purpose of flight until such time as it comes to rest at the end of the flight and the primary propulsion system is shut down, in which:

(a) a person is fatally or seriously injured as a result of:

- being in the aircraft, or
- direct contact with any part of the aircraft, including parts which have become detached from the aircraft, or
- direct exposure to jet blast, except when the injuries are from natural causes, self-inflicted or inflicted by other persons, or when the injuries are to stowaways hiding outside the areas normally available to the passengers and crew; or

(b) the aircraft sustains damage or structural failure which:

- adversely affects the structural strength, performance or flight characteristics of the aircraft, and
- would normally require major repair or replacement of the affected component, except for engine failure or damage, when the damage is limited to a single engine (including its cowlings or accessories), to propellers, wing tips, antennas, probes, vanes, tires, brakes, wheels, fairings, panels, landing gear doors, windscreens, the aircraft skin (such as small dents or puncture holes), or for minor damages to main rotor blades, tail rotor blades, landing gear, and those resulting from hail or bird strike (including holes in the radome); or

(c) the aircraft is missing or is completely inaccessible.

Note 1. For statistical uniformity only, an injury resulting in death within thirty days of the date of the accident is classified, by ICAO, as a fatal injury.
Note 2. An aircraft is considered to be missing when the official search has been terminated and the wreckage has not been located.
Note 3. The type of unmanned aircraft system to be investigated is addressed in Annex 13 Section 5.1.
Note 4. Guidance for the determination of aircraft damage can be found in Attachment E.

Serious incident. An incident involving circumstances indicating that there was a high probability of an accident and associated with the operation of an aircraft which, in the case of a manned aircraft, takes place between the time any person boards the aircraft with the intention of flight until such time as all such persons have disembarked, or in the case of an unmanned aircraft, takes place between the time the aircraft is ready to move with the purpose of flight until such time as it comes to rest at the end of the flight and the primary propulsion system is shut down.

- Note 1. The difference between an accident and a serious incident lies only in the result.
- Note 2. Examples of serious incidents can be found in Attachment C.
- Incident. An occurrence, other than an accident, associated with the operation of an aircraft which affects or could affect the safety of operation.
- Note. The types of incidents which are of main interest to the International Civil Aviation Organization for accident prevention studies are listed in Attachment C."

References

1. EASA. *Annual Safety Review 2019*; European Aviation Safety Agency: Cologne, Germany, 2019. Available online: https://www.easa.europa.eu/sites/default/files/dfu/Annual%20Safety%20Review%202019.pdf (accessed on 27 November 2019). [CrossRef]
2. UK CAA. *CAP 1367–Aircraft Maintenance Incident Analysis*; Civil Aviation Authority: West Sussex, UK, 2015.
3. UK CAA. *CAA Paper 2009/05: Aircraft Maintenance Incident Analysis*; Civil Aviation Authority: West Sussex, UK, 2009.
4. EASA. *Annual Safety Review 2018*; European Aviation Safety Agency: Cologne, Germany, 2018.
5. EASA. *Annual Safety Review 2014*; European Aviation Safety Agency: Cologne, Germany, 2015.
6. EASA. *Annual Safety Review 2016*; European Aviation Safety Agency: Cologne, Germany, 2016.
7. EASA. *Annual Safety Review 2017*; European Aviation Safety Agency: Cologne, Germany, 2017.
8. Hieminga, J. Identification of Safety Improvements in Aviation Maintenance Using EU-Wide Incident Reports. Master's Thesis, Cranfield University, Cranfield, UK, 2018.
9. ICAO. State of Global Aviation Safety (ICAO Safety Report 2019 Edition), Montreal, International Civil Aviation Organisation. 2019. Available online: https://www.icao.int/safety/Documents/ICAO_SR_2019_final_web.pdf (accessed on 16 March 2020).
10. Eriksson and Kovalainen. *Qualitative Methods in Business Research*; SAGE Publications Ltd.: Thousand Oaks, CA, USA, 2016.
11. Braun, V.; Clarke, V. Using thematic analysis in psychology. *Qual. Res. Psychol.* **2006**, *3*, 77–101. [CrossRef]
12. Brooks, J.; McCluskey, S.; Turley, E.; King, N. The Utility of Template Analysis in Qualitative Psychology Research. *Qual. Res. Psychol.* **2015**, *12*, 202–222. [CrossRef] [PubMed]
13. University of Sheffield. *Template Analysis*. 2014. Available online: https://www.sheffield.ac.uk/lets/strategy/resources/evaluate/general/data-analysis/template-analysis (accessed on 11 August 2018).
14. CGE Risk Management Solutions. 2017. Available online: https://www.cgerisk.com/knowledgebase/The_history_of_bowtie (accessed on 14 June 2020).
15. Liamputtong, P.; Ezzy, D. *Qualitative Research Methods*, 2nd ed.; Oxford University Press: Oxford, UK, 2005.
16. McHugh, M.L. Interrater reliability; the kappa statistic. *Biochem. Med.* **2012**, *22*, 276–282. [CrossRef]
17. Cohen's. A Coefficient of Agreement for Nominal Scales. *Educ. Psychol. Meas.* **1960**, *20*, 37–46. [CrossRef]
18. Altman. *Practical Statistics for Medical Research*; Chapman and Hall: Boca Raton, FL, USA, 1991.
19. IATA. *IATA Annual Review 2017*; International Air Transport Association: Montréal, QC, Canada, 2018.
20. IATA. *Safety Report 2017*; International Air Transport Association: Montréal, QC, Canada, 2018.
21. UK CAA. In *CAA PAPER 2011/03—CAA PAPER 2011/03CAA 'Significant Seven' Task Force Reports*; Civil Aviation Authority: West Sussex, UK, 2011.
22. Boeing. *Maintenance Error Decision Aid*; Boeing Commercial Airplane Group: Seattle, WA, USA, 1994.
23. UK CAA. *CAP 718–Human Factors in Aircraft Maintenance and Inspection*; Civil Aviation Authority: West Sussex, UK, 2002.
24. Reason, J. *Comprehensive Error Management in Aircraft Engineering: A Manager's Guide*; British Airways Engineering: Heathrow, UK, 1995.
25. Reason, J. *Hobbs, A. Managing Maintenance Error: A Practical Guide*; Ashgate: Aldershot, UK, 2003.
26. Burban, C. Human Factors in Air Accident Investigation: A Training Needs Analysis. Ph.D. Thesis, Cranfield University, Cranfield, UK, 2016.

27. JTSB (Japan Transport Safety Board). Aircraft Serious Incident Investigation Report, Vietnam Airlines, VN-A146. 2010. Available online: https://www.skybrary.aero/bookshelf/books/1223.pdf (accessed on 10 June 2020).
28. Nisula, J. *"ARMS Methodology for Operational Risk Assessment"* [PowerPoint Presentation]. Aviation Risk Management Solutions (ARMS) Group under the Aegis of ECAST. 2009. Available online: https://www.easa.europa.eu/document-library/general-publications/arms-methodology-operational-risk-assessment-presentation (accessed on 14 June 2020).
29. Maurino, D.E. Human factors and aviation safety: What the industry has, what the industry needs. *Ergonomics* **2000**, *43*, 952–959.
30. Reason, J. *The Human Contribution: Unsafe Acts, Accidents and Heroic Recoveries*; Routledge: Abingdon, UK, 2008.

Article

A Preliminary Investigation of Maintenance Contributions to Commercial Air Transport Accidents

Fatima Najeeb Khan [1,†], Ayiei Ayiei [2,†], John Murray [3,4,†], Glenn Baxter [5,†] and Graham Wild [4,*]

1 Institute of Aviation Studies, The University of Management and Technology, Johar Town, Lahore, Punjab 54770, Pakistan; fatima.najeebk@gmail.com
2 School of Engineering, RMIT University, Melbourne 3000, Australia; ayiei.ayiei@student.rmit.edu.au
3 School of Engineering, Edith Cowan University, Joondalup 6027, Australia; john.murray@ecu.edu.au
4 School of Engineering and Information Technology, University of New South Wales, Canberra 2612, Australia
5 School of Tourism and Hospitality Management, Suan Dusit University, Bangkok 77110, Thailand; g_glennbax@dusit.ac.th
* Correspondence: g.wild@adfa.edu.au; Tel.: +61-2-6268-8672
† These authors contributed equally to this work.

Received: 14 August 2020; Accepted: 26 August 2020; Published: 2 September 2020

Abstract: Aircraft maintenance includes all the tasks needed to ensure an aircraft's continuing airworthiness. Accidents that result from these maintenance activities can be used to assess safety. This research seeks to undertake a preliminary investigation of accidents that have maintenance contributions. An exploratory design was utilized, which commenced with a content analysis of the accidents with maintenance contributions (n = 35) in the official ICAO accident data set (N = 1277), followed by a quantitative ex-post facto study. Results showed that maintenance contributions are involved in 2.8 ± 0.9% of ICAO official accidents. Maintenance accidents were also found to be more likely to have one or more fatalities (20%), compared to all ICAO official accidents (14.7%). The number of accidents with maintenance contributions per year was also found to have reduced over the period of the study; this rate was statistically significantly greater than for all accidents (5%/year, relative to 2%/year). Results showed that aircraft between 10 and 20 years old were most commonly involved in accidents with maintenance contributions, while aircraft older than 18 years were more likely to result in a hull loss, and aircraft older than 34 years were more likely to result in a fatality.

Keywords: accidents; aircraft; airworthiness; aviation; maintenance; safety

---

1. Introduction

Aircraft maintenance can be described as activities performed to maintain an aircraft in a serviceable and airworthy condition. 'Maintenance' includes activities involving component or aircraft repair, inspection, overhauling, troubleshooting, and modifications [1–3]. Aircraft maintenance is a critical task that is essential for warranting aviation safety in relation to the life-cycle of an aircraft [4–6]. Even though maintenance is regarded as one of aviation's many high-risk areas due to its direct impact on aviation safety, it still has a considerable contribution in aircraft maintenance accident and incident occurrences [7–10]. Hobbs and Williamson [11] and Floyd [12] highlight the importance of understanding maintenance errors along with promoting a culture of identifying, reporting, and learning from maintenance errors for improving work quality and safety.

PeriyarSelvam, et al. [13] reports that maintenance costs typically make up between 10% and 20% of aircraft operational costs. The Airline Maintenance Cost Executive Commentary is published by International Air Transport Association's Maintenance Cost Technical Group, contains annual data acquired from airlines around the world. These data are based on the airline's maintenance cost data. Data collected from 54 airlines for 2018 shows that the airlines spent USD 69 billion on aircraft maintenance, repair, and overhaul. This represented approximately 9% of the total airline operational costs [14].

The general aim of this work is to help improve safety in scheduled commercial air transport; this will be achieved by understanding how accidents with maintenance contributions are unique in contrast to all other scheduled commercial air transport accidents. The knowledge gained from understanding these features of accidents with maintenance contributions will hopefully prevent further accidents, and hence, save lives and reduce the cost (both direct and indirect) to the aviation industry. The research questions addressed by this work include:

1. How many accidents in the ICAO official accident dataset are contributed to by maintenance factors, and by extension, what proportion of ICAO official accidents have a maintenance contribution? (RQ1)
2. How does the distribution of accidents that show a contribution from maintenance activities differ to all scheduled commercial air transport accidents, reported by ICAO? (RQ2)
3. Has the number of 'official' accidents with maintenance contributions reduced over time? (RQ3)
4. How does the age of an aircraft in an 'official' accident with maintenance contributions influence the outcome (fatalness and aircraft damage) of the occurrence? (RQ4)

In response to these research questions, the following research hypotheses were proposed:

1. There is currently no reported number for the proportion of accidents that have maintenance as a contributing factor. Estimates for air traffic management accidents are on the order of 8%, hence a similar single digit percentage would be reasonable to expect.
2. Accidents with maintenance contributions will show unique features in comparison to all aviation accidents; occurrences will typically be categorized as system component failures, and they will be more common in earlier phases of flight.
3. Given the short timeframe of the ICAO official accident dataset (since 2008), it is anticipated that the number of events will have remained constant over time, potentially with a slight reduction.
4. Older aircraft will be more likely to result in fatalities and hull losses.

2. Literature Review

2.1. Aircraft Maintenance Related Safety Occurrences in General Aviation

Nelson and Goldman [15] presented research on maintenance related accident investigations for general aviation aircrafts and home built aircrafts, reports for a period between 1983 and 2001 obtained from the National Transportation Safety Board (NTSB). All maintenance related accidents were divided into two databases one for amateur built aircraft and the other for GA. The databases were further analyzed as per human factors taxonomy of maintenance related casual factors. The reports were analyzed to establish the frequencies of fatalities and injuries, airframe time, phase of operation, and time since last inspection. The report further compared the taxonomy results of maintenance related casual factors for amateur built aircraft to general aviation maintenance related accidents. The research findings showed that the main cause of the maintenance related accidents in both amateur built aircraft and general aviation aircraft are due to installation of aircraft parts, at 32% and 17% respectively. Hence, maintenance considered as a causal accident factor, is approximately six times more likely to result in a fatal outcome in amateur built aircrafts as compared to general aircraft accidents.

Studies for analyzing maintenance related safety occurrences were carried out by Rashid et al. [16] and Saleh et al. [17]. The human factor analysis of both studies revealed that the most contributing factor towards maintenance incidents are associated with improper procedures being followed for inspections and installation of components, along with casual factors that are deeply rooted within the organizational and managerial levels. Rashid, Place, and Braithwaite [16] statistically and on a human factor basis analyzed 58 helicopter maintenance-induced safety occurrences. Data was acquired from incidents reports obtained from the Australia Transport Safety Bureau, TSB of Canada, CAA of New Zealand, the UK's AAIB, and the USA's NTSB. Saleh, Tikayat Ray, Zhang, and Churchwell [17], presented risk factor-based research findings focused on maintenance and inspection of helicopter accidents tracing the time for when the error was committed to the actual time when the accident took place. The study showed that about 31% of maintenance related accidents occurred within the first 10 flight hours. It also revealed that most of the preventive maintenance activities errors occurred due to nonconformance with published regulations or maintenance plans. The study recommended the providence of better training; emphasis on the development, use, and implementation of checklists; and strong awareness of the importance of safety along with isolation of workload from maintainers.

2.2. Aircraft Maintenance Related Safety Occurrences in Commercial Air Transport

Several studies have been conducted to understand the reasons behind maintenance related safety occurrences for promoting better safety culture in the commercial air transport sector. Research findings based on studies carried out by Suzuki et al. [18], Insley and Turkoglu [7] and Geibel et al. [19] revealed that the main causes of maintenance occurrences are due to inadequate maintenance procedures, lack of responsibility, and incorrect installations.

Insley and Turkoglu [7] presented a study based on enhancing the understanding of the safety critical functions related to the nature of aircraft maintenance-related accidents and serious incidents, between 2003 and 2017. For the selected time period it was found that runway excursions, air turnback's, and on-ground gear-related events were the most common maintenance events while most of the system/component failures related to engine, landing gear, and flight controls. The data related to fatal accidents revealed collision events, engine-related events, and inadequate maintenance procedures were the most concerning maintenance factor. Similar results were observed by Suzuki, von Thaden, and Geibel [18] based on data collected for 1000 incidents centered on coordination problems in commercial aircraft maintenance from NASA's Aviation Safety Report System (ARSR) for a period of two years from August 2004–July 2006. This study revealed that three problematic behaviors—not delivering information, sending wrong information, and lack of responsibility—are potential sources of impairment for safety procedures in aircraft maintenance. Geibel, von Thaden, and Suzuki [19], presented study results by analyzing issues that cause errors in airline maintenance. Technician qualifications, inspections, parts installation, contract maintenance issues, and log book documentation were the five main categories identified as high-profile performance-based error categories for the study based on 1000 incident reports identified for aircraft maintenance related issues from NASA ASRS for a time period from July 1997 through August 2006. The study concluded over that over 53% of the undesirable outcomes analyzed in the ASRS data were attributed to skill-based errors, such as slips, lapses, and perceptual errors, followed by routine violations (15%), and decision-making errors (9%).

A study based on the impact of human factors training for maintenance personnel to reduce maintenance incidents in the European Union and the United States was carried out by Reynolds et al. [20]. Data regarding the subject training was compared prior to the implementation of the human factors training, 1991–1998 and after the implementation of the subject training, 2000–2006. The study revealed that following the introduction of human factors training the mechanical incidents in the EU dropped from 33% to 22% while the percentage of mechanical incidents in the United States increased resulting in a significant statistical difference in rates for the US relative to the EU. Further a study conducted by Ng and Li [21] provides theoretically supported concepts aimed to provide assistance in analyzing aircraft maintenance incidents. The study investigated the causes of 109 aircraft maintenance incidents for which data was acquired from various airline companies. The results revealed that more than 60% of the aircraft maintenance tasks could be categorized as rule-based tasks as per the Rasmussen's SRK framework while almost "50% of the incidents can be explained by well-known error types or work factors in the psychological literature".

2.3. Aircraft Maintenance Related Safety Occurrences in Military Aviation

Human Factor Analysis and Classification System (HFACS) was used by Schmidt et al. [22] and Illankoon, Tretten, and Kumar [8] to analyze maintenance mishaps in military aviation. Schmidt, Schmorrow, and Figlock [22] carried out analyses of the influence of human factors on naval aviation maintenance mishaps. Information for a total of 470 maintenance related mishaps was gained from the Naval Safety Center's Information Management Systems for the fiscal years, 1990–1997. The study revealed that supervisory, maintainer, and working latent conditions are present that can impact maintainers in the performance of their jobs. Illankoon, Tretten, and Kumar [8] analyzed data acquired from a fighter aircraft fleet looking for reported maintenance deviations over a period of 38 months beginning from January 2013, using HFAC-ME (Maintenance Extension) taxonomy to find and mark hidden causal factors. The study identifies attention, memory errors, inadequacy of processes, and documentation as key causal factors. The study also provides insight on how situational awareness (SA) interventions may contribute to the reduction of maintenance deviations while at the same time capture hidden causal factors.

2.4. Research Gap

There is a clear difference in specific results in the previous studies, along with similarities. These differences depending on data sets used and are likely real features of the corresponding populations. Previous work has looked at collections of all safety occurrences (accidents and incidents), subsets (just incidents etc.), and applications to specific segments of the aviation industry (GA, etc.). Given that the annual ICAO Safety Reports are published by the international body for air transport, it is important to understand accidents with maintenance contributions in this specific data set. Not only can statistically significant features be identified, trends over time can be analyzed to show how accidents with maintenance contributions in scheduled commercial air transport operations, utilizing large transport category aircraft, have evolved.

3. Materials and Methods

3.1. Research Design

This study utilized a mixed method approach, specifically an exploratory design, commencing with a qualitative content analysis to provide data for a quantitative ex-post facto study [23]. In this approach, the categorical data from the accidents with maintenance contributions were extracted, and then the narratives were coded further to generate additional categorical variables. Once the data was coded, it was then analyzed in an ex-post facto study, to analyze the distributions in comparison to all aviation accidents to assess if any observed differences were statistically significant.

3.2. Data Collection, Coding, and Cleaning

The primary source of data for this work was the set of ICAO official accidents [24]. This list of official accidents includes those used in the ICAO annual safety reports and is made up of all safety occurrences that are accidents, in a scheduled commercial operation, and investigated by the relevant national authority. Entries in the ICAO official accident dataset were cross referenced with the Aviation Safety Network (ASN) to provide narratives for the accidents with maintenance contributions. While the ASN database (provided by the Flight Safety Foundation [25]) includes maintenance as one of the contributing/causal factors (of which there are almost 100 cases), full text narrative searches of all entries in the ASN were undertaken. Advanced searches in both Google and Bing were conducted to the extent of the search engines, including omitted duplicates. The searches used criteria that limited the website searched (using "site:", to the ASN database) and the title of the results (using "intitle:", and "ASN Aircraft accident"). The additional search terms then used were "maintenance", "mechanic", "technician", electrician", "AME" (aircraft maintenance engineer), "LAME" (licensed aircraft maintenance engineer), "incorrect installation", "incorrectly installed", "inadequate inspection", "airworthiness directive", "service bulletin", and "inadequate maintenance". The general term "maintenance" was used last, as it returned the most results, and many of the cases were returned with the other more specific search terms, and hence if already selected were identifiable due to the hyperlink's color change.

Using the details and narratives from the ASN, the 35 ICAO official accidents were coded with:

- Maintenance issue,
- Type of operation,
- Operator's business model,
- Phase of flight (in which the maintenance issue first appeared)
- Age (difference between year of the aircraft's first flight and the year of the accident),
- Accident category (A1 a hull loss, or A2 repairable), and
- ICAO occurrence categories.

The maintenance issues used by ASN to code these accidents include:

- Repair of previous damage,
- Engine issue,
- Failure to follow airworthiness directives or service bulletins,
- Wrong or incorrect installation of parts, and
- General issues (substandard practices etc).

The ICAO data already includes the number and type of engines. The range of mass categories used in the ICAO official accident dataset was insufficient, so a lookup-table was created for all the aircraft (1277, removing duplicates), including the weight (MTOW) and manufacturer (manufacturer was coded as the current active company responsible for the type, e.g., the DHC-8 is coded as Bombardier). These codes were then added to the records for all the accidents (not just the maintenance accidents). Another lookup-table was created for all the three-letter country codes, to give continent and ICAO region. These were used to code the region of the accident and the region of the operator.

The final data set utilized the downloaded spreadsheet from ICAO [24]. Additional columns were added to this spreadsheet; including: MTOW, manufacturer, the operator's business model and the type of operation, the two regions (operator and occurrence), the phase of flight where the maintenance issue occurred, the date separated (day, month, and year), the year of first flight (and then the difference between that and year of accident giving the aircraft age), the identified maintenance issue (and associated system and sub-system), and the separated occurrence categories (downloaded as a single comma separated text string for each accident) giving a sample size of 60 due to the fact that accidents can be coded with multiple categories.

3.3. Data Analysis

3.3.1. Non-Parametric Analysis

For RQ2, given the small sample size, 35 accidents with maintenance contributions, testing was limited to utilizing Fisher's exact test [26]. This type of testing with small sample sizes has previously been utilized in other post-accident analyses [27]; as with that work, Fisher's exact tests were undertaken in MATLAB to determine the *p*-values. Specifically, in the testing of the various characteristics or features coded, the observed (O) data were the 35 accidents with maintenance contributions, while the expected (E) are those of the complete set of 1277 official accidents. The statistical hypotheses to be tested are given as

$$H_0: \quad P_{O,n} = P_{E,n}$$
$$H_A: \quad P_{O,n} \neq P_{E,n}$$

where P is the proportion of the n'th category. That is, n separate Fisher's exact tests were conducted.

3.3.2. Longitudinal Analysis

For RQ3, correlation was used to determine if any statistically significant trend existed over the 12 years of the study. Pearson's correlation coefficient (r) is a measure of association between two interval or ratio variables. In this case, the number of accidents and the year. The statistical hypotheses to be tested are given as

$$H_0: \quad r = 0$$

$$H_A: \quad r \neq 0$$

This test is assessing if there is an association between the number of accidents in each year and the year. A secondary question can be posed here, and that is, if the accident counts have reduced, has the rate of reduction for maintenance accidents reduced at the same rate or a different rate to all accidents? This requires two simple linear models

$$n_{all} = \alpha_1 + \beta_1 t \tag{1}$$

and

$$n_m = \alpha_2 + \beta_2 t \tag{2}$$

where α_1 and β_1 are the model coefficients to predict the count for all official ICAO accidents (n_{all}), while α_2 and β_2 are the model coefficients to predict the counts for accidents with maintenance contributions (n_m), and t is time in years. Assessing the difference in the rate of change requires a combined model, given as

$$n_\Delta = n_m - pn_{all} = \alpha_3 + \beta_3 t \tag{3}$$

where the new 'dependent variable' is given as the relative difference in the two accident groups (p is the proportion of all accidents which are maintenance related and approximately 35/1277). That is, multiplying n_{all} by p gives a count relative to 35 instead of 1277, which means they are on the same scale. If this new count, or more correctly, the difference in the count (n_Δ), diverges ($\beta_3 \neq 0$), then the rates are not equal ($\beta_1 \neq \beta_2$). These can be expressed as

$$H_0: \quad \beta_3 = 0 \quad \text{or} \quad \beta_1 = \beta_2$$

$$H_A: \quad \beta_3 \neq 0 \quad \text{or} \quad \beta_1 \neq \beta_2$$

A subtle difference which needs to be mentioned is that since two dependent variables are being regressed against the independent variable, then the degrees of freedom are the number of observations (12) subtract 3, not the 2 associated with simple linear regression.

Logistic regression is needed to answer RQ4. This is because both fatalness and fate of the airframe (accident category) are both dichotomous variables (fatal or not, hull lost, or not), while the aircraft age is a continuous variable. Logistic regression is the ideal tool to measure association between a dichotomous dependent variable and a continuous independent variable. The statistical hypotheses to be tested are given as

$$H_0: \quad \beta = 0$$

$$H_A: \quad \beta \neq 0$$

where β is the variable in the fitted logit function, that relates the continuous variable to the dichotomous output. The logit has the form,

$$\pi(x) = \frac{e^{\alpha+\beta x}}{1 + e^{\alpha+\beta x}}. \tag{4}$$

Here, π is the estimated probability of the dichotomous outcome at the given predictor level (x), in this case, age of the aircraft.

4. Non-Parametric Results and Analysis

4.1. Summary of Results

Table 1 shows the results for the Fisher's exact tests comparing accidents with maintenance contributions to all accidents in the ICAO official accident data set. The characteristics that are statistically significant include region of occurrence (ROc), phase of flight (PoF), occurrence categories (OC), and manufacturer (Manu). The characteristics which were not statistically significant are the fatalness (Fat), region of operator (ROp), maximum takeoff mass (Mass), engine types (ET), and number of engines (nE). Table 2 shows the Fisher's exact test results for those characteristics which were tested against a uniform expected distribution. Here, all but the operator's business model (BM) were statistically significant; including, maintenance issue (MI), system/component issue (SC), the type of operation (Op), and the age of the aircraft at the time of the accident in decades (Age). Each of these characteristics will be discussed further below.

Table 1. Fisher's exact test, comparing characteristics of accidents with maintenance contributions to all accidents, in the ICAO official accident data set.

	Fat	ROp	ROc	PoF	OC	Manu	Mass	ET	nE
p	0.28	0.58	0.04	0.03	~0	0.02	0.93	0.84	0.59
Conc	N	N	Y	Y	Y	Y	N	N	N

Table 2. Fisher's exact test, to determine if characteristics of accidents with maintenance contributions are different to an expected uniform distribution.

	MI	SC	Gear	BM	Op	Age
p	0.01	0.01	0.17	0.31	0.02	0.03
Conc	Y	Y	N	N	Y	Y

4.2. Comparative Data

4.2.1. Fatalness

Figure 1 shows the distribution of fatal to non-fatal accidents. For all the official ICAO accidents, 14.7% resulted in at least one fatality. For accidents with maintenance contributions this increased to 20%. As noted in Section 4.1, this difference is not statistically significant. The lack of statistical significance is due to the comparatively small sample size. That is, more than 1277 accidents have occurred in the aviation industry since 2008. However, specific criteria need to be met for an accident such that it is included in the ICAO official statistics. Of this population, the resultant sample of accidents with maintenance contributions that are fatal is too small to draw definitive conclusions. The conclusion of interest being that accidents with maintenance contributions appear to be more fatal than the 'average' accident. Further work is needed to confirm this conclusion.

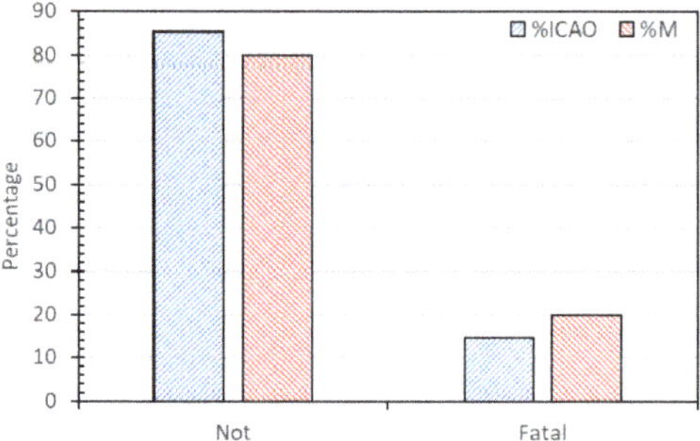

Figure 1. Distribution of fatalness for accidents with maintenance contributions.

4.2.2. World Region

While it might be worth considering an expected distribution for a region based on traffic numbers, comparing the distribution of regions for accidents with a maintenance contribution to the distribution of regions for all accidents removes sampling bias. That is, different countries may have different levels of reliability when it comes to ensuring relevant aviation accidents are shared with ICAO. By comparing a distribution of accidents to another distribution of accidents, this potential bias is removed. This is because in principle a country should be as likely to report an accident with a maintenance contribution as any other accident. Region is coded as both the region of the operator, and the region in which the accident occurred. Figure 2a shows the region of the operator (ROp) which was not statistically significant in Section 4.1. There is, however, a noticeable lack of North American maintenance accidents, with the Middle East and South and Latin America having slightly more than expected. Figure 2b shows the distribution of accidents for the region in which those accidents occurred (ROc). A similar trend can be seen, but more pronounced, hence the reason this test in Section 4.1 was statistically significant. The interesting result here is the spike for the Middle East. Looking at these cases there are no accidents associated with the big three Middle Eastern airlines (Emirates, Etihad, and Qatar); the cases are two Iranian, a Sudanese, and a charted aircraft operating for Saudi Arabia. The other noteworthy difference is the lack of accidents in Africa; previous work looking at human factors (HF) related accidents showed a greater prevalence of these accidents in Africa [28].

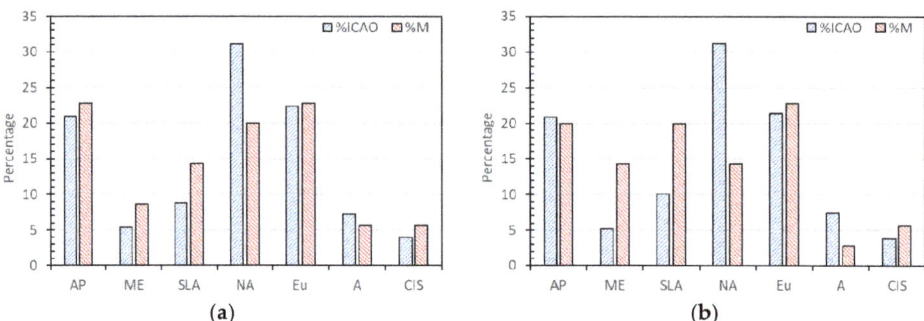

Figure 2. Distribution of world region for accidents with maintenance contributions: (**a**) region of the operator (ROp); (**b**) region where the accident occurred (ROc).

4.2.3. Phase of Flight

Figure 3 shows the distribution of accidents with maintenance contributions by phase of flight. The reason for recoding the phase of flight from the original ICAO codes is that a number of maintenance related issues manifest themselves prior to causing 'the accident'. For example, gear related issues manifest themselves at takeoff (gear up) or approach (gear down), but the 'accident' is considered to have occurred upon landing. When the phase of flight as coded by ICAO was utilized (Figure 3a), no statistically significant result is observed (not included in Section 4.1), although there is a slight excess of landing accidents, again expected by the fact that maintenance issues are most commonly associated with the landing gear. When recoded (Figure 3b), the results are statistically significantly different (as shown in Section 4.1). As hypothesized, there are more accidents during climb; however, there is no increase during takeoff. Again, during climb (ICL) is when the gear is retracted, and this spike corresponds to these failures occurring at that time.

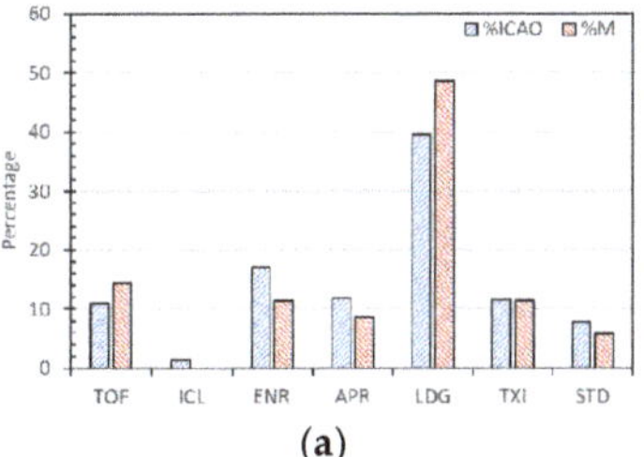

(a)

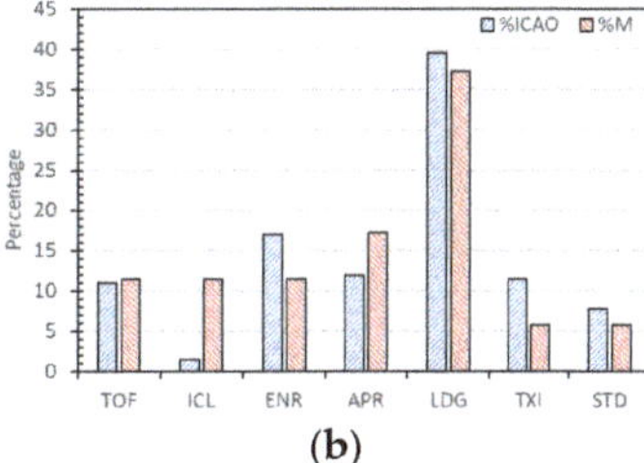

(b)

Figure 3. Distribution of phase of flight for accidents with maintenance contributions: (**a**) the phase in which the accident occurred (end result); (**b**) the phase in which the maintenance issue manifested.

4.2.4. Occurrence Category

Any occurrence category that resulted in less than 5% of the total were grouped into an 'other' category. The results showing the associated occurrence categories in Figure 4 are not surprising, with system component failure non-powerplant (SCF-NP) the most common, followed by system component failure powerplant (SCF-PP). While in the ICAO official accident statistics SCF are common, they are more common in accidents with maintenance contributions. The increase in SCF occurrences is balanced by slightly less than expected abnormal runway contact (ARC) and runway excursions (RE); more significant is the lack of 'other' for accidents with maintenance contributions, and the fact that 'other' for ICAO official accidents includes turbulence (TURB) which is second most common cause in the ICAO official accident statistics.

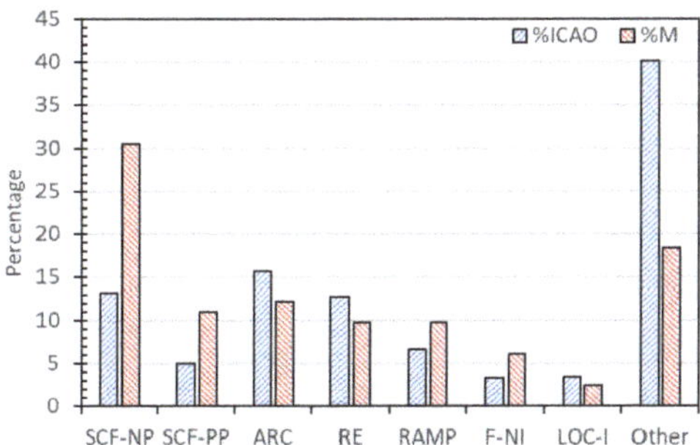

Figure 4. Distribution of occurrence categories for accidents with maintenance contributions.

4.2.5. Manufacturer

The distribution of accidents by manufacturer is shown in Figure 5. The mode for both distributions is clearly Boeing, which is expected based on the number of Boeing, McDonnell, and Douglas aircraft that have been in operation between 2008 and 2019. There are, however, fewer Airbus accidents with maintenance contributions relative to all official accidents. The key significant difference is for Ilyushin aircraft, where the percentage of maintenance accidents is 10 times the percentage of all accidents.

This result mirrors previous research into human factors accidents [28], where accidents with Russian aircraft were more likely to be involved in those accidents. Interestingly, Antonov had no maintenance accidents while it had 43 other accidents in the ICAO official data set.

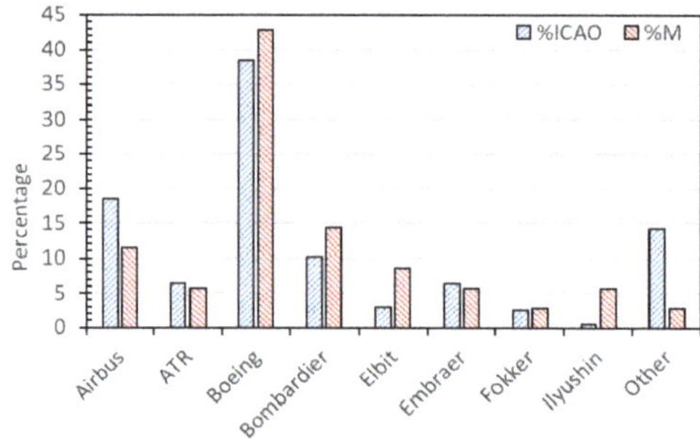

Figure 5. Distribution of aircraft manufacturer for accidents with maintenance contributions.

4.2.6. Mass Category

The ICAO Accident/Incident Data Reporting (ADREP) taxonomies code aircraft size as the mass category. These include maximum takeoff masses of (1) less than 2.25 tonnes, (2) 2.25 tonnes to 5.7 tonnes, (3) 5.7 tonnes to 27 tonnes, (4) 27 tonnes to 272 tonnes, and (5) above 272 tonnes. Given ICAO Official Accidents only include aircraft above 5.7 tonnes, the first two categories were omitted in the results. The distribution of accidents by mass category is shown in Figure 6. The mode is 3 (medium aircraft), and this mass category includes the majority of narrow body (single isle) large transport category aircraft, and the smaller wide body (twin isle) large transport category aircraft. The distributions are almost identical, and the test result was not statistically significant.

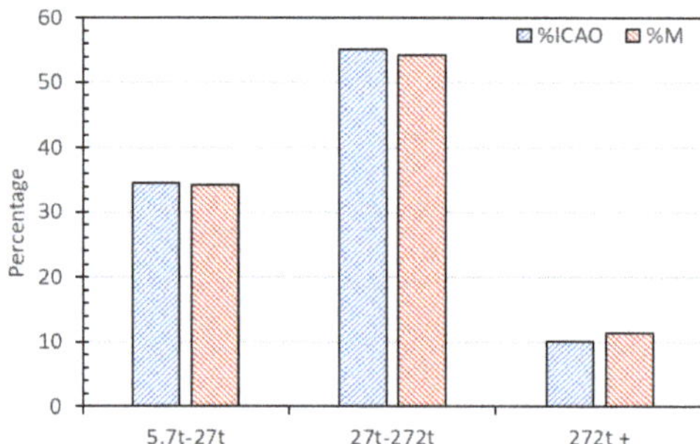

Figure 6. Distribution of aircraft mass category for accidents with maintenance contributions.

4.2.7. Engines

The ICAO ADREP taxonomies use multiple categories to differentiate the most common engine types used on aircraft. However, the ICAO official accidents code engines as reciprocating (piston), turboprop, and jet (which captures turbofan and turbojet engines). From Figure 7a, the most common engine in accidents (both maintenance and all) is jet. For the number of engines, Figure 7b, a twin engine aircraft is typically involved in accidents. This is because twin engine aircraft are the most common. For both engine type and engine count, there was no statistically significant difference between all accidents and those with maintenance contributions.

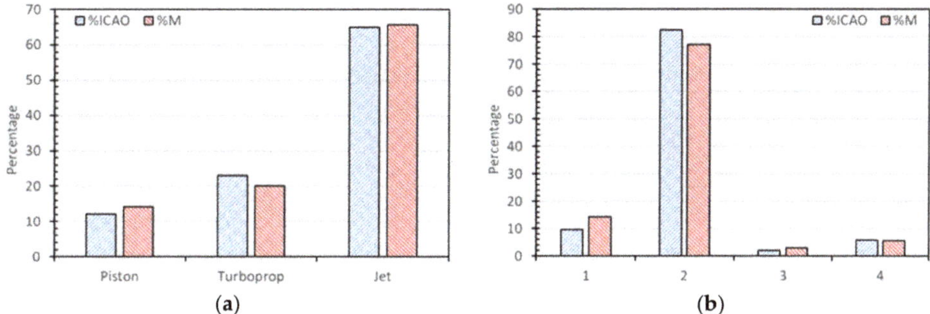

Figure 7. Distribution of aircraft engines for accidents with maintenance contributions: (**a**) the type of engine; (**b**) the number of engines.

4.2.8. Maintenance Issue

From Figure 8a, 'general' (substandard practices, insufficient maintenance, qualification, training, etc.) has the highest count, and the observed variation is statistically significant. Engine and part issues are as expected, while 'AD/SB' (failure to follow airworthiness directives or service bulletins) and 'PD' (repaired previous damage), have low counts. Figure 8b shows the distribution of system/component involved in accidents with maintenance contributions. The statistically significant result indicates that the landing gear is involved in 47% of accidents, with engine is in 26%. Given the unexpectedly high count for gear related issues, a further sub-code of gear issue was created for these 16 accidents (Figure 8c) where mechanical/structural/physical issues were identified as the most common.

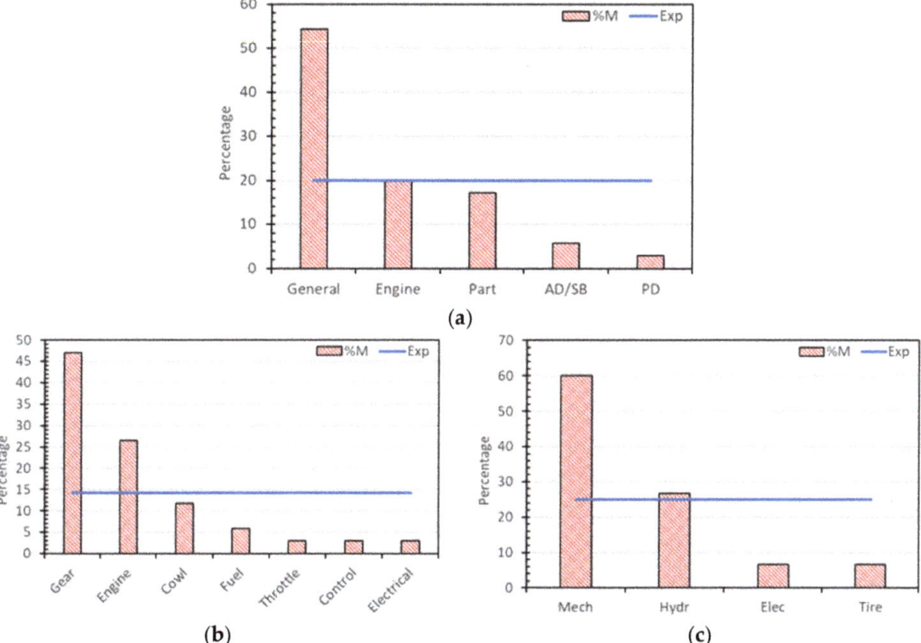

Figure 8. Results for the maintenance and technical issues for accidents with maintenance contributions: (**a**) the broad maintenance issue; (**b**) the system or component involved in the maintenance issue; (**c**) the type of gear fault.

4.2.9. Operator and Operation

While in Figure 9a, there is a spike for FSNCs, this is not statistically significant. In fact, if we compare to a previous study that investigated business models in HF accidents [28], the distribution of accidents with maintenance contributions has a similar shape, and hence with a limited sample size, this would also not produce a statistically significant result. For the type of operation or service, the result is statistically significant, with the peak value for domestic and the minimum value for charter responsible for this. It should be noted that ICAO indicates globally that 60% of revenue passenger kilometers are currently international [29], leaving 40% for domestic (including regional). However, we can clearly see that accidents with maintenance contributions are more likely to occur in domestic operations, so relative to traffic, this becomes even more significant.

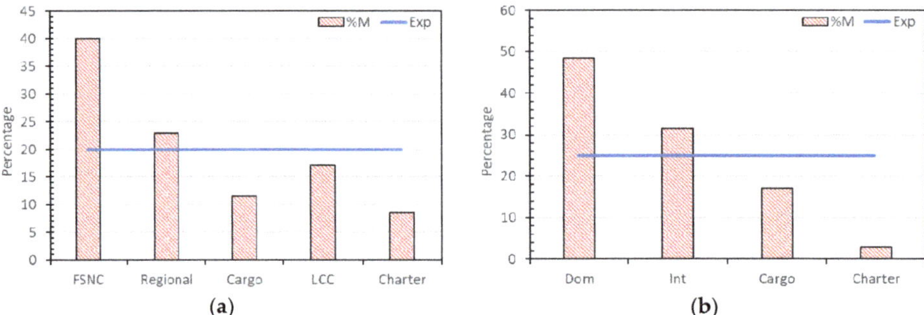

Figure 9. Results for aviation aspect of the accidents with maintenance contributions: (**a**) the operators business model; (**b**) the type of operation/service.

4.2.10. Age

Figure 10 shows the distribution of accidents with maintenance contributions by age. The mode is 10 to 20 years old. The interesting feature is the small count for aircraft that are 30 to 40 years old, and the data set included no accidents with maintenance contributions that were older than 40 years. Looking at the airlines involved, and consulting Airfleets.net [30], the average fleet age was (17.5 ± 4) years, which agrees with the average age for the 35 aircraft given as (19 ± 3) years. As such, the likely distribution coincides with the distribution of raw aircraft age (the ages of all aircraft in service).

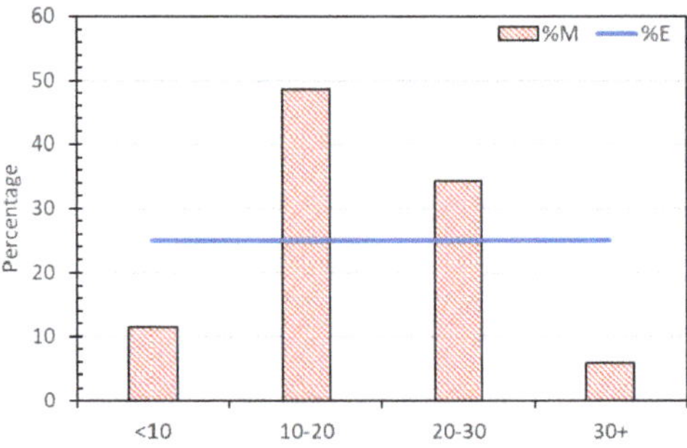

Figure 10. Results for the age of aircraft involved in aircraft maintenance related accidents.

5. Parametric Results and Analysis

5.1. Longitudinal Study

Table 3 shows the results for the various correlation and regression tests. The variables tested included:

1. ICAO, the total number of ICAO official accidents,
2. M, the number of ICAO official accidents with maintenance contributions,
3. M%, the percentage of accidents with maintenance contributions relative to all ICAO official accidents,
4. pICAO, the proportion of ICAO official accidents (relative to 2008),
5. pM, the proportion of accidents with maintenance contributions (relative to 2008), and
6. The model given by (3) above.

Table 3. Correlation and regression test results for the various longitudinal factors.

	ICAO	M	M%	pICAO	pM	Model
β	−2.77	−0.29	−0.14	−0.02	−0.05	−0.21
r^2	0.24	0.42	0.10	0.24	0.42	0.22
n	12	12	12	12	12	12
v	10	10	10	10	10	9
t	5.58	8.45	3.29	5.58	8.45	4.73
p	<0.01	<0.01	0.01	<0.01	<0.01	<0.01
Sig	Y	Y	Y	Y	Y	Y

All of the tests were statistically significant. The results for the proportions match the respective raw count results, which is to be expected. However, the proportion results enable the β values to be directly compared, since they are relative counts. We note that maintenance accidents appear to have reduced over time 2.5 times faster than the rate of reduction in all ICAO official accidents. The statistical significance of this is supported by the results for the model given by (3). Figure 11 shows the annual accident counts for the accidents with maintenance contributions and all of the ICAO official accidents.

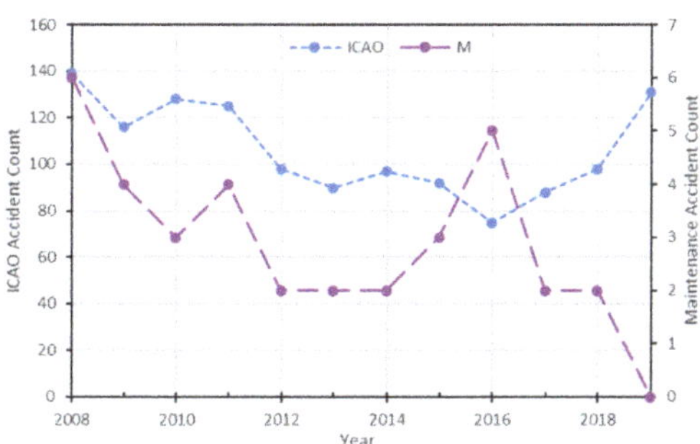

Figure 11. The trends over time for accidents with maintenance contributions (secondary axis, purple line with large dashes) and all ICAO official accident (blue line with small dashes).

5.2. Logistic Regression Results

Research question 4 (RQ4) was posed to test the hypothesis that older aircraft are 'more dangerous' than newer aircraft. Clearly the context of this statement is only with regards to those accidents with a maintenance contribution. The two aspects here (fatalness and aircraft damage) and their relationship to age can be assessed with logistic regression. Figure 12a shows the resultant logit fitted to the fatalness data (0 = non-fatal, 1 = fatal), and the logit predicts that older aircraft are statistically significantly more likely to be associated with a fatality (McFadden's pseudo $R^2 = 0.11$, $\chi^2 = 3.97$, $p = 0.046$); to be clear, this is just statistically significant, and the R^2 suggests only 11% in the variation of fatalness is due to the age of the aircraft, and the odds are even for an aircraft that is 34 years old. The limitation here is the small sample size, giving only a few fatal outcomes. Looking at the outcome for the aircraft (Figure 12b), the logit now much more clearly increases with age as hypothesized (McFadden's pseudo $R^2 = 0.30$, $\chi^2 = 14.6$, $p < 0.01$). That is, about 30% of the variation in the aircraft damage outcome is predicted by the aircraft's age, and the odds are even for an aircraft that is 18 years old.

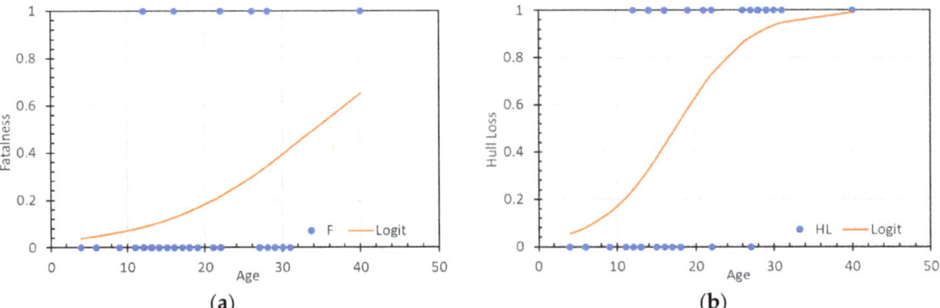

Figure 12. Results for the logistic regression investigating the effect of aircraft age on: (**a**) fatalness of maintenance accidents (0 = no fatalities, 1 = fatalities); (**b**) aircraft damage from maintenance accidents (0 = aircraft repaired, 1 = aircraft damaged beyond repair = hull loss).

5.3. Aircraft Maintenance Fraction

The Australian Transport Safety Bureau's report into HF in aircraft maintenance [31] states that "precise statistics are unavailable" for the proportion of aviation accidents and incidents that are the result of improper maintenance. This is the inspiration for RQ1. While it would also be interesting to know the proportion of aviation incidents that have a maintenance contribution or causation, the ICAO official accident data is exactly that, only accident data. We have a database that contains n number of accidents each year, and we have a dataset that contains m number of accidents with maintenance contributions each year. Figure 13 shows this annual percentage of ICAO official accidents which have maintenance contributions.

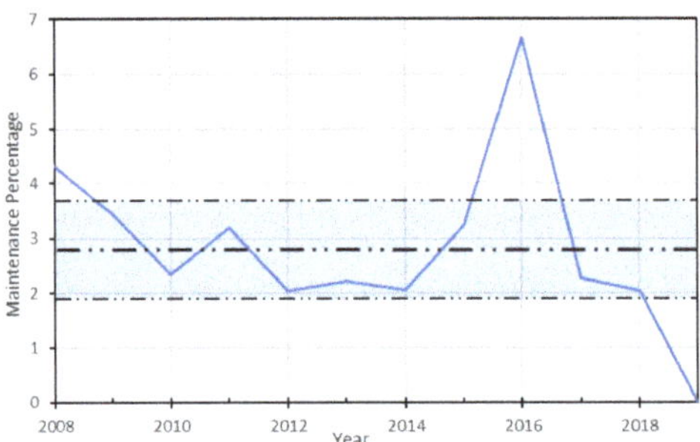

Figure 13. Percentage of all accidents in the ICAO official accident data set that have been identified with a maintenance contribution, 2008 to 2019. The shaded band illustrates the confidence interval of the average, (2.8 ± 0.9%).

Annually, the average percentage of accidents with maintenance contributions is given as (2.8 ± 0.9%). However, the result of the regression test for this was statistically significant, in terms of a reduction by 0.14% per year. That is, on average, this percentage is reducing by 1 every 7 years.

The peak in 2016, and then the lack of an accident in 2019 suggests this reducing trend is an artifact of the small sample size. To conclude, approximately 3% of accidents are the result of aircraft maintenance.

6. Case Study

6.1. Rational

The results of the quantitative analysis produce findings that lack of context. While a trend over time can be shown to be statistically significant, further questions involving question of 'why' cannot be answered. Using the results of the quantitative analysis, a 'typical' case, however, can be identified. A case study of 'typical' accidents provides contextualization and helps to explain why the observed quantitative results exist. While a perfectly 'average' case would be ideal (embodying all modes), finding such as case is unrealistic. Even a study of "the average man" [32], with normally distributed anthropometry characteristics allowing average to mean within 0.3 standard deviations, after 10 characteristics out of 132 [33], there was no "average man" [34]. As such, we consider an indicative case here; this case is non-fatal, occurred in Europe with a European FSNC operator, during climb, that was a system component failure non-powerplant, for a twin engine jet aircraft in mass category 4, and is a general maintenance issue. The case is not Boeing (it is Airbus), not domestic (it is international), and is to a gear issue. That is, there are 10 'average' or important characteristics and 3 that are not. The specific case involves a failure to follow Aircraft Maintenance Manual (AMM) procedures and an inadvertent aircraft swap by maintenance technicians led to the A319 aircraft, shortly after take-off, experiencing a detachment of fan cowl doors leading to hydraulic loss, fuel leak, and an engine fire.

6.2. The Flight

On 24, May 2013 the A319 aircraft departed from Heathrow for Oslo, Norway. As the aircraft rotated the fan cowl doors on both engines detached. The left engine was not damaged by the resulting debris; however, the right engine suffered damage from its detaching cowl door (Figure 14a). There was a subsequent loss of one hydraulic system and fuel also began leaking from the right engine. The loss of fuel from this engine was such that if the crew had not shut-down the engine on approach to land, the fuel supply to it would have been exhausted [35].

The crew requested and was cleared to return to Heathrow with an initial PAN call. On approach, the crew made the decision to shut-down the right engine. Shortly after this decision was made a fire warning for the right engine sounded. The crew's response to this warning was to, "quickly shut down the right engine and discharged the first fire extinguisher bottle" [35]. A mayday call was made and after the required wait time, the second fire bottle was fired into the right engine.

After landing and coming to a halt on the runway, ATC informed the flight crew that there were flames visible in the right engine. Fire services arrived quickly to the aircraft and commenced fighting the fire, and an initial decision was made not to evacuate the aircraft. However, this was quickly changed to a request for the left engine to be shut down and a cabin to be evacuated from the left side of the aircraft. This resulted in no fatal or serious injuries to crew or passengers [35].

6.3. Maintenance of the Accident Aircraft

The aircraft had been inspected by two maintenance technicians, through the night, prior to the accident flight. One of the required tasks was to lift the fan cowl doors and inspect the oil level of the integrated drive generator (IDG). The inspection revealed the oil for the IDG on each engine needed to be replenished. As the technicians did not have the requisite equipment or oil for this task they decided to continue to complete the checks for the aircraft, move to inspect other aircraft, and return to the accident aircraft later in the shift after they had retrieved the correct equipment and oil.

In contravention of the AMM procedures, the technicians lowered the fan cowl doors and left them "open, resting on the hold-open device (depicted in Figure 14b). (With the) hook re-engaged with the latch handle, (therefore the) cowling (was) not locked" [35]. In effect, the cowl doors were like the Grand Old Duke of York's men, neither up on telescopic struts nor down and locked, rather they were held somewhere in between. The technicians saw this as a common practice, albeit one that was not in the promulgated procedures. The investigators found that other technicians in the company saw this as a common practice. The rationale was that by leaving the doors open it was a hazard to staff.

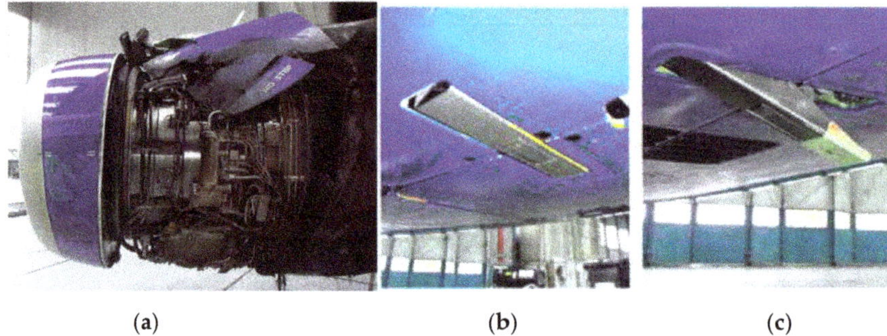

(**a**) (**b**) (**c**)

Figure 14. (**a**) Resultant damage to the right engine, with the cowling detached; The difference in the cowl fully locked (**b**) and just latch but not locked (**c**) [35].

6.4. *Drift into Failure*

This process is described by Dekker [36] as drift into failure, where workers make changes to practices so as to help them achieve the end goals. The new practice gets accepted by the workforce, it becomes "normal" and as the practices are the new normal, the workers do not see a need to report their diverging actions [37]. This drift of practice away from promulgated procedures is not necessarily a precursor to an accident. As Dekker [38] writes,

> *"not following procedures does not necessarily lead to trouble, and safe outcomes may be preceded by just as (relatively) many procedural deviations as those that precede accidents".*

The aircraft manufacturer, after the event, surveyed operators of the A32X series of aircraft and found that 69% of fan cowl loss events followed the opening of the cowl for the checking or replenishing of the IDG oil levels [35].

The investigators of the accident flight discovered how widespread the practice of leaving the cowl doors on hold-open device was when, in interviews with other technicians, it was suggested that 70% of staff followed this non-promulgated procedure [35]. The normality of this practice was further reflected in an examination of the operator's internal audits over the years 2011–2013 which found no reports of this practice deviation of not complying with the AMM procedures.

6.5. *Aircraft Swap*

After progressing with their planned course of action to complete checks on other aircraft, the maintenance crew had a break from work then proceeded to obtain the required equipment and supplies to replenish the IDG oil levels of the accident aircraft. The two technicians then went to another aircraft believing it was the aircraft they intended to service; it was, however, the wrong aircraft. They were surprised to note that the cowl doors were closed and latched and thought that perhaps other staff had locked the cowl doors. They decided to check all was well and opened the cowl doors and noted that the IDG oil levels on both engines were at acceptable levels. They rationalized that the oil levels had risen to acceptable levels as the engine had cooled. So they correctly closed up the cowl doors and verified each other's work [35], on the wrong aircraft.

This is an example of an aircraft swap error—that is, "required maintenance being carried out on an incorrect aircraft" (p. 78) [35]. The investigators found aircraft swap errors were "an occasional, infrequent occurrence" (p. 78) [35]. In having the correct plan of returning to the accident aircraft to complete the service but incorrectly carrying out the plan by going to the wrong aircraft the technicians unsafe act was a slip, rather than a mistake [39]. The outcome of the normalized changes to promulgated procedures when opening cowl doors, along with the aircraft swap error was that the accident aircraft was dispatched to service with the cowl doors still on the hold-open device and not fully latched.

6.6. Active Failures and Latent Conditions

Reason's [39] first description and illustration of the now ubiquitous "Swiss-Cheese" model, had five "planes". These planes were defenses against accidents occurring. The first four planes did not have holes through them. The last two planes interacted with local events and along with the limited window of opportunity afforded by a hole in the last planes an accident could occur. A further refinement by Reason of his model was published in 1997. He saw that to understand accidents there needed to be three elements, hazards, defenses, and loses. The defenses had latent conditions arising from organization factors as well as local workplace factors.

The defenses the accident organization has in place against an event such as the cowl doors not being fully latched involved a member of the flight crew and a ground crew worker conducting separate visual checks of the aircraft including the latching devices. Unfortunately, workplace failures punched holes in this localized defense. Neither the co-pilot on his walk around inspection nor the tug driver on his inspection noticed the locking devices protruding below the cowls. The latent condition that contributed to the lack of visual recognition of the unsafe condition of the cowl doors was the positioning of the latching devices being close to the ground and not easy to see. To visually check these devices, the workers were required to be on hands and knees on the ground [35], an option that obviously did not appeal to the co-pilot or tug driver.

One of the 'planes' in Reason's (1990) original iteration of the Swiss Cheese model was labeled "fallible decisions", Kourousis, et al. [40] identified the increased defenses the manufacturer of the A32X series of aircraft inserted into the safety system in the aftermath to this accident in an effort to reduce or eliminate the occurrences of cowl doors not being fully latched at take-off. Mandated modifications included new hardware and new procedures. However, the authors noted the potential latent failings that could arise from these newly implemented modifications [40]. These latent conditions arising from fallible decisions made by people who are not proximal to the accident may hinder or even work directly against the desired effect of reducing the number of cowl door incidents the mandated modifications are seeking.

7. Discussion

7.1. Findings

For RQ1 we note that on average aircraft maintenance contributes to (2.8 ± 0.9%) of all accidents. This rate appears to be showing a slight but statistically significant decrease with time, which if it was to continue suggests in 20 years maintenance could consistently contribute to no accidents; although this is an ambition statement being well outside the predictive capability of this simple longitudinal analysis. Of these maintenance accidents, the properties of note are the prevalence of general maintenance issues, such as inadequate maintenance and slips and lapses (like those associated with failing to latch engine cowls). The small number of AD/SB and previous damage accidents suggests when mandatory and completed, maintenance activities are effective in ensuring an aircraft is safe and airworthy. Looking at the systems involved, it was noted that issues with the landing gear accounted for almost half (47%) of cases; of these, 60% were due to structural/physical issues (not hydraulic or electrical issues). This is not surprising given the large mechanical loads placed on the landing gear.

There are a number of interesting characteristics of accidents with maintenance contributions relative to all ICAO official accidents. While the fatalness was slightly higher, 20% relative to 14.7%, this difference was not statistically significant. In terms of world regions, there was no statistically significant difference for the region of operator. However, for the region of the occurrence, there is a statistically significant difference. This is driven by a spike in accidents in the Middle East and South/Latin America, and a lower than expected count in North America. For phase of flight, a spike of engine related maintenance issues during initial climb resulted in a statistically significantly different distribution, one of which is further highlighted in the case study presented. The results for occurrence categories are as expected; an excess of system component failures, and a lack of 'environment' related occurrences. Of note here is that runway related occurrences are as expected. In terms of aircraft properties (number of engines, type of engine, and MTOW), there was no difference between accidents with maintenance contributions and all ICAO official accidents. However, for the manufacturer, the relatively accident free Ilyushin aircraft showed a spike in accidents with maintenance contributions that resulted in a statistically significant difference in the distribution of accidents by manufacturer. The final aircraft property considered was age, the results for which suggest that an aircraft between 10 and 20 years old is more likely to have an accident with a maintenance contribution, and very old aircraft (those over 30 years) were the least observed to be involved in accidents with maintenance contributions. In terms of the commercial operation and operator, there was no statistically significant difference for the operator, but domestic scheduled services did show a statistically significant count, again likely due to the volume of domestic traffic.

In response to RQ3, it is noted that all accident counts reduced over the period of the study. The relatively significant reduction of accidents with maintenance contributions since 2008 is promising. The fact that these accidents have reduced 'faster' than 'average' is also promising. Finally for RQ4, it was noted that while accidents with maintenance contributions occur less often as an aircraft becomes older (given older aircraft are less likely to be utilized in commercial air transport for economic reasons), the outcomes of those limited accidents tends to be worse, with the odds of both a fatal outcome and the aircraft being written-off increasing with age.

The undertaken case study highlights that maintenance issues are not exclusively an issue of budget conscious LCCs in commercial air transport, operating third tier aircraft. The fact that an indicative case involves a legacy airline (British Airways), and one of the two work horses of high capacity narrow body operations (Airbus A32X), highlights that maintenance can contribute to accidents across the aviation industry. The case also highlights that simple slips and lapses can result in an accident at great cost to the operator. So while significant improvements were implemented in the 1990s, there is still a lot of work that needs to be done to eliminate maintenance contributions to accidents in scheduled commercial air transport, and if not eliminate, consistently result in zero cases per year.

7.2. Assumptions and Limitations

The key limitation of this work is the small size of the data set. There is limited statistical power associated with small data sets. Of note here is the just statistically significant result of the odds of a fatality based on aircraft age. A larger dataset would help to either confirm of disprove this. Similarly, the data set size has influenced the trend in percentage of maintenance accidents over time, with significant 'noise' in the last three years.

It could also be argued that the lack of information about incidents, which are far more common than accidents, is a limitation. It should, however, be noted that accidents result in significant damage or injury, even hull loss or death, unlike incidents. As such, research in accidents is arguably more important in aviation safety.

The use of uniform expected distributions for some of the categorical variables is also a limitation. For the maintenance issue and the systems/components involved the goal is to simply assess if one of these is statistically speaking more likely than the others. In contrast, for the operator (business model) and operation (type of service), these would benefit from a non-uniform expected distribution. With time and effort, these codes could be created for all 1277 accidents in the ICAO official dataset. The additional insight gained from this could be useful or limited.

It would be ideal to analyze the dataset looking for covariances between the categorical variables considered. This would ideally help identify latent classes in the data (combinations of variable values that are more likely to occur together, and hence present a greater safety risk, e.g., an Ilyushin cargo aircraft in the Middle East). However, the limited number of accidents with maintenance contributions means that performing cross tabulations for these would result in such small counts that the associated Fisher's exact test would likely yield no statistically significant results.

7.3. Future Work

Future work will utilize all the collated maintenance accidents from the ASN database (of which there are approximately 360, spanning 1940 to 2019), to see if all accidents with maintenance contributions are different to those captured in the official ICAO accident dataset. This will also require a comparison to the ASN dataset as a population, rather than the ICAO dataset. This will enable all of the properties in this study to be expanded beyond the scheduled commercial large transport category aircraft operations captured by the ICAO official accident statistics. Many of the limitations will be overcome with this larger data set, presenting even further opportunities for future work.

The other follow up question that remains unanswered is the proportion of incidents with a maintenance contribution or causation. The hypothesis is that a much higher percentage of incidents could be the result of maintenance issues, which are rectified before they become accidents. Many readers will have experienced either personally or professionally a delay at the gate because of 'technical issues'. While maintenance personnel are clearly fixing technical issues, the topic of this work demonstrates that they also cause other technical issues, but how many? To answer this question requires an all-encompassing dataset with narrative information available to search for maintenance issues.

8. Conclusions

This work has investigated official ICAO accidents with maintenance contributions. The use of an exploratory research design enabled interesting features of these accidents to be identified (during the qualitative phase) and studied further (in the quantitative phase) to understand how the values of these variables are distributed and vary relative to all accidents. Maintenance was found to have contributed to approximately 3% of all accidents and resulted in slightly more accidents with a fatality (at least 1); that is, 20% of accidents with maintenance contributions ended with a fatality, while in the ICAO official accident data, only 15% were fatal. Relative to all accidents, the number of accidents with maintenance contributions was also found to be reducing at a greater rate over the 12-year span of the data used in this study. That is, accidents with maintenance contributions are reducing at a rate of 5% per year, while the rate at which all accidents is reducing is 2% per year. Finally, the effect of aircraft age was quantified. Accidents with maintenance contributions typically occur with an age of 10 to 20 years, and an average age of 19 years. Based on age, the outcome of the accident for crew and passengers (in terms of fatalness) and for the aircraft (in terms of a hull loss) were more likely to end badly. Specifically, for whether or not the aircraft was written off (damaged beyond repair), there was a significant trend showing that even odds occurred at 18 years; that is if an aircraft is older than 18 years when in an accident it is more likely than not to be written-off. For fatalness, even odds occurred at an age of 34 years, which means that if the aircraft involved in the accident was more than 34 years old, it was more likely than not to involve a fatality.

Author Contributions: Conceptualization, F.N.K., J.M. and G.W.; Data curation, G.W.; Formal analysis, J.M., G.B. and G.W.; Investigation, F.N.K., A.A., J.M., G.B. and G.W.; Methodology, G.W.; Project administration, G.B.; Resources, A.A. and G.W.; Supervision, G.B.; Visualization, G.W.; Writing—original draft, F.N.K., J.M. and G.W.; Writing—review & editing, A.A., G.B. and G.W. All authors have read and agreed to the published version of the manuscript.

Funding: This research received no external funding.

Conflicts of Interest: The authors declare no conflict of interest.

References

1. EASA. *Official Journal of the European Union: Commission Regulation (EU) No 1321/2014*; European Aviation Safety Agency: Cologne, Germany, 2014.
2. FAA. *Maintenance Programs for U.S.-Registered Aircraft Operated Under 14 CFR Part 129 FAA*; Federak Aviation Authority: Washington, DC, USA, 2009.
3. PCAA. *Approved Maintenance Organisations-Air Navigation Order*, 3rd ed.; Pakistan Civil Aviation Authority: Karachi, Pakistan, 2019.
4. Wang, L.; Sun, R.; Yang, Z. Analysis and evaluation of human factors in aviation maintenance based on fuzzy and AHP method. In Proceedings of the 2009 IEEE International Conference on Industrial Engineering and Engineering Management, Hong Kong, China, 8–11 December 2009; pp. 876–880.
5. Rajee Olaganathan, D.; Miller, M.; Mrusek, B.M. Managing Safety Risks in Airline Maintenance Outsourcing. *Int. J. Aviat. Aeronaut. Aerosp.* **2020**, *7*, 7. [CrossRef]
6. Bağan, H.; Gerede, E. Use of a nominal group technique in the exploration of safety hazards arising from the outsourcing of aircraft maintenance. *Saf. Sci.* **2019**, *118*, 795–804. [CrossRef]
7. Insley, J.; Turkoglu, C. A Contemporary Analysis of Aircraft Maintenance-Related Accidents and Serious Incidents. *Aerospace* **2020**, *7*, 81. [CrossRef]
8. Illankoon, P.; Tretten, P.; Kumar, U. A prospective study of maintenance deviations using HFACS-ME. *Int. J. Ind. Ergon.* **2019**, *74*, 102852. [CrossRef]
9. Lestiani, M.E.; Yudoko, G.; Purboyo, H. Developing a conceptual model of organizational safety risk: Case studies of aircraft maintenance organizations in Indonesia. *Transp. Res. Procedia* **2017**, *25*, 136–148. [CrossRef]
10. Dalkilic, S. Improving aircraft safety and reliability by aircraft maintenance technician training. *Eng. Fail. Anal.* **2017**, *82*, 687–694. [CrossRef]
11. Hobbs, A.; Williamson, A. Skills, rules and knowledge in aircraft maintenance: Errors in context. *Ergonomics* **2002**, *45*, 290–308. [CrossRef] [PubMed]
12. Floyd, H.L. Maintenance Errors as Cause for Electrical Injuries-What We Can Learn from Aviation Safety. In Proceedings of the 2019 IEEE IAS Electrical Safety Workshop (ESW), Jacksonville, FL, USA, 4–8 March 2019; pp. 1–6.
13. PeriyarSelvam, U.; Tamilselvan, T.; Thilakan, S.; Shanmugaraja, M. Analysis on costs for aircraft maintenance. *Adv. Aerosp. Sci. Appl.* **2013**, *3*, 177–182.
14. IATA MCTG. *Airline Maintenance Cost Executive Commentary*; International Air Transport Association: Montreal, QC, Canada, 2019.
15. Nelson, N.L.; Goldman, S.M. Maintenance-Related Accidents: A Comparison of Amateur-Built Aircraft to all Other General Aviation. *Hum. Factors Ergon. Soc. Annu. Meet.* **2003**, *47*, 191–193. [CrossRef]
16. Rashid, H.; Place, C.; Braithwaite, G. Helicopter maintenance error analysis: Beyond the third order of the HFACS-ME. *Int. J. Ind. Ergon.* **2010**, *40*, 636–647. [CrossRef]
17. Saleh, J.H.; Tikayat Ray, A.; Zhang, K.S.; Churchwell, J.S. Maintenance and inspection as risk factors in helicopter accidents: Analysis and recommendations. *PLoS ONE* **2019**, *14*, e0211424. [CrossRef]
18. Suzuki, T.; Von Thaden, T.L.; Geibel, W.D. Coordination and safety behaviors in commercial aircraft maintenance. In Proceedings of the Human Factors and Ergonomics Society Annual Meeting, New York, NY, USA, 22–26 September 2008; pp. 89–93.
19. Geibel, W.D.; Von Thaden, T.L.; Suzuki, T. Issues that precipitate errors in airline maintenance. In Proceedings of the Human Factors and Ergonomics Society Annual Meeting, New York, NY, USA, 22–26 September 2008; pp. 94–98.
20. Reynolds, R.; Blickensderfer, E.; Martin, A.; Rossignon, K.; Maleski, V. Human Factors Training in Aviation Maintenance: Impact on Incident Rates. In Proceedings of the Human Factors and Ergonomics Society Annual Meeting, San Francisco, CA, USA, 27 September 2010; pp. 1518–1520.
21. Ng, M.-W.; Li, S.Y. An analysis of aircraft maintenance incidents using psychological and cognitive engineering knowledge. In Proceedings of the Human Factors and Ergonomics Society Annual Meeting, Washington, DC, USA, 19–23 September 2016; pp. 1676–1680.
22. Schmidt, J.; Schmorrow, D.; Figlock, R. Human factors analysis of naval aviation maintenance related mishaps. In Proceedings of the Human Factors and Ergonomics Society Annual Meeting, San Diego, CA, USA, 30 July 2000; pp. 775–778.

23. Wild, G.; Murray, J.; Baxter, G. Exploring Civil Drone Accidents and Incidents to Help Prevent Potential Air Disasters. *Aerospace* **2016**, *3*, 22. [CrossRef]
24. ICAO. ICAO iSTARS API Data Service. Available online: https://www.icao.int/safety/istars/pages/api-data-service.aspx (accessed on 18 July 2020).
25. Aviation Safety Network. ASN Aviation Safety Database. Available online: https://aviation-safety.net/database/ (accessed on 18 July 2020).
26. Raymond, M.; Rousset, F. An Exact Test for Population Differentiation. *Evolution* **1995**, *49*, 1280–1283. [CrossRef] [PubMed]
27. Ayiei, A.; Murray, J.; Wild, G. Visual Flight into Instrument Meteorological Condition: A Post Accident Analysis. *Safety* **2020**, *6*, 19. [CrossRef]
28. Kharoufah, H.; Murray, J.; Baxter, G.; Wild, G. A review of human factors causations in commercial air transport accidents and incidents: From to 2000–2016. *Prog. Aerosp. Sci.* **2018**, *99*, 1–13. [CrossRef]
29. ICAO. *ICAO Long-Term Traffic Forecasts*; International Civil Aviation Organization: Montreal QC, Canada, 2016.
30. Airfleet.net. Airline Fleet Age. Available online: https://www.airfleets.net/ageflotte/fleet-age.htm (accessed on 7 July 2020).
31. ATSB. *An Overview of Human Factors in Aviation Maintenance*; Australian Transport Safety Bureau: Canberra, Australia, 2018.
32. Daniels, G.S. *The Average Man?* Air Force Aerospace Medical Research Lab Wright-Patterson Air Force Base: Dayton, OH, USA, 1952.
33. Hertzberg, H.; Daniels, G.S.; Churchill, E. *Anthropometry of Flying Personnel-1950*; Antioch College: Yellow Springs, OH, USA, 1954.
34. Parker, M. *Humble Pi: A Comedy of Maths Errors*; Penguin Books Limited: London, UK, 2019.
35. AAIB. *Report on the Accident to Airbus A319-13, G-EUOE London Heathrow Airport 24 May 2013*; AAIB: Farnborough, UK, 2015.
36. Dekker, S. Why we need new accident models. *Hum. Factors Aerosp. Saf.* **2004**, *2*. Available online: https://www.diva-portal.org/smash/record.jsf?pid=diva2%3A244497&dswid=-3453 (accessed on 2 August 2020).
37. Dekker, S. *Drift into Failure: From Hunting Broken Components to Understanding Complex Systems*; CRC Press: Boca Raton, FL, USA, 2016.
38. Dekker, S. Failure to adapt or adaptations that fail: Contrasting models on procedures and safety. *Appl. Ergon.* **2003**, *34*, 233–238. [CrossRef]
39. Reason, J. *Managing the Risks of Organizational Accidents*; Routledge: New York, NY, USA, 2016.
40. Kourousis, K.I.; Chatzi, A.V.; Giannopoulos, I.K. The airbus A320 family fan cowl door safety modification: A human factors scenario analysis. *Aircr. Eng. Aerosp. Technol.* **2018**, *90*, 967–972. [CrossRef]

Article

Analysis of Aircraft Maintenance Related Accidents and Serious Incidents in Nigeria

Khadijah Abdullahi Habib [1] and Cengiz Turkoglu [2,*]

1 Civil Aviation Authority, Abuja 900108, Nigeria; khabdullahi@yahoo.com
2 Cranfield Safety and Accident Investigation Centre, Cranfield University, Cranfield MK43 0AL, UK
* Correspondence: cengiz.turkoglu@cranfield.ac.uk; Tel.: +44-1234-754-019

Received: 1 September 2020; Accepted: 26 November 2020; Published: 11 December 2020

Abstract: The maintenance of aircraft presents considerable challenges to the personnel that maintain them. Challenges such as time pressure, system complexity, sparse feedback, cramped workspaces, etc., are being faced by these personnel on a daily basis. Some of these challenges cause aircraft-maintenance-related accidents and serious incidents. However, there is little formal empirical work that describes the influence of aircraft maintenance to aircraft accidents and incidents in Nigeria. This study, therefore, sets out to explore the contributory factors to aircraft-maintenance-related incidents from 2006 to 2019 and accidents from 2009 to 2019 in Nigeria, to achieve a deeper understanding of this safety critical aspect of the aviation industry, create awareness amongst the relevant stakeholders and seek possible mitigating factors. To attain this, a content analysis of accident reports and mandatory occurrence reports, which occurred in Nigeria, was carried out using the Maintenance Factors and Analysis Classification System (MxFACS) and Hieminga's maintenance incidents taxonomy. An inter-rater concordance value was used to ascertain research accuracy after evaluation of the data output by subject matter experts. The highest occurring maintenance-related incidents and accidents were attributed to "removal/installation", working practices such as "accumulation of dirt and contamination", "inspection/testing", "inadequate oversight from operator and regulator", "failure to follow procedures" and "incorrect maintenance". To identify the root cause of these results, maintenance engineers were consulted via a survey to understand the root causes of these contributory factors. The results of the study revealed that the most common maintenance-related accidents and serious incidents in the last decade are "collision with terrain" and "landing gear events". The most frequent failures at systems level resulting in accidents are the "engines" and "airframe structure". The maintenance factors with the highest contribution to these accidents are "operator and regulatory oversight", "inadequate inspection" and "failure to follow procedures". The research also highlights that the highest causal and contributory factors to aviation incidents in Nigeria from 2006 to 2019 are "installation/removal issues", "inspection/testing issues", "working practices", "job close up", "lubrication and servicing", all of which corresponds to studies by other researchers in other countries.

Keywords: flight safety; aviation accidents; airworthiness; aircraft maintenance; Nigerian aviation accidents

1. Introduction

Aircraft maintenance is an important part of the global aviation industry. It sometimes entails complicated tasks to be carried out by Aircraft Maintenance Engineers (AME) often with considerable time constraints [1]. In recent times, the aviation industry has experienced a rapid advancement in technology such as highly automated and integrated systems, which increases the burden on the AMEs to maintain both old and new fleets. AMEs also need to constantly improve their knowledge compared to the AMEs in previous times [1]. These advancements in technology have the tendency to introduce new types of maintenance errors, this is because hindsight cannot always be used to assess the system safety of new and complex designs. However, it should also be noted that other advancements such as fail-safe systems, enhanced hardware and in recent decades the use of health monitoring technologies on engines, systems and even structures have contributed to the reduction in maintenance and inspection workload [2].

1.1. Study Background, Accidents and Maintenance-Related Events

The Nigerian Civil Aviation Regulations (Nig. CARs) [3] define commercial air transport and general aviation as: "an aircraft operation involving the public transportation of passengers, cargo or mail for remuneration or hire" and "an aircraft operation other than a commercial air transport operation or an aerial work operation", respectively.

Aviation activities in Nigeria increased post-independence in 1963 after the Federal government of Nigeria obtained all Nigerian Airways shares [4] and the increase was especially observed in commercial air transport. In 2006, by the Civil Aviation Act, the Nigerian Government established Nigerian Civil Aviation Authority (NCAA) as an autonomous regulatory body and the Accident Investigation Bureau which has the sole responsibility of independently investigating accidents in Nigeria in accordance with International Civil Aviation Organisation (ICAO) Annex 13.

Nigeria has experienced various accidents and serious incidents over the years. The most recent fatal accident occurred when a Sikorsky S–76C + operated by Bristow Helicopters (Nigeria) Limited crashed at Oworonsoki area of Lagos in 2015 and unfortunately claimed all 12 souls on board [5]. Daramola [6] analysed all accidents that occurred in Nigeria from 1985 to 2010 using the Human Factors Analysis and Classification System (HFACS) which is a taxonomy framework widely used in aviation. Findings from Daramola's research showed that skill-based error, inadequate supervision and environment were the three most contributory factors.

While this study focuses on the analysis of accidents, serious incidents and occurrences that must be reported to the regulatory authority, e.g., "Mandatory Occurrence Reporting" in Nigeria, the previous studies focusing on European and global datasets enabled us to use the previously developed taxonomies based on the analysis of such data. Therefore, understanding the trends outside of Nigeria is also important. The European Aviation Safety Agency [7] Annual Safety Review (ASR) identified aircraft maintenance as a safety issue affecting Key Risk Areas (KRA). The KRAs affected are aircraft upset, runway excursion, and aircraft environment in commercial and non-commercial operations. Aircraft maintenance also had higher risk occurrence in comparison to other safety issues identified [7]. Analysis of 120 accidents and incidents which occurred from 2014 to 2018 revealed that aircraft maintenance contributed to 17 of them [7]. Unfortunately, maintenance errors were not considered in European Aviation Safety Agency (EASA)'s ASR prior to 2014.

The UK Air Accidents Investigation Branch (AAIB) includes aircraft maintenance and inspection as one of the safety recommendation topics in their ASR. The maintenance- and inspection-related recommendations are shown in Table 1.

Table 1. UK Air Accidents Investigation Branch (AAIB) maintenance- and inspection-related safety recommendations, adapted from AAIB [5,8,9].

Year	Aircraft Mx/Inspection Recommendation	Total Number of Recommendations	Maintenance Recs/Total Recommendations
2018	4	37	10.80%
2017	3	66	4.55%
2016	1	125	0.80%

Top findings from the International Air Transport Association [10] safety report for 2018 show that maintenance events were the sixth most significant threat by 13% contribution to the total number of accidents between 2014 and 2018 but they contributed to only 7% of the fatal accidents during the same period.

The same report also showed an interesting discrepancy about the contribution of maintenance events to total number of accidents between IOSA-registered operators (17%) and non-IOSA-registered operators (10%) revealed an interesting fact. This is rather contradictory to the difference in overall safety performance between the IOSA-registered and non-IOSA-registered operators as the latter's performance is significantly poorer and this is used by International Air Transport Association (IATA) to promote the IATA Operational Safety Audit (IOSA) programme.

A total of 586 lives out of 4.3 billion passengers were lost in 2018, due to events in commercial air transport. This is in contrast to the 67 fatalities recorded out of 4.1 billion passengers in the year 2017—tagged the "safest year ever" [7]. In the general aviation category, EU-registered aircraft with Maximum Certificated Take-Off Mass (MCTOM) above 2250 kg experienced events that saw the loss of 12, while EU-registered aircraft with MCTOM under 2250 kg experienced the loss of 159 lives [11]. Maintenance contribution to flight safety has been identified as one of the current and emerging safety risks by IATA in the 2018 annual review [12]. Details can be found in Appendix A.

In Africa, the average fleet age for different operators is in the range of 6 to 28 years and Nigerian based airlines were found to operate the oldest aircraft in comparison to air operators based in Ethiopia, South Africa and Rwanda [13]. One of the key challenges the airlines operating aging fleet face is the additional maintenance tasks such as corrosion prevention and control tasks which aim to ensure the airworthiness of the aircraft [14].

African carriers are consistently banned by the European Commission partly due to inadequate authority oversight [15]. In spite of this, a review of various academic journals via Google Scholar, Scopus, Science Direct, Elsevier and Emerald found very limited literature presenting the analysis of aircraft accidents and incidents in Nigeria or Africa. Additionally, Table 2 shows the departures and accidents rate by the International Civil Aviation Organisation (ICAO) Regional Aviation Safety Group (RASG) region of occurrence.

Table 2. Departures and accidents rates by Regional Aviation Safety Group (RASG) region of occurrence. Adapted from International Civil Aviation Organisation (ICAO) (2018).

Regional Aviation Safety Group	Estimated Departures (Millions)	Number of Accidents	Accident Rate (per Million Departures)
Africa	1.3	7	5.3
Asia Pacific	11.8	20	1.7
Europe	8.7	12	1.4
Middle East	1.3	2	1.6
Pan America	13.5	47	3.5
Worldwide	36.6	88	2.4

The data for the African continent are in bold to highlight higher than average accident rate, in spite of the lower estimated departures compared to other regions.

1.2. Aim and Objectives

The aim of this research is to identify the most significant aircraft-maintenance-related causal and contributory factors to accidents and serious incidents in Nigeria. Additionally, the study also aims to highlight the importance of utilising taxonomies for data analysis and in order to identify mitigation strategies.

To achieve this aim, the following objectives were developed:

1. Identify and validate maintenance-related accidents in commercial aircraft category and general aviation category aeroplanes, which were published in the last 10 years, i.e., 2009–2019.
2. Identify and validate all maintenance-related occurrences in commercial aircraft category and general aviation category aeroplanes that occurred from 2006 to 2019.

3. Qualitative analysis of the data using Insleys's [16] MxFACS taxonomy (Appendix B) for the accidents and Hieminga's [17] taxonomy (Appendix C) for the serious incidents.
4. Collect and analyse data from Subject Matter Experts (SMEs) in Nigerian Accident Investigation Bureau (AIB), Nigerian Civil Aviation Authority (NCAA) and the maintenance engineers practising in Nigeria.
5. Identify root causes of the analysed accidents, serious incidents and occurrences via survey capturing the views of Aircraft Maintenance Engineers in Nigeria about the potential mitigating measures to prevent recurrence.

1.3. Research Structure

A research structure as demonstrated in Figure 1 was set out with all the challenges and previous studies listed above. It details the steps that were followed in order to achieve clarification on different issues previously raised and to possibly reveal new information from incidents analysed.

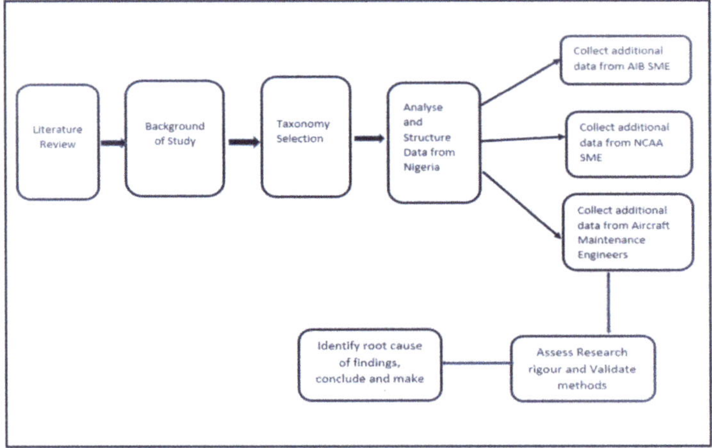

Figure 1. Research structure.

This structure shows the process necessary to be followed in order to achieve the research aim and objectives. Following a literature review, a background study was carried out on previous studies showing existing taxonomies, their advantages and disadvantages. The choice to select taxonomies developed by Insley [16] and Hieminga [17] was due to their acceptable level of inter-rater concordance. It was also to check if the results would coincide with research previously carried out in other countries.

Finally, it was to determine if developing a taxonomy in Nigeria for the identification of maintenance-related incidents and accidents, with the contribution of stakeholders in the industry would aid in identifying and predicting future events.

Responses from SMEs were collected via online questionnaires (Appendix D) in order to further understand the results of the accident/serious incident and occurrence data analysis and discover the root causes behind some discrepancies noted. The responses from the SMEs aided in gathering more detailed information of the method used and what is currently occurring in the industry. After the analysis of SME data, a survey was carried out to capture the views of maintenance engineers to further understand the issues raised by SMEs.

2. Literature Review

2.1. Errors, Classifications and Taxonomies

The term "Error" has been defined in various ways and while there is a general understanding of the term, there is no universal definition [18]. For the purpose of the present study, one applicable definition of error as defined by Reason (1995) is:

"An error is the failure of planned actions to achieve their desired goal".

Errors do not occur randomly and can be controlled effectively. Deviations of different kinds are involved in all errors. These deviations may either be connected to an adequate plan with unintended associated actions or adequate actions with inadequate planning for the outcome intended [19]. The starting point of an investigation is human error, where the investigation highlights what errors should be focused on [20]. Nevertheless, the ultimate focus of any investigation should be not to apportion blame but to identify the organisational, environmental causes of errors so that mitigation measures can be put in place.

The term classification can be defined as a "spatial, temporal or spatio-temporal segmentation of the world". A classification system can be described as a set of metaphorical or other kinds of boxes that things can be put in, to do some kind of work, either bureaucratic or knowledge production [21].

Lambe [22] defines the term taxonomy as the rules or conventions of order or arrangement, where an effective taxonomy has key attributes of being a classification scheme, semantic and a knowledge map.

Various error classifications are designed based on what is in need [19]. Classification systems are expected to meet criteria such as classifying according to origin, mutually exclusive categories and completeness. No working classification as accurately met all these requirements at once [23].

According to Dekker [20], the intent of error classification tools is as uncomplicated in principle as the tools are laborious in implementation. Their simplicity is due to the ease at which humans can manipulate them as they are basic to the consciousness of humans, their complexity can be attributed to the various ways in which they can be used, making the outcomes subjective and inconsistent. The main purpose of these tools should go beyond focusing on the peripheral error and further probe the system for root cause of the occurrence [20]. Although the intent of error classification is understood, Dekker argues that there is limited clarification as regards to reasons behind choices made by an investigator when using error classification tools to analyse accidents or incidents.

To develop an extensive accident or incident reporting system, a taxonomy that takes various causes of human errors into consideration must be provided [24]. The context in which these events occur should also be taken into consideration. In this case, aviation-maintenance-related events are the areas of application.

2.2. Taxonomies Currently in Use

According to ICAO, development of common terms, definitions and taxonomies for safety reporting systems in aviation would bring about worldwide coordination. It would also remove the constraints of aviation safety analysis that are caused by lack of common global descriptors. Without data standards, the value of safety information would be diminished, and different descriptions would result in creation of various contrasting efforts [25].

The most widely used taxonomies in the aviation industry are ICAO Accident/Incident Data Reporting (ADREP), Maintenance Error Decision Aid (MEDA), and Human Factors Analysis and Classification System Maintenance Extension (HFACS-ME). The validity of the categorisation using these taxonomies is highly dependent on gathering key information in detail.

2.2.1. ICAO ADREP

The Accident/Incident Data Reporting program (ADREP) taxonomy was developed by ICAO in 2000 and combined with European Coordination Centre for Accident Incident Reporting Systems

(ECCAIRS) taxonomy in 2013. It is used globally for safety related events categorisation and description [26].

It is a combination of several taxonomies, some of which are descriptive factors, events, phases, occurrence category, organisation, category of aircraft, etc. However, there are seven basic categories in ADREP [26].

In Europe, the combined taxonomy is used for the Mandatory incident reporting which is ultimately managed by ECCAIRS. However, it does not contain a structure for initial and continuing airworthiness data collection [27].

It is very broad and known to be complex and sometimes, difficult to use when categorising data, however, Cheng et al. argue that it is the most complete aviation safety event taxonomy ICAO has developed [28].

2.2.2. HFACS

The framework for the Human Factors Analysis and Classification System (HFACS) is used for the classification and identification of contributory factors to accident occurrences. It has been widely used in various industries such as marine, rail, road and healthcare [29].

The HFACS-ME taxonomy includes the maintenance extension which is used to classify causal factors contributing to maintenance-related aviation occurrences [30]. It is a very useful taxonomy for identifying maintenance-related incidents, especially where there is sufficient human factors details.

2.2.3. MEDA

The Maintenance Error Decision Aid (MEDA), though reactive in nature, is widely accepted by aviation personnel and used to investigate factors contributing to maintenance-related accidents and incidents [30].

It was developed by Boeing along with stakeholders in the industry in 1992 to further understand issues related to maintenance [31].

2.2.4. MxFACS Taxonomy

In 2018, Insley reviewed and analysed 112 aircraft maintenance-related accidents and serious incidents which occurred between 2003 and 2017. The data were obtained from the Aviation Safety Network database and aimed to provide a better understanding of the causal and contributory factors. This study enabled the development of a new taxonomy (MXFACS) and the structure of the taxonomy was based on Bowtie methodology and included three levels. The first level descriptors indicated the outcome of the event, the second level focused on system/component failures while the third level aimed to identify causal factors in the maintenance environment. Some accident reports contained obvious maintenance-related errors [16].

2.2.5. Hieminga's Taxonomy

In 2018, 1232 mandatory occurrence reports from the European Central Repository were analysed to develop a "two-level taxonomy". The structure of the taxonomy included two levels. Level 1 was based on high-level overview of the maintenance processes while the level 2 descriptors included more granular coding to aim to determine potential causal and contributing factors [17].

2.3. Previous Studies Trends

The most noticeable and reoccurring issue throughout the previous studies is the presence of omission errors in all events that have been analysed. In particular, installation error has been highlighted frequently throughout historical studies [10,16,17,32–34].

Although different phraseologies have been used in the various studies, the installation errors are attributed to incomplete maintenance, incorrect maintenance or inadequate installations. According to Johnston et al., installation errors are the most prevalent type of maintenance error [35].

2.4. Research Rationale

Although aircraft maintenance errors do not account for a large portion of aviation-related accidents and fatalities, there is still a highly visible contribution to various events. Some of these events have been fatal, leading to the loss of lives, property and confidence in the industry [36]. Taxonomies are used to highlight and categorise event outcomes as well as causal factors to identify trends and focus on key areas to prevent recurrence [37].

We believe this study provides new knowledge about the potential measures to prevent aircraft maintenance-related accidents, serious incidents and occurrences in the future for the following reasons. Firstly, there is no publicly available literature that specifically focuses on the analysis of aircraft maintenance-related events in Nigeria. Secondly, the most recent analysis of accidents in Nigeria, which was carried out by Daramola [6], covered events from 1985 to 2010. Therefore, this study offers an up to date analysis of events since 2010 as well. Thirdly, the study also offers the analysis of occurrence reports which are received by the Nigerian CAA and are not publicly available. Therefore, this study would also help future researchers gather information on occurrence reports in Nigeria, which are not publicly available.

The most common taxonomies being used globally can sometimes be complicated when trying to analyse events in Nigeria. Furthermore, it is beneficial for the Nigerian aviation industry and relevant stakeholders to be aware about the importance of analysing events over a period of time in the most suitable way they can. This would aid in identifying trends and preventing future events.

3. Method

The data used in this study were gathered from three sources, they are:

(a) Accident investigation reports available to the public via the NAIB website.
(b) Mandatory occurrence reports which are only available to NCAA staff.
(c) Survey responses from Subject Matter Experts (SMEs).

3.1. Accident Analysis with MxFACS

Insley's MxFACS was selected to analyse the accident reports. The taxonomy consists of a three-level hierarchy:

(a) First level—Event Outcome
(b) Second level—System/Component Failure
(c) Third level—The maintenance contributing factors that led to the system/component failure and the ultimate event.

The MxFACS taxonomy makes use of the Bowtie Risk Assessment Model to identify risks, causal and contributory factors. The three levels are derived from the "top event" element, "consequence" element and "threats" element, respectively, as shown in Figure 2. The maintenance error is taken as an equivalent of the "hazard" in each accident. It aids in identifying the action, the outcome and the context which are three of four basic elements of error [16].

The data contained in the Accident Investigation Bureau (AIB) publications were in PDF. Therefore, each report published in the last decade was downloaded. The documents were thoroughly reviewed in order to identify the maintenance-related accidents. After identification, each accident report was analysed and coded by using the MxFACS taxonomy structure.

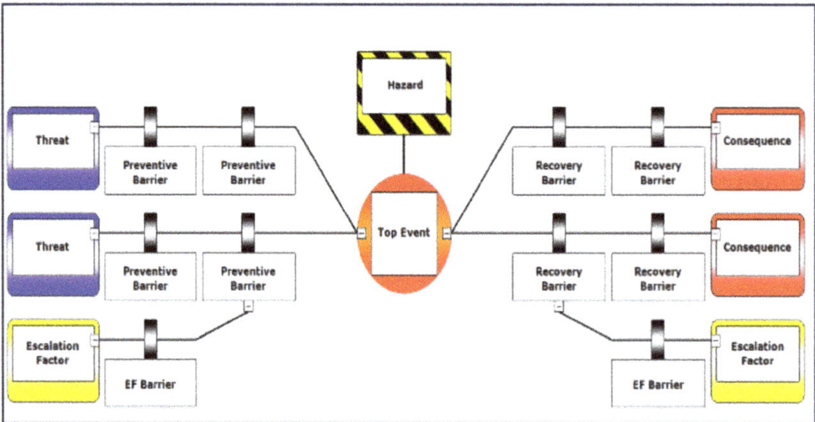

Figure 2. Bowtie Risk Assessment Model, CGE Risk Management Solutions (2017).

3.2. *Incident Analysis with Hieminga's Maintenance Incident Taxonomy*

Hieminga's taxonomy was selected to analyse the reports which were sourced from the Safety Deficiency Incidence Analysis (SDIA) mandatory occurrence report database. The taxonomy consists of a two-level hierarchy:

- First level: This level follows a logical maintenance process, i.e., from planning to preparation. It also considers general category issues which are excluded from other categories. It follows through to different tasks and concludes with a job close up.
- Second level: This level is also a logical practical maintenance process and is comprised of as many substructures present as possible.

Although it gives the reporter the opportunity to select a descriptor that is as precise as possible due to its broad spectrum [17], it is a very broad coding system and could confuse maintenance personnel. This is because one incident could be coded into more than one first and second level category. This taxonomy utilises familiar words for classification which makes it easier for maintenance personnel to report incidents appropriately.

The data contained in the SDIA dataset were thoroughly analysed. All maintenance-related incidents were identified and classified in accordance with Hieminga's template.

3.3. *Collection of Data: Accident Investigation Reports Available to the Public*

The process of collecting maintenance-related accident data involved downloading all accident reports available to the public. All commercial aircraft category and general aviation category accident reports published in the last decade, i.e., 2009–2019, were identified and downloaded.

There was a total of 70 accidents published in Portable Document Format (PDF). In order to identify the maintenance-related accidents, each document was analysed by the lead author. The next step was to identify the maintenance-related accidents and compile them in a dataset. All accident investigation reports that occurred in Nigeria were sourced from Nigeria's Accident Investigation Bureau's (AIB) publication database.

The AIB's database contains official documented reports of all previous civil accidents in Nigeria. It is the only aviation parastatal in Nigeria with the principal responsibility of performing independent investigations into aviation accidents and making safety recommendations to the relevant agencies [38] and this satisfies the standards and recommended practices defined in ICAO Annex 13.

3.4. Collection of Data: Mandatory Occurrence Reports (MORs)

The MORs were sourced from the Nigerian Civil Aviation Authority (NCAA). In particular, it was received from the Safety Deficiency Incidence Analysis (SDIA) unit under the Directorate of Airworthiness Standards (DAWS).

The NCAA was established in 1999 to comply with ICAO's requirements. ICAO required every member state to set up an organisation with the responsibility of ensuring compliance with air navigation rules [6]. NCAA is responsible for the safety oversight of the aviation industry in Nigeria.

All civil aviation events that meet the criteria for reportable occurrences defined in the NCARs are reported to the NCAA through an MOR form. These incidents must be reported within 72 h of its occurrence [3]. The incidents are then analysed, uploaded to SDIA's Google Drive database and monitored till closure.

A total of 2530 incidents from 2006 to 2019 were stored in the SDIA MOR database. The data were de-identified before analysis for the purpose of confidentiality. All the incidents were analysed to discern maintenance-related incidents. They were compiled in a dataset and reviewed to ensure that only maintenance-related events were considered.

3.5. Collection of Data: Subject Matter Experts (SMEs)

Feedback, clarification and recommendation were sought from three different groups of SME. This was to acquire relevant information for recommending next steps of the data analysis. Another reason was to create awareness of the process of developing customised taxonomy for maintenance-related accidents and incidents in Nigeria. The three different groups of SMEs were;

(a) Four Aviation Safety Inspectors at the NCAA SDIA unit;
(b) Seven Air Safety Investigators from the NAIB;
(c) Twenty-five Aircraft Maintenance Engineers (AMEs) practising in Nigeria.

The data were collected by sending links of the three different surveys to three different groups. This was distributed via online questionnaires (shown in Appendix D). These were administered by the Qualtrics software.

The first questionnaire contained four open-ended questions. This was responded to by NCAA SME participants. Their experience of collecting and analysing the MORs in Nigeria made them adequately qualified to contribute to this study. The questions covered the following topics:

1. Experience and view on current taxonomy being used;
2. Assessment of the study's methodology and data output;
3. Suggestions of other taxonomies to be used and possibility of developing customised taxonomy in Nigeria;
4. Recommendation of adequate methods to further predict and make adequate safety plans by using the results of this data.

The second questionnaire contained three open-ended questions. This was responded to by AIB SME participants. Their wealth of experience in carrying out accident investigations, developing the reports and publishing them made them adequately qualified to contribute to this study. The questions covered the following topics:

1. Experience on carrying out long-term reviews of previous accident reports and views about the benefit of such reviews.
2. The depth of human factors training received and the availability of human factors experts within the AIB.
3. Assessment of the study's methodology, data output and recommendation for improvements.

The third questionnaire contained three open-ended and five multiple-choice questions. It was distributed to AMEs practising in Nigeria. A total of 25 responses were received. The question covered topics related to:

1. Type and years of experience;
2. Prioritising the identified maintenance contributory factors likely to cause future accident, by using a scale;
3. Experience, challenges faced and opinion of following maintenance instructions;
4. Experience, challenges faced and opinion of inspection instructions adequacy.

All the questionnaires were sent along with a PowerPoint presentation containing the aims, objectives and methodology. The first two surveys were accompanied by an Excel worksheet containing the taxonomies used and sample data to enable them to code the data using the taxonomies. The data output of this study was also presented to the first two SME groups.

These questionnaires were also sent out due to the limited data available on Nigeria aviation industry. An inductive approach was used to gather the information for this qualitative aspect [39].

3.6. Data Analysis

Although there are various well-known taxonomies currently being used to code accidents and incidents datasets, Insley's MxFACS taxonomy and Hieminga's taxonomy were used to code the accidents and serious incidents dataset. According to Hieminga (2018), having reviewed the taxonomies used in "Maintenance Error Decision Aid" (MEDA), which is an investigation tool developed by Boeing, HFACS-ME Framework, CAA Paper 2009/05 [33] and CAP 1367 [32]—none of these appeared to inhibit the two scales of adequate "usability" and "comprehensiveness" at the same time. This led to the solution of developing a different taxonomy to aid in coding incident events.

Another reason why a new taxonomy was not developed for the analysis of accidents and incidents in Nigeria was due to inadequate standard phraseology present in the dataset analysed. It can be argued that although the thematic analysis provides flexibility for the researcher, that same flexibility could lead to inconsistency and a lack of coherence when developing themes from a dataset.

Thematic analysis is a suitable qualitative research method that can be used in a wide range of analysing large qualitative datasets. Its advantage is that trustworthy and insightful findings can be produced using this method [40]. It has also been described as a method used to identify, organise and describe reporting themes within a dataset [41]. This was the main method of qualitative analysis used for the SME's survey.

3.7. Analysis of SME Survey

The NVivo 12 plus is a qualitative data analysis software and it was the primary software used for survey analysis. All the survey questions were downloaded from Qualtrics in Microsoft Word format. These were then uploaded to the software as separate projects. These projects were analysed individually for identification of reoccurring words and themes in each response to the question. The responses were to aid in the methodology used, provide in-depth information and guidance in the study. The analysis did not focus only on responses tallied with the questions asked. This was to avoid missing out on new and important information. This is an inductive analysis method which allows researchers to code data without bias [41].

3.8. Evaluation of Research Rigour

According to Brink, a valid study demonstrates what is in existence, a valid measurement and measures what it was created to measure. A reliable study should produce the same results consistently when repeated by a different researcher [42]. However, these terms are difficult to apply to qualitative research methods when compared to quantitative research methods. Rigour is a more suitable term to measure validity and reliability of a quantitative research method [43].

Although Liamputtong and Ezzy argue that it does not perfectly verify the reliability and validity of qualitative research as stated above, coherence between different researchers is important to provide meaningful information. Inter-rater reliability or inter-rater concordance can be used to assess the level of coherence between two or more researchers. Cohen's kappa is the most commonly used measure for assessing this match [44].

To assess the proportion of coherence and corrected for chance, Cohen's Kappa was used in this study. Equation (1) shows how it is derived, Equations (2) and (3) show how formula components are determined.

$$\kappa = \frac{P_o - P_e}{1 - P_e} \tag{1}$$

$$P_o = \frac{\sum_{i=1}^{n} R}{n} \tag{2}$$

$$P_e = \frac{\sum_{i=1}^{n} \frac{c_i \times r_i}{n}}{n} \tag{3}$$

where κ = Cohen's Kappa, P_o joint probability of agreement, P_e = chance agreement, R = rater agreements, n = total number of ratings, c = column marginal and r = row marginal.

To evaluate the research rigour, SMEs from the NCAA coded a sample dataset using Hieminga's maintenance incident taxonomy. This dataset was selected from one year and all the information was cleaned. The SMEs from the AIB coded all the maintenance error accidents identified using the MxFACS taxonomy. The researcher and SME's coding were compared to determine Cohen's Kappa using IBM SPSS V.25 statistics software.

3.9. Ethical Considerations

To protect the rights of participants, especially when conducting a qualitative research, certain moral and ethical problems could be encountered [45]. The study included collection of data from subject matter experts. It was therefore crucial to initially seek participants consent, inform them about their participation being voluntary and ensure their anonymity and confidentiality. Only relevant components were assessed. The university granted research ethical approval after it was requested for.

4. Results

4.1. Reliability of Taxonomies Used

McHugh (2012) interprets the level of agreement of Cohen's Kappa [46]. This is shown in Table 3 below.

Table 3. Cohen's Kappa Level of Agreement. Adapted from: McHugh (2012).

Value of Kappa	Level of Agreement	Percentage of Data that Are Reliable
0–0.20	None	0–4%
0.21–0.39	Minimal	4–15%
0.40–0.59	Weak	15–35%
0.60–0.79	Moderate	35–63%
0.80–0.90	Strong	64–81%
Above 0.90	Almost Perfect	82–100%

Table 4 below shows the Kappa value of this research's inter-rater concordance with the AIB SME. Table 5 shows the Kappa value of this research's inter-rater concordance with the NCAA SME.

The values showed a moderate to strong level of coherence between the researcher and SME when the MxFACS taxonomy was used to categorise maintenance-related accidents. This suggests that there is a strong research rigour when compared to the levels as shown by McHugh [46].

Table 4. Derived agreement statistics in all levels for researcher and AIB Subject Matter Expert (SME).

Researcher and SMEs	κ	P_o	P_e
Level 1	0.70	0.70	0.001
Level 2	0.80	0.80	0.002
Level 3	1	1	0.0001

Table 5. Derived agreement statistics in all levels for researcher and AIB SME.

Researcher and SME	κ	P_o	P_e
Level 1	0.70	0.70	0.001
Level 2	0.80	0.80	0.002

The values also show a moderate to strong level of coherence between the researcher and SME when Hieminga's maintenance incidents taxonomy is used to categorise maintenance-related incidents. This suggests that there is a strong research rigour when compared to the levels as shown by McHugh [46].

4.2. MxFACS Level 1—Event Outcome

A total of 70 accident reports were analysed, however, only 11 of them were identified with maintenance-related causal or contributory factors. These 11 events were then categorised in accordance with the MxFACS taxonomy to highlight the nature of the event. A new coding of "other" was created and added to the taxonomy to categorise maintenance-related events that did not match any of the themes. Table 6 Shows level 1 (Event Outcome) in detail

Table 6. Level 1 (Event Outcome).

Event Outcome	n	% Fatality
Cabin fume event		
Insulation blanket fumes		
Collision		
Collision with building		
Collision with terrain	2	18
Collision with water		
Collision with another aircraft		
Depressurisation		
Progressive depressurisation		
Rapid depressurisation		
Diversion or air turn back		
Air turn back		
Diversion	1	
Fire		
In-flight fire	1	
On-ground fire	1	
In-flight shut down		
In-flight engine shut down		
Landing-related event		
Approach and landing without auto flight assistance		
Degraded hydraulic system functionality during landing		
Ditching		
Engine failure upon landing		
Forced landing	3	
Hard landing		
Landing short of runway		
Wheels-up landing	2	

Table 6. *Cont.*

Event Outcome	*n*	% Fatality
LandingGear-related event		
In-flight LG-related event		
On-ground LG-related event		
Runway-related event		
Runway excursion		
Structural damage		
Empennage damage		
Lower fuselage structural damage		
Wing structural damage		
Other	1	

The data highlight that in all maintenance-related Event Outcomes, only terrain collision led to a fatal accident, which was 18% of all maintenance-related accidents in the last decade [47]. This correlates with EASA [7] which identified terrain collision as one of the KRAs, further supported by the global accident/serious incident review in 2018 [36].

Surprisingly, runway excursion and ground collision, which have a high potential to be caused by maintenance error, were not affected in the last decade [7]. Only 16 percent of all accidents in the last decade are related to maintenance error. This could be due to inadequate focus given to this safety critical aspect of aviation.

4.3. MxFACS Level 2—System/Component Failures

All 11 accident reports were further coded to identify the top-level system/component affected. This is shown in Table 7.

Table 7. Level 2 System/Component Failures.

System/Component Failures	*n*
Electrical power	0
Engine	3
Flight controls	0
Fuel	0
Instrumentation and indication	0
Insulation	0
Landing gear	0
Pressurisation	0
Steering	0
Structure	6
Windscreen	0
Other	2

A new coding of "other" was also created in level 2 and added to the taxonomy to categorise other maintenance-error-related events that did not match any of the themes. Level 2 (System/Component Failures) shows that engine, aircraft structure and "other" were directly affected by maintenance issues.

The UK Civil Aviation Authority [32] identified Air Transport Association (ATA) Specification 100 Chapter 32 "Landing Gear" as the second most affected system out of the top six events related to maintenance error. However, no landing gear system was identified as being affected by maintenance error in the last decade. This may be caused by the limited attention being given to maintenance errors in Nigeria. This can be improved upon by the AIB targeting maintenance, airworthiness and human factors causal and contributory factors when carrying out investigations. It also highlights why NCAA should include maintenance-related causal and contributory factors in data analysis. It may also be due to adequate inspection and testing carried out on such a visible area as the landing gear.

The accidents were further coded in accordance with MxFACS level 3 to identify the nature of aircraft maintenance errors that contribute to accidents in Nigeria.

4.4. Level 3—Maintenance Errors

All 11 accident reports were further coded to identify the top-level system/component affected and the codes can be seen in Table 8.

Table 8. Level 3 Maintenance Errors.

Maintenance Factors	n
Airworthiness directive	1
AMM (Aircraft Maintenance Manual)	3
Check	0
FOD (Foreign Object Debris)	1
Human Factors	1
Inadequate maintenance	0
Incorrect maintenance	0
Inspection	3
Organisational	1
Overhaul	0
Oversight	7

Inadequate and incorrect maintenance were not identified as causal or contributory factors in any of the accidents analysed. However, research has highlighted that these types of omission and commission errors are common in aviation maintenance [19,48]. The absence of these errors may be due to inadequate information provided to the AIB by the personnel involved. Another plausible reason may be due to the difference in phraseology used in taxonomies.

The results show a high presence of Failure to follow procedures which can be classified under "AMM" and "Airworthiness Directive". Failure to follow procedure is a maintenance issue as well as an organisational issue. This is because every team member, including managers who mount pressure on AMEs, are involved in the maintenance chain [49]. This is also in agreement with organisational factors being one of the maintenance contributory factors. This is a growing concern in the aviation maintenance area.

There was a surprisingly low number of human-factor-related errors in the accident events. During the categorisation, no assumptions were made, hence where there were no human factors mentioned, it was not categorised. According to Sarter and Alexander, human error contributed to 70% of major aviation accidents [50]. The analysis highlights discrepancies in limited human factors considerations during investigations.

Oversight by operators and authorities was identified as the highest causal/contributory factor to maintenance-related events. Research by Dhillon showed that operators and regulator's oversight led to some fatal maintenance-related accidents [51]. Oversight should be a critical aspect of aviation as it aids all personnel involved to obtain second views and opinions.

4.5. Analysis of Mandatory Occurrence Report by Using Hieminga's Taxonomy

All the incidents in the SDIA MOR database were coded in accordance with Hieminga maintenance incidents taxonomy. The results show that a total of 588 incidents were related to maintenance error. This is about 23% of all incidents which occurred from 2006 to 2019. The breakdown of total number of MORs, total number of maintenance-related MORs and the number of most frequent Level 1 category events is shown in Figure 3. There are several findings which are related to the scope of this study.

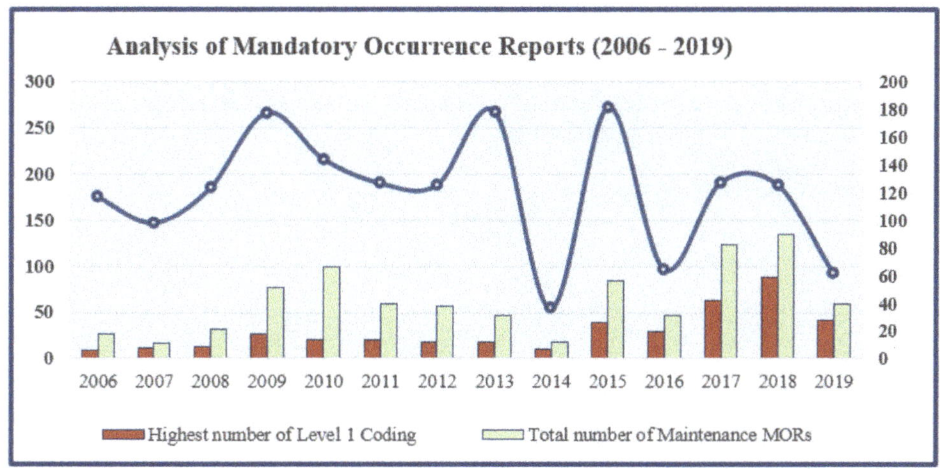

Figure 3. Analysis of maintenance-related MORs (Mandatory Occurrence Reports) between 2006 and 2019.

Firstly, the total number of MORs fluctuates and does not follow a pattern. When such statistics are presented, one important consideration is to normalise the numbers by presenting them as rate of occurrences based on traffic numbers. Nevertheless, it is not always possible to determine the cause of certain data points such as 2014 being the lowest number of mandatory occurrence reports the Nigerian CAA has received.

Secondly, the rate of the total number of maintenance-related MORs has increased in recent years. Particularly, as shown in Table 9, the maintenance-related MORs were more than 40% of the total number of MORs within the last three years the data set covered. While such statistics may be concerned, it is important to understand the further analysis of the data and identify the causal and contributory factors.

Table 9. Analysis of Mandatory Occurrence Reports (2006–2019).

Year	Total Number of MORs	Total Number of Mx MORs	The Rate of Mx MORs	Highest Number of Level 1 Coding	Highest Number of Level 1 Coding
2006	175	18	10%	6	Inspection/testing issue
2007	147	11	7%	8	Inspection/testing issue
2008	185	21	11%	9	Removal/Installation issue
2009	266	51	19%	18	Inspection/testing issue
2010	216	66	31%	14	Removal/Installation issue
2011	190	40	21%	14	Working practices
2012	188	38	20%	12	Removal/Installation issue
2013	267	31	12%	12	Removal/Installation issue
2014	55	12	22%	7	Removal/Installation issue
2015	272	56	21%	26	Removal/Installation issue
2016	97	31	32%	20	Removal/Installation issue
2017	191	82	43%	42	Removal/Installation issue
2018	188	90	48%	59	Removal/Installation issue
2019	93	40	43%	28	Removal/Installation issue

Thirdly, one clear finding from this analysis was that "removal/installation issues" have been the most frequent event category. For example, out of 588 maintenance-related MORs in total, 259 of these events were caused by "removal/installation issues". Furthermore, this category appeared as the "most frequent" event within the last 8 years and ten times within the 14 years the dataset covered. This is certainly in alignment with several other studies including the ones which led to the development of two taxonomies used in this study [34].

The "removal/installation issues" category was followed by working practices which contained most errors related to accumulation of dirt and contamination. This may be caused by the dry season weather conditions in Nigeria. Studies have shown that dust would contribute to aviation safety through corrosion, blockage of Pitot-static tube, etc. [52]. Inspection and testing issues also had a high contribution to maintenance-related incidents, followed by job close up, lubricating and servicing. This correlates with previous studies carried out by Latorella and Prabhu [53].

4.6. Results of the SME Survey

Information was gathered from the responses to the questionnaires. Themes were identified and nodes created to extract the most important elements of the survey. The information identified from three different groups is written below along with brief discussions on them.

4.6.1. NCAA SDIA SME RESPONSE

1. *Experience and view of current taxonomies*

The respondents gave information related to the process of receiving, uploading, investigating, monitoring and closing incidents reported to them.

All the respondents stated that the ADREP taxonomy is currently being used to analyse mandatory occurrence reports. It is not suitable for capturing maintenance errors because it is very broad.

2. *Assessment of study methodology*

The respondents were in support of the methodology used, however, some of them suggested that the taxonomy be narrowed down to avoid a cumbersome process of classifying incidents

3. *Suggestions on other taxonomies or developing one and recommendations for safety plans*

Some of the respondents suggested that MEDA or bowtie analysis with a risk matrix be used to identify maintenance errors. Some respondents agreed to the development of maintenance error taxonomies for Nigeria.

There was information related to authenticity of reports sent to the NCAA, the phraseology being used by the maintenance engineers in the industry, insufficient data provided to the SMEs and reluctance to report incidents.

The respondents recommended a standardised online reporting system, organising safety symposiums, workshops and creating awareness of the presence of just culture.

4.6.2. AIB SME RESPONSE

1. *Experience of analysis previous reports within a period,* e.g., *a decade*

All the AIB respondents agreed that analysis was carried out on previous reports, however, they have not experienced such analysis being carried out since 2009. The previous analysis carried out are not available to the public.

2. *Depth of human factors knowledge and availability of human factors personnel*

The respondents stated that although their AIB commissioner is focused on human capacity building and ensures they are adequately trained, there is a need for human factors training for the AIB investigators. There is currently one human factors expert who was recruited in July 2019.

3. *Assessment of study methodology and recommendation for improvements*

4.6.3. AME SME RESPONSE

1. *Experience and challenges faced following maintenance instructions*

Seventy-five percent of the respondents suggested maintenance instructions were not always practicable to follow. Some of the challenges faced by them were:

Some manufacturer's manuals and some aircraft type were more complicated than normal.

Time constraint, human factors involvement, inadequate special tools and inadequate access to internet to update manuals.

2. *Experience and challenges with inspection instruction adequacy*

Eighty-five percent of the respondent stated that some maintenance instructions are inadequate. They face challenges such as ambiguity and incomplete instructions.

They recommended developing a process of sending feedback to the manufacturers through the most appropriate means.

5. Discussion

This research has demonstrated the nature of aircraft maintenance errors that significantly contribute to or the cause accidents and serious incidents in Nigeria. Factors such as failure to follow procedures in manuals, human factors, foreign object damage, inadequate inspection and operator and regulatory oversight were identified as the nature of maintenance errors that lead to accidents, while job close up, installation/removal, lubrication/servicing, working practice and maintenance control were identified as the nature of events leading to serious incidents in Nigeria. The SME surveys highlighted various challenges faced and recommendations. Considering that the scope of this study is the nature of maintenance errors, these are hereby analysed critically, arguments are developed and comparison with previous studies carried out.

5.1. MxFACS—All Levels Discussion

In the accident analysis, 16 percent of all the accidents in the last decade are maintenance-error-related. This is slightly higher than the previous studies carried out that show that 12% of major aircraft accidents are caused by maintenance discrepancies [54–57].

A probable explanation for this could be due to the type of maintenance culture that exists in Nigeria. A study carried out by Olufunke revealed that there is a need to emphasise the importance of maintenance culture to Nigerians in every industry [58]. She also highlighted how maintenance personnel should be highly valued and importance be given to them to motivate them.

Another probable reason for this could be the high rate of traffic in Nigeria in the last decade [55]. This corresponds with findings of studies carried out by Saleh et al. which revealed that from 2005 to 2015, 14–21% of helicopter accidents in the US were related to flawed maintenance and inspection [59].

Airworthiness directive and AMM were identified as contributory factors. These could be classified under failure to follow procedures in manuals. A probable explanation for this which is taken from the SME survey, could be due to some of these documents being complex or cumbersome. According to Drury and Johnson, "procedures not followed" is now frequent in incident and accident reports in aviation. Findings from an Federal Aviation Administration (FAA) study by Johnson and Watson revealed that during a heavy maintenance check carried out within 90 days, the number one factor that caused major malfunctions was failing to comply with maintenance documents [60]. This corresponds with the findings of this study.

Another probable cause could be loss of confidence in the document as any error found by the end user would decrease the user's confidence in the document [61]. It could also be over-confidence of highly experienced maintenance engineers in carrying out simpler tasks.

Inadequate maintenance and incorrect maintenance had no contribution to any of the accidents analysed. This is surprisingly one of the common commission errors in aircraft maintenance (Reason and Hobbs, 2003) which does not correspond to this study. A probable cause could be the terms and phrases used in the reports. Another probable cause could be that aircraft maintenance errors did not receive adequate attention [53].

Inadequate inspection being identified as one of the causal or contributory factors to the accidents is not surprising. A probable cause for this could be lack of required special tools as suggested by one of the SME respondents or improper use of tools provided. Another probable cause could be fatigue [62]. This correlates with this study's findings. Without a fatigue risk management system in place for maintenance engineers in Nigeria, some organisations may tweak the laws regarding rest and duty limitations which do not consider commute time. This fatigue can be classified under human factors which was identified as one of the factors contributing to accidents.

Foreign Object Damage (FOD) was identified as one of the factors contributing to accidents. Studies carried out by Hussin et al. revealed FOD is a rising concern in the aviation industry [63]. An analysis of events that occurred from 1998 to 2008 was carried out by Australian Transport Safety Board (ATSB); the results show that 116 events were caused by FOD. This corresponds with the findings of this research that FOD can contribute to events.

Operator and regulatory oversight were identified as the nature of maintenance errors that lead to accidents. A plausible explanation may be due to the low attention given to aircraft maintenance errors by both operator and regulatory body. Accident Investigation Bodies around the world regularly issue recommendations for the regulatory authorities to consider taking action in many different areas. Effective oversight can be one of these areas when the investigation identifies clear evidence of ineffective oversight the regulator or the operator. According to Drury, a report by the Federal Aviation Administration (FAA) recommended increased regulatory oversight for repair stations [64]. This corresponds with the findings of this study. Another plausible explanation for operator oversight could be that the management is not balancing safety goals with production goals, which could lead to events [25].

5.2. Hieminga Maintenance Incidents Taxonomy Discussion—All Levels

In all the serious incidents analysed, 23% were attributed to maintenance errors. An analysis by Marais and Robichaud of 3242 incident reports showed that 10% can be attributed to maintenance error, which has remained constant in the past decade [65]. This correlates with the findings of this study about how maintenance-related errors cause or contribute to serious incidents.

Job close up, i.e., close up not performed correctly, was identified as one of the natures of maintenance errors. An example of this type of error identified was engine cowls not latched. This is an omission error which is common as stated earlier. An analysis of accidents cause by Cowan et al. revealed that maintenance errors such as leaving engine cowl unlatched could lead to separation during flight, this causes structural failures. This corresponds with the findings of this study. A probable cause for this could also be fatigue as explained earlier [66].

Installation and removal was also identified as one of the natures of maintenance errors contributing to incidents. A plausible explanation for this could be carrying out tasks without the approved document as discussed earlier. In recent years, "failing to follow procedures" has been identified by the FAA as a consistent causal factor and as a result an "online training programme" was developed to look at this issue holistically. Another plausible explanation could be the presence of an aging aircraft being operated as discussed earlier, which would require additional maintenance. This corresponds to the result of this finding. Inadequate oversight from operators could also lead to this type of error because quality control on aircraft maintenance helps to highlight discrepancies during audits [53].

Inspection testing was identified as a nature of maintenance-related error leading to accidents. A plausible cause for this could be inspection overdue or inadequate tools to carry out inspections. A study by Boeing revealed that 16% of hull loss and 20% of accidents that occurred from 1982 to 1991 could have been prevented by a change in maintenance inspection [54].

Working practice. A probable reason for working practice being a maintenance-related contributory factor may be due to organisation culture. According to Pettersen and Aase, operational work practice is part of the safety and regulatory systems of the industry but can be highly influenced by organisational framework [67]. This means that the personnel tend to formulate how tasks should be carried out and formulate grey zones within themselves. This eventually becomes normal especially during time pressure.

A probable cause for identifying lubrication and servicing as one of the natures of maintenance error could be using the wrong fluid, insufficient lubricant or servicing overdue. An example of a fatal accident related to this took place on the Alaska Airlines Flight 261. Insufficient lubrication of the jackscrew assembly led to thread failure; further contributing to this was extending the lubrication interval, which was approved by the FAA [40].

6. Conclusions

Prior to concluding this study, a number of challenges and limitations faced are addressed.

6.1. *Challenges and Limitations of the Study*

One key challenge faced during the study was gathering the mandatory occurrence report data because it was not available in the public domain and furthermore the protecting the identity of individuals and organisations involved are vitally important.

Another challenge was inadequate information in the incidents and accident reports. So many incidents had to be omitted due to insufficient information related to the scope of this study.

Gathering all three questionnaires in a short time was very challenging, however, it was possible to finish most of the analysis in time. Some parts of the data from the SMEs were not analysed due to time constraint.

The most important challenges faced during the study was the unavailability of adequate literature related to the aviation industry in Nigeria. This is one of the problems this study aims to solve.

The most significant limitation of this study is the fact that the coding of accidents/serious incidents as well as MORs was only validated by one SME for each category of data. Therefore statistical significance cannot be claimed; however, the nature of the study never aimed to be a quantitative approach and it is believed that the results can still provide real insight into the maintenance-related events and their potential causes and contributing factors.

6.2. *Conclusion of the Study*

The study shows the nature of aircraft maintenance errors that contribute to or cause accidents and incidents in scheduled commercial, non-scheduled commercial and general aviation category. The utilisation of the MxFACS and Hieminga's maintenance incidence taxonomy yielded similar results with existing global research. It also highlighted how one maintenance error could be caused by another maintenance error within the same taxonomy. It also shows that utilising taxonomy can aid in predicting some future accidents. It may, however, not predict some occurrences as these taxonomies were created using hindsight.

The results of the study revealed that the most common maintenance-related "Event Outcomes" in the last decade are "collision with terrain" and "landing gear events". The systems of components that were affected the most during accidents are the aircraft engine and structure. The maintenance factors with the highest contribution to these accidents are operator and regulatory oversight, inadequate inspection and failure to follow procedures.

The research also highlights the highest causal and contributory factors to aviation incidents in Nigeria from 2006 to 2019 are installation/removal issues, inspection/testing issues, working practices, job close up, lubrication and servicing. All of which corresponds to studies by other researchers in other countries.

The trend over the years revealed that an increase in air traffic in Nigeria led to an increase in the number of maintenance-related incidents, however, it is worthy to note that data on air traffic from 2018 to 2019 was not available for comparison.

The study's findings could contribute to the limited literature related to maintenance errors and incidents in Nigeria. It would also aid all relevant stakeholders in understanding the nature of errors that pose a threat to the safety performance in Nigeria.

7. Recommendations

With a deeper understanding of the challenges and suggestions provided by the SMEs, it can be recommended that a less complex taxonomy be developed for the identification and categorisation of maintenance-related events in Nigeria.

Oversight should target human error as much as they do for technical failures. This can help in predicting possible events and identifying trends to aid in implementing a risk-based oversight approach by the regulatory authority.

Additional Human Factors training would help particularly the inspectors and data analysts in the regulatory authorities and the accident investigators to focus on key human performance issues during the performance of their duties.

Author Contributions: Conceptualization, K.A.H. and C.T.; methodology, K.A.H. and C.T.; formal analysis, K.A.H.; investigation, K.A.H.; resources, K.A.H. and C.T.; data curation, K.A.H.; writing—original draft preparation, K.A.H.; writing—review and editing, K.A.H. and C.T.; visualization, K.A.H. and C.T.; supervision, C.T.; project administration, K.A.H. and C.T. All authors have read and agreed to the published version of the manuscript.

Funding: This research received no external funding.

Acknowledgments: This paper is the result of Khadijah Habib's individual research project as part of her studies at Cranfield University to pursue the MSc Safety and Human Factors in Aviation award. We appreciate the Commonwealth Scholarship she received from the UK government. We are also grateful for all the contribution we received from the subject matter experts in Nigerian Regulatory Authority, Nigerian Accident Investigation Bureau and other professionals working in the Nigerian Aviation Sector.

Conflicts of Interest: The authors declare that there is no conflict of interest regarding the publication of this paper. Khadijah Abdullahi Habib has received a Commonwealth Scholarship for her studies at Cranfield University and this paper is the outcome of her individual research project, which is the partial fulfilment of the requirements for the MSc degree award.

Appendix A

Table A1. IATA 2019 Safety Report.

Accident Type	Maintenance Operations	Maintenance Operations: SOPs and Checking	Maintenance Operations: Training Systems	Maintenance Events
Aircraft Accidents	9%	8%	2%	13%
Fatal Aircraft Accidents	4%	4%	4%	7%
Non-Fatal Aircraft Accidents	9%	9%	2%	14%
IOSA Aircraft Accidents	11%	10%	2%	17%
Non-IOSA Aircraft Accidents	7%	6%	2%	10%
LOC-I	5%	5%	5%	11%
RWY/TWY EXC				2%
IN-F DAMAGE	9%	9%		22%
GND DAMAGE	13%	9%		17%
G UP LDG/CLPSE	34%	32%	7%	49%
RWY COLL	10%	10%		
Jet Aircraft	10%	9%	1%	17%
Turboprop	6%	6%	4%	5%
Cargo	7%	7%	2%	7%
Africa	10%	10%		15%
Asia/Pacific	7%	6%	1%	10%
CIS	4%			9%
Europe	7%	7%		7%
Latin America and the Caribbean	11%	11%	4%	29%
Middle East and North Africa	16%	16%	5%	32%
North America	13%	13%	4%	13%
North Asia	0%	0%	0%	0%

Appendix B

Table A2. MxFACS Taxonomy (Insley, 2018).

Level 1 Occurrence	Level 2 System/Component Failures	Level 3 Maintenance Factor(s)
Cabin fume event	Electrical power	Airworthiness directive
Insulation blanket fumes	Electrical fire	Not followed
	Electrical interruption	
Collision	Significant loss of function	AMM
Collision with building	Engine	Incorrect information
Collision with terrain	Cowling separation	Missing information
Collision with water	Engine fire	Failure to follow procedure
	Engine icing	Procedures difficult to follow
Depressurisation	Engine separation from aircraft	
Progressive depressurisation	Engine surge	Check
Rapid depressurisation	Engine wash contamination	Check not undertaken
	Flameout	Inadequate check
Diversion or Air Turnback	Fuel starvation	
Air turnback	Loss of thrust	
Diversion	Propeller separation	FOD
	Throttle stagger	Tool left in aircraft
Fire	TR cowling separation	Contaminants in aircraft
In-flight fire	TR not deploying	
On-ground fire	Uncontained engine failure	Human Factors
	Flight controls	Maintainer fatigue
In-flight shut down	Elevator detachment	Time pressure
In-flight engine shut down	Loss of flap control	Unqualified maintenance personnel
	Loss of pitch control	
Landing-related events	Uncommanded roll	Inadequate maintenance
Approach and landing without autoflight assistance	Fuel	Inadequate instructions
Degraded hydraulic system functionality during landing	Fuel leak	Inadequate maintenance
Ditching	Fuel tank rupture	Procedures
Engine failure upon landing	Instrumentation and indication	Non-airworthy component released into service
Forced landing	Blocked pitot tube (ASI error)	Part missing
Hard landing	False engine fire indication	Part not reattached
Landing short of runway	IRS incorrect	Part or latch not secured
Wheels-up landing	Insulation	
	Insulation blanket collapse onto high temperature component	Incorrect maintenance
1.8 LG-related event	Landing gear	Incorrect adjustment
In-flight LG-related event	LG assembly damage	Incorrect assembly
On-ground LG-related event	LG collapse	Incorrect component installed
	LG fire	Incorrect installation
Runway-related event	LG not fully extended	Incorrect procedure
Runway excursion	Loss of braking	Incorrect rigging
	Shock absorber separation	
Structural damage	Tyre failure	Inspection
Empennage damage	Violent vibration	Inspection does not identify defect
Lower fuselage structural damage	Wheel(s) lost	Inspection not undertaken
Wing structural damage	Pressurisation	Insufficient inspection
	Outflow valve opening in error	No fault found
Other	Steering	Organisational
Other event	Loss of nose wheel steering	Inadequate maintenance documentation
	Structure	Inadequate reporting
	Fuselage damage	Inadequate training
	Hole in bulkhead	Lack of training
	Skin crack	Misleading paperwork
	Wing separation	Poor resource planning
	Total aircraft damage	Overhaul
	Windscreen	Overhaul not undertaken
	Improperly maintained windscreen obstructing vision during visual approach	Part used past expiry
	Other	(a) Deliberate exceedance of service lifetime

Appendix C

Table A3. Hieminga (2018) Maintenance error Taxonomy.

Level 1	Level 2
Maintenance Control	Work orders not carried out Mismatch between logs/work order and work carried out/actual configuration Scheduled tasks overdue Mismatch between MX forecast and actual times/cycles Action not signed off AD not embodied/not in compliance on a/c Action sign off/explanation incorrect or unclear Instructions/limitations to other team/shift/department not communicated/unclear/incorrect Additional inspections not planned/carried out Defect deferred with incorrect procedure/reference/follow-up Work orders/task not in planning, or planned with incorrect interval
Maintenance documentation	Authorisation does not cover work carried out/authorisation issued Instructions or references incorrect/unclear Incorrect or incomplete documentation present/used
Parts supply/tracking/life limits	Incorrect part supplied Parts supplied with incomplete/incorrect repair, modification, configuration or condition. Parts supplied with FOD/damage/corrosion present Parts supplied with incorrect/incompatible life remaining Mismatch between parts installed and tracking system Incorrect life recorded in tracking system Time expired parts (found to be) fitted Uncertainty about part documentation
Tool issue	Incorrect tool used/available Tools used had incorrect calibration status
Job access/job set-up issue	Incomplete/incorrect job set up Damage caused by access equipment Damage caused by lifting equipment/special tools
Working practices	Created opportunity for damage/contamination/FOD Accumulation of dirt/fluids/grease/water/other contamination present Damage present, or damage caused by work carried out Incorrect procedure used or procedure applied incorrectly
Troubleshooting issues	Results incorrect Results unclear Previous troubleshooting did not clear the issue
Lubrication/servicing issue	Lubrication not (correctly) carried out Wrong type lubricant used Lubrication overdue Servicing not (correctly) carried out Refill task incomplete/incorrect Servicing overdue
Inspection/testing issue	Inspection or test not carried out or not complete Inspection or test carried out incorrectly Inspection or test results not carried forward Inspection or test did not identify an existing issue
Installation/removal issue	Clearance issue Part missing Part incorrect Part unserviceable Installation/removal incomplete Damage present, caused by installation/removal Installation/removal incorrect Wrong (consumable) material used Wrong fastener used Wrong software version loaded or wrong config/setup Incorrect/incomplete follow-up after installation/removal

Table A3. *Cont.*

Level 1	Level 2
Modification/repair issue	Modification not carried out IAW AMM/SRM/other instructions AMM/SRM/other instructions for modification not clear Modification completed but technical issues still present Modification completed, incorrect follow-up Repair not carried out IAW AMM/SRM/other instructions AMM/SRM/other instructions for repair not clear Repair completed but technical issues still present Repair completed, incorrect follow-up Uncertainty about status/certification basis for modification/repair
Activation/deactivation issue	Activation/deactivation incorrect Deactivated system/component, but no fault found
Job close-up	Close up not performed correctly Tools/parts/FOD left behind Job not completed

Appendix D

Table A4. Surveys—Subject Matter Experts.

Regulatory Authority Subject Matter Expert Questionnaire
Dear Participant, This study is about identifying and understanding the contributory factors to aircraft maintenance related accidents and incidents in Nigeria. All relevant information regarding the methods used would be made available to you. This survey has been prepared for Aviation Safety Inspectors (ASI) at the Safety Deficiency Incidents Analysis (SDIA) unit of the Nigerian Civil Aviation Authority (NCAA). A total of five open ended questions would be presented to you and your responses/ideas would be highly beneficial to this study.
Q1 What taxonomy do you use in analysing occurrence data? Does this taxonomy support coding of maintenance error or maintenance related occurrences? What other taxonomy/taxonomies would you prefer to use? Please describe your experience and process of analysing the mandatory occurrence reports
Q2 With respect to the data output of this research, please evaluate and discuss your opinion of the methodology used and the output. What could have been done better?
Q3 Please discuss other methods that can be used to identify and prioritise aircraft maintenance related high risk areas. Do you think developing customised taxonomies for maintenance related events would help identify high risk areas in Nigeria?
Q4 In order to further predict incidents, make adequate plans (such as new rule making, safety promotion, training, workshops, increase/targeted oversight etc.) using the results of this data analysis, what methods can you recommend for aviation regulatory authorities and all relevant stakeholders?
Q5 Please discuss the main challenges in terms of data integrity or quality. Is there sufficient detail and information available within the MORs submitted/dataset to determine human factors related causal and contributory factors?
Q2 As an air Accident Investigator with the Accident Investigation Bureau, are you satisfied with the depth of human factors included in your training? Do you have a separate department which focuses on Human Factors related issues such as human factors in aircraft maintenance?
Q3 With respect to the data outputs of this research, please evaluate and discuss your opinion of the methodology used and the output. What could have been done better?
Accident Investigation Bureau Subject Matter Expert Questionnaire
Dear Participant, The aim of this study is to explore the contributory factors to aircraft maintenance-related accidents and incidents in Nigeria in order to achieve a deeper understanding to this safety critical aspect of the aviation industry. To achieve this aim, one of the objectives was to qualitatively analyse the accident investigation reports published by the Accident Investigation Bureau in the last 10 years. This was achieved by using Insley's (2018) Maintenance Factors Analysis and Classification System (MxFACS) taxonomy to code the data.

Table A4. *Cont.*

The results of the analysis showed that the aircraft maintenance-related accidents were attributed to the following contributory factors.

A. Human Factors
B. Operator's oversight
C. Inadequate inspection
D. Incorrect maintenance
E. Failing to follow procedures
F. Noncompliance with Airworthiness Directives

This questionnaire is designed for Air Safety Investigators of the Accident Investigation Bureau (Nigeria). A total of three open ended questions would be presented to you and your responses/ideas would be highly beneficial to this study.

Q1 Does the Accident Investigation Bureau carry out long-term (e.g., last 10 years) reviews of previous accident trends? Do you think that such reviews (e.g., the one carried out in this study focusing on airworthiness and maintenance) may help to identify and prioritise high risk areas and plan mitigation actions such as targeted oversight, rulemaking or safety promotion?

Q2 As an air Accident Investigator with the Accident Investigation Bureau, are you satisfied with the depth of human factors included in your training?
Do you have a separate department which focuses on Human Factors related issues such as human factors in aircraft maintenance?

Q3 With respect to the data outputs of this research, please evaluate and discuss your opinion of the methodology used and the output. What could have been done better?

Aircraft Maintenance Engineers Questionnaire

Dear Participant,
The aim of this study is to explore the contributory factors to aircraft maintenance-related accidents and serious incidents in Nigeria in order to achieve a deeper understanding to this safety critical aspect of the aviation industry.
To achieve this aim, one of the objectives was to qualitatively analyse the accident investigation reports published by the Accident Investigation Bureau in the last 10 years.
This was achieved by using Insley's (2018) Maintenance Factors Analysis and Classification System (MxFACS) taxonomy to code the accidents and Hieminga's (2018) taxonomy to code the serious incidents.
The results of the analysis showed that the aircraft maintenance-related accidents were attributed to the following contributory factors listed in alphabetical order.

A. Failing to follow procedures
B. Human Factors
C. Inadequate inspection
D. Incorrect maintenance
E. Noncompliance with Airworthiness Directives
F. Operator's oversight

This questionnaire is designed for Aircraft Maintenance Engineers (AME) in Nigeria to identify root cause of study findings the output of the data analysis.

Q1 Are you an aircraft maintenance engineer?

Q2 How many years of experience do you have in the aircraft maintenance industry in Nigeria?

Q3 From your experience and in your view, which of the following contributory factors to maintenance related accidents are more likely to cause future accidents? Where (one) 1 indicates least likely and (seven) 7 indicates most likely

1. Not following AMM (Incorrect information, missing information, failure to follow procedure, procedures difficult to follow)
2. Non-compliance with Airworthiness directive
3. Human Factors (Maintenance engineer fatigue, time pressure, unqualified personnel
4. FOD (Tool left in aircraft, Contaminants in aircraft)
5. Inspection (Inspection does not identify defect, Inspection not undertaken, Insufficient inspection)
6. Organisational (Inadequate maintenance documentation, Inadequate reporting, Inadequate training, Lack of training, Misleading paperwork, Poor resource planning)
7. Incorrect maintenance (Incorrect adjustment, Incorrect assembly, Incorrect component installed, Incorrect installation, Incorrect procedure, Incorrect rigging)

Table A4. *Cont.*

Q4 Following maintenance instructions in the AMM/SRM etc. is not always practical/possible. Strongly agree/Agree/Somewhat agree/Neither agree nor disagree/Somewhat disagree/Disagree Strongly disagree
Q5 Please briefly elaborate your experience with regards to following procedures. If they are not always practical/possible to follow, what are the main reasons/challenges which prevents you from following them to the letter? Please recommend possible solutions
Q6 Some inspection instructions are not sufficient enough to identify defects Strongly agree/Agree/Somewhat agree/Neither agree nor disagree/Somewhat disagree/Disagree/Strongly disagree
Q7 Please briefly elaborate your experience with regards to inadequate inspection instructions. What are the main challenges faced with inspection instructions? Please recommend possible solutions

References

1. Chang, Y.H.; Wang, Y.C. Significant human risk factors in aircraft maintenance technicians. *Saf. Sci.* **2010**, *48*, 54–62. [CrossRef]
2. Zhong-Ji, T.; Zhiqiang, Z.; Yan-Bin, S. The Overview of the Health Monitoring Management System. *Phys. Procedia* **2012**, *33*, 1323–1329. [CrossRef]
3. Nigeria Civil Aviation Regulations Part 18 (Nig. CARs). Available online: https://ncaa.gov.ng/media/qkqodgnd/aviation-part-18.pdf (accessed on 20 May 2020).
4. Ladan, S.I. An analysis of air transportation in Nigeria'. *JORIND* **2012**, *10*, 230–237.
5. Air Accidents Investigation Branch. *Annual Safety Review*; Air Accidents Investigation Branch: Farnborough, UK, 2017. Available online: https://assets.publishing.service.gov.uk/media/594a62dfed915d0baa00001a/AAIB_Annual_Safety_Review_2016.pdf (accessed on 20 August 2019).
6. Daramola, A.Y. An investigation of air accidents in Nigeria using the Human Factors Analysis and Classification System (HFACS) framework'. *J. Air Transp. Manag.* **2014**, *35*, 39–50. [CrossRef]
7. EASA. *Annual Safety Review 2019*; European Aviation Safety Agency: Cologne, Germany, 2019; Available online: https://airportcreators.com/compliance-certification/?gclid=Cj0KCQiAwf39BRCCARIsALXWETyus3uqDmz-B5cl0I4SLT8qUPXXFKuO0nefs-WZIPXwrpHA8A2gyYcaAn7bEALw_wcB (accessed on 10 December 2020).
8. Air Accidents Investigation Branch. *Annual Safety Review 2017*; Air Accidents Investigation Branch: Farnborough, UK, 2018. Available online: https://assets.publishing.service.gov.uk/media/5a9ff04a40f0b64d821c4005/AAIB_Annual_Safety_Review_2017_Hi_res.pdf (accessed on 20 August 2019).
9. AAIB. *Annual Safety Review 2018*; Air Accidents Investigation Branch: Farnborough, UK, 2019. Available online: https://assets.publishing.service.gov.uk/media/5cb5ad2ded915d3f4c2eabe5/AAIB_Annual_Safety_Review_Hi_Res.pdf (accessed on 20 August 2019).
10. IATA. *Safety Report 2018*; International Air Transport Association: Montréal, QC, Canada, 2019; Available online: https://libraryonline.erau.edu/online-full-text/iata-safety-reports/IATA-Safety-Report-2018.pdf (accessed on 25 August 2020).
11. Eurostat, Statistics Explained; Air Statistics in the EU. Available online: https://ec.europa.eu/eurostat/statistics-explained/index.php/Air_safety_statistics_in_the_EU (accessed on 21 July 2019).
12. IATA. *IATA Annual Review 2018*; International Air Transport Association: Montréal, QC, Canada, 2018; Available online: https://www.iata.org/contentassets/c81222d96c9a4e0bb4ff6ced0126f0bb/iata-annual-review-2018.pdf (accessed on 1 June 2018).
13. Vanguard, NWAFOR. Nigeria's Airlines Fly Africa's Oldest Planes. Vanguard. Available online: https://www.vanguardngr.com/2018/11/nigerias-airlines-fly-africas-oldestplanesdanas-aircraft-28-years-old/ (accessed on 21 July 2019).
14. Yonggang, W.; Honglang, L. Summary and Analysis of the Aging Aircrafts' Failure'. *Procedia Eng.* **2011**, *17*, 303–309. [CrossRef]
15. European Commission. The EU Air Safety List. Available online: https://ec.europa.eu/transport/modes/air/safety/air-ban_en (accessed on 2 June 2020).

16. Insley, J. A contemporary Analysis of Aircraft Maintenance-Related Accidents and Serious Incidents. Master's Thesis, Cranfield University, Cranfield, UK, 2018.
17. Hieminga, J. Identification of safety improvements in aviation maintenance using EU-wide incident reports. Master's Thesis, Cranfield University, Cranfield, UK, 2018.
18. Reason, J. *A Life in Error: From Little Slips to Big Disasters*; Surrey: Ashgate, UK, 2013.
19. Reason, J. Safety in the operating theatre—Part 2: Human error and organisational failure. *Curr. Anaesth. Crit. Care* **1995**, *6*, 21–126. [CrossRef]
20. Dekker, S. *The Field Guide to Human Error Investigations*; Ashgate Publishing International: Aldershot, UK, 2002; p. 63.
21. Bowker, G.C.; Star, S.L. Building Information Infrastructures for Social Worlds—The Role of Classifications and Standards. In *Community Computing and Support Systems*; Springer: Berlin/Heidelberg, Germany, 1998; pp. 231–248.
22. Lambe, P. *Organising Knowledge: Taxonomies, Knowledge and Organisational Effectiveness*; Chandos Publishing: Oxford/London, UK, 2007.
23. International Civil Aviation Organisation. *ICAO Safety Report*; ICAO: Montréal, QC, Canada, 2018.
24. Shappell, S.A.; Wiegmann, D.A. A human error approach to accident investigation: The taxonomy of unsafe operations. *Int. J. Aviat. Psychol.* **1997**, *7*, 269–291. [CrossRef]
25. ICAO. (2003) Doc. 9824 AN/450: Human Factors Guidelines for Aircraft Maintenance Manual. In *Commercial Aviation Safety Team. International Civil Aviation Organisation Common Taxonomy Team History*; International Civil Aviation Organisation (ICAO): Montréal, QC, Canada, 2014; Available online: http://www.intlaviationstandards.org/Documents/CICTTStandardBriefing.pdf (accessed on 20 August 2019).
26. Stojić, S.; Vittek, P.; Plos, V.; Lališ, A. Taxonomies and their role in the aviation Safety Management Systems. *e-Xclusive J.* **2015**, *1*, 1–8.
27. Whiting, B.D. Improved technical airworthiness taxonomy: Capturing business intelligence to support an effective safety management system. In Proceedings of the AIAC18: 18th Australian International Aerospace Congress (2019): HUMS-11th Defence Science and Technology (DST) International Conference on Health and Usage Monitoring (HUMS 2019): ISSFD-27th International Symposium on Space Flight Dynamics (ISSFD), Melbourne, Australia, 24–25 February 2019; Engineers Australia, Royal Aeronautical Society: Canberra, Australia, 2019; p. 93.
28. Cheng, L.; Schon, Z.Y.; Arnaldo Valdés, R.M.; Gómez Comendador, V.F.; Sáez Nieto, F.J. A Case Study of Fishbone Sequential Diagram Application and ADREP Taxonomy Codification in Conventional ATM Incident Investigation. *Symmetry* **2019**, *11*, 491. [CrossRef]
29. Zhang, M.; Zhang, D.; Goerlandt, F.; Yan, X.; Kujala, P. Use of HFACS and fault tree model for collision risk factors analysis of icebreaker assistance in ice-covered waters. *Saf. Sci.* **2019**, *111*, 128–143. [CrossRef]
30. Rashid, H.S.J.; Place, C.S.; Braithwaite, G.R. Helicopter maintenance error analysis: Beyond the third order of the HFACS-ME. *Int. J. Ind. Ergon.* **2010**, *40*, 636–647. [CrossRef]
31. Rankin, W. MEdA investigation processes. In *Boeing Commercial Aero Magazine*; 2007; pp. 15–21.
32. Erickson. *UK CAA (2015) CAP 1367–Aircraft Maintenance Incident Analysis*; Civil Aviation Authority: WEST Sussex, UK, 2015; Available online: http://publicapps.caa.co.uk/docs/33/CAP%201367%20template%20w%20charts.pdf (accessed on 20 August 2019).
33. Erickson. *UK CAA Paper 2009/05: Aircraft Maintenance Incident Analysis*; Civil Aviation Authority: West Sussex, UK, 2009.
34. Hobbs, A.; Williamson, A. Associations between errors and contributing factors in aircraft maintenance. *Hum. Factors* **2003**, *45*, 186–201. [CrossRef]
35. Johnston, N.; McDonald, N. *Aviation Psychology in Practice*, 1st ed.; Routledge: London, UK, 1994; pp. 1–390.
36. Insley, J.; Turkoglu, C.; A Comprehensive Review of Accidents and Serious Incidents. Royal Aeronautical Society Human Factors Engineering Group Conference. 2019. Available online: https://www.youtube.com/watch?v=yIlJRfd8F0g (accessed on 22 May 2020).
37. Drury, C.G. Errors in aviation maintenance: Taxonomy and control. *Proc. Hum. Factors Soc. Annu. Meet.* **1991**, *35*, 42–46. [CrossRef]
38. Dempsey, P.S. Independence of aviation safety investigation authorities: Keeping the foxes from the henhouse. *J. Air L. Com.* **2010**, *75*, 223.
39. Bryman, A.; Bell, E. *Business Research Methods*; Bell & Bain Ltd.: Glasgow, UK, 2007; Volume 4.

40. Nowell, L.S.; Norris, J.M.; White, D.E.; Moules, N.J. Thematic analysis: Striving to meet the trustworthiness criteria. *Int. J. Qual. Methods* **2017**, *16*, 1609406917733847. [CrossRef]
41. Holloway, C.M.; Johnson, C.W. *Distribution of Causes in Selected US Aviation Accident Reports between 1996 and 2003*; Nasa Technical Reports Server 2004; CiteSeerX: Washington, DC, USA, 2014.
42. Braun, V.; Clarke, V. Using thematic analysis in psychology. *Qual. Res. Psychol.* **2006**, *3*, 77–101. [CrossRef]
43. Brink, H.I. Validity and reliability in qualitative research. *Curationis* **1993**, *16*, 35–38. [CrossRef] [PubMed]
44. Liamputtong, P.; Ezzy, D. *Qualitative Research Methods*; Oxford University Press: Melbourne, Australia, 2005; Volume 2.
45. Wirtz, M.; Kutschmann, M. Analyzing interrater agreement for categorical data using Cohen's kappa and alternative coefficients. *Die Rehabil.* **2007**, *46*, 370–377. [CrossRef]
46. Klopper, H. The qualitative research proposal. *Curationis* **2008**, *31*, 62–72. [CrossRef]
47. McHugh, M.L. Interrater reliability: The kappa statistic. *Biochem. Med.* **2012**, *22*, 276–282. [CrossRef]
48. Nigerian Accident Investigation Bureau (NAIB). Aircraft Accident Report Dana/2012/06/03/F. Available online: https://reports.aviation-safety.net/2012/20120603-0_MD83_5N-RAM.pdf (accessed on 22 May 2020).
49. Reason, J.; Hobbs, A. *Managing Maintenance Error: A Practical Guide*, 1st ed.; Ashgate CRC Press: Aldershot, UK, 2003.
50. Siebenmark, J.; FAA Adviser: Following procedure not just AMT problem. AINONLINE. Business Aviation. 7 May 2019. Available online: https://www.ainonline.com/aviation-news/business-aviation/2019-05-07/faa-adviser-following-procedure-not-just-amt-problem (accessed on 20 August 2019).
51. Sarter, N.B.; Alexander, H.M. Error types and related error detection mechanisms in the aviation domain: An analysis of aviation safety reporting system incident reports. *Int. J. Aviat. Psychol.* **2000**, *10*, 189–206. [CrossRef]
52. Dhillon, B.S. *Human Reliability, Error, and Human Factors in Engineering Maintenance: With Reference to Aviation and Power Generation*; CRC Press: Boca Raton, FL, USA, 2009; pp. 1–109.
53. Kallos, G. Dust impact on aviation. In Proceedings of the 6th International Workshop on Sand/Dust Storms and Associated Dust Fall, Athens, Greece, 7–9 September 2011.
54. Latorella, K.A.; Prabhu, P.V. A review of human error in aviation maintenance and inspection. *Int. J. Ind. Ergon.* **2000**, *26*, 133–161. [CrossRef]
55. Boeing Airplane Company. *Accident Prevention Strategies: Removing Links in the Accident Chain*; Boeing Commercial Airplane Group: Seattle, WA, USA, 1993.
56. Rogers, B.L.; Hamblin, C.J.; Chaparro, A. Classification and analysis of errors reported in aircraft maintenance manuals. *Int. J. Appl. Aviat. Stud.* **2008**, *8*, 295.
57. Hobbs, A.; Williamson, A. *Aircraft Maintenance Safety Survey: Results*; Department of Transport and Regional Services: Sydney, Australia, 1994.
58. Olufunke, A.M. Education for maintenance culture in Nigeria: Implications for community development. *Int. J. Sociol. Anthropol.* **2011**, *3*, 290–294.
59. Saleh, J.H.; Ray, A.T.; Zhang, K.S.; Churchwell, J.S. Maintenance and inspection as risk factors in helicopter accidents: Analysis and recommendations. *PLoS ONE* **2019**, *14*, e0211424. [CrossRef]
60. Johnson, W.B.; Watson, J. *Reducing Installation Error in Airline Maintenance*; Federal Aviation Administration/ Office of Aviation Medicine: Washington, DC, USA, 2001.
61. Drury, C.G.; Johnson, W.B. Writing Aviation Maintenance Procedures that People Can/Will Follow. *Proc. Hum. Factors Ergon. Soc. Annu. Meet.* **2013**, *57*, 997–1001. [CrossRef]
62. Sumwalt, R.L. *The Role of Maintenance and Inspection in Accident Prevention*; National Transportation Safety Board, Inspection Authorization Renewal Seminar: Washington, DC, USA, 2014. Available online: https://www.ntsb.gov/news/speeches/RSumwalt/Documents/Sumwalt_140208 (accessed on 10 December 2020).
63. Hussin, R.; Ismail, N.; Mustapa, S. A Study of Foreign Object Damage (FOD) and Prevention Method at the Airport and Aircraft Maintenance Area. *IOP Conf. Ser. Mater. Sci. Eng.* **2016**, *152*, 012038. [CrossRef]
64. Drury, C.G.; Guy, K.P.; Wenner, C.A. Outsourcing aviation maintenance: Human factors implications, specifically for communications. *Int. J. Aviat. Psychol.* **2010**, *20*, 124–143. [CrossRef]
65. Marais, K.B.; Robichaud, M.R. Analysis of trends in aviation maintenance risk: An empirical approach. *Reliab. Eng. Syst. Saf.* **2012**, *106*, 104–118. [CrossRef]

66. Cowan, T.; Acar, E.; Francolin, C. *Analysis of Causes and Statistics of Commercial Jet Plane Accidents between 1983 and 2003*; Mechanical and Aerospace Engineering Department, University of Florida: Gainesville, FL, USA, 2006.
67. Pettersen, K.A.; Aase, K. Explaining safe work practices in aviation line maintenance. *Saf. Sci.* **2008**, *46*, 510–519. [CrossRef]

Publisher's Note: MDPI stays neutral with regard to jurisdictional claims in published maps and institutional affiliations.

Article

Learning from Incidents: A Qualitative Study in the Continuing Airworthiness Sector

James Clare [1] and Kyriakos I. Kourousis [1,2,*]

[1] School of Engineering, University of Limerick, V94 T9PX Limerick, Ireland; james.clare@ul.ie
[2] School of Engineering, RMIT University, Melbourne, VIC 3001, Australia
* Correspondence: kyriakos.kourousis@ul.ie

Received: 16 December 2020; Accepted: 18 January 2021; Published: 22 January 2021

Abstract: Learning from incidents (LFI) is a useful approach when examining past events and developing measures to prevent ensuing recurrence. Although the reporting of incidents in the aircraft maintenance and continuing airworthiness domain is well appointed, it is often unclear how the maximum effect of safety data can be efficaciously applied in support of LFI in the area. From semi-structured interviews, with thirty-four participants, the gathered data were thematically analyzed with the support of NVivo software. This study establishes a relationship between an incident in its lifecycle and the learning process. The main aim of this work is to elucidate factors that enable LFI. The analysis of the data revealed, for example, the benefits of a just culture and the use of formal continuation training programs in this respect. Moreover, it identified limitations inherent in current processes such as poor event causation and poorly designed learning syllabi. Additionally, aspects such as a lack of regulatory requirements for competence in the areas of learning for managers and accountable persons currently exist. This thematic analysis could be used in support of organizations examining their own processes for learning from incidents. Additionally, it can support the development of terms of reference for a continuing airworthiness regulatory working group to examine, strengthen and better apply LFI in the aviation industry.

Keywords: learning form incidents; airworthiness; aircraft maintenance; safety management

1. Introduction

If it were possible for all organizations to learn effective lessons from the past, the effects of future unwelcome events might be limited [1]. Aviation safety depends to a large extent on the efficacious efforts of all involved in the system [2]. Research has acknowledged the importance of event information when it comes to learning and preventing recurrence [3]. Thankfully, major events such as accidents are becoming less frequent and generate less points for learning [4]. In contrast, there are numerous incidents with less severe consequences and if appropriately considered, these could offer an earlier insight into the circumstances that enable unwelcome events. Predefined and relevant information harvested from incident reporting systems is a major element of learning and preserving acceptable levels of safety. Hobbs and Williamson [5] highlight the importance of aircraft maintenance staff being aware of the cumulative effect of "seemingly insignificant" incidents as this amplifies the need to be proactive when it comes to learning from incidents. This research undertook a qualitative examination of staff involved in aircraft maintenance and continuing airworthiness operations in order to identify factors that could augment learning from incidents within this industry sector.

In the areas of continuing airworthiness and aircraft maintenance, safety management systems include incident and occurrence reporting [6] as an obligation. It is common for incidents to be discovered within organizations and reported with the assistance of such "systems of systems" [7]. On an operational level, initial training on human factors and company procedures is intended to specify and re-affirm the category and type of occurrence and incident that should be reported. Recent developments in European Union (EU) regulations [8] empower voluntary and confidential reporting and are independent of all other individual obligations. Detecting and identifying hazards highlighted through incident reporting systems is also recommended by the International Civil Aviation Organization (ICAO) standards and recommended practices as an effective means of augmenting levels of safety. However, Gerede [9] strongly suggests that a failure to foster a just culture is considered to have a negative impact upon effective data collection (reporting), organizational learning and the subsequent ability to learn from incidents.

Drupsteen and Wybo [10] reaffirm organizations use experience gained from past events in order to improve safety. Effective learning can be considered as a successful translation of safety information into knowledge. Utilizing information from events with learning potential can actively improve the operating environment and help prevent recurrence. Learning in this context can often be experienced as modifying or implementing new knowledge where cultural, technical or procedural elements are integrated. Therefore, when learning is transformed into measures to prevent re-occurrence, an organization often has a reasonable means of mitigating future similar events. Argyris and Schön [11] highlight the importance of learning to detect and address effective responses to errors. Their "theory in action" concept is the focal point for this determination. The first of its two components, "theory in use" is one that guides a person's behavior. It is often "tacit" and is how people behave routinely. Very often these observed "habits" are unknown to the specimen. The second element is known as "espoused theory", namely what people say or think they do. Drupsteen and Guldenmund [12] mention that espoused theory comprises of "the words we use to convey what we do, or what we like others to think we do".

However, it is important to re-affirm the linkages that exist between individuals and organizational learning. The introduction of safety management systems (SMS) has initiated a shift in how organizational errors are viewed. Firstly, equipment has become increasingly more reliable, but the human form has not displayed the same response. In the second instance, the impact of complexities associated with an increasing cognitive load for staff is just beginning to be realized. The existence of a potential for blaming an individual is now being aligned with organizational responsibilities. Prior to this, event causation was often misrepresented or even over quantified the human input as organizational factors were not always considered. They offer an insight into the connection between individual actions and organizational initiatives designed to secure the best safety outcomes. Fogarty et al. [13] also recognize the role that both individual factors have on human error and the inputs both can have on preventing recurrence.

ICAO Doc 9859 ICAO [14] defines a template for aviation operators and regulators to support the application of a variety of proactive, predictive and reactive oversight methodologies. In addition to routine monitoring schemes, voluntary and mandatory reporting, post incident follow-up, there are also regular safety oversight audits. These audits and inspections often set out to establish if there is a difference between espoused theory and the theory in use (e.g., is the task being correctly performed in accordance with the documented procedure/work instruction or is there a deviation from approved data and practice?). However, Drupsteen and Guldenmund [12] caution auditors not to "focus too much on the documentation of procedures" alone. In such cases, the oversight audit may be ineffective because of its sole focus on espoused theories of the organization only and not the theory-in-use. These authors translate this idea of poor focus on theory in action, into a valid learning component arising from incidents. They also highlight the "espoused" aspect where those attempting to learn from incidents often fail to experience the desired learning because outcomes are not fully aligned with the practical objectives of a learning from incidents (LFI) initiative. For learning to be most effective, espoused theory and theory in use should be reasonably well aligned. Ward et al. [15] propose it is necessary to further develop an operational model that can account for "what is meant to happen and what actually happens".

Continuing airworthiness and aircraft maintenance and activities performed in EU member states are subjected to rules that mandate reporting of defined issues. Repositories of reported data tend to be populated by sources that are predominantly the subject of mandatory reporting requirements. Conventional safety oversight models also only verify the presence of reporting media and repositories in this segment of the industry. Jacobsson et al. [16] avow the degree of interest invested in learning from incidents but question its efficiency in some organizations. Although unwelcome events are less prevalent, less severe events still provide learning opportunities. There is often only a primary focus for organizations upon reporting in line with each state's own reporting obligations. Unfortunately, a narrow focus on this single element of an incident in its lifecycle can negate the potential benefits of learning from incidents at an organizational level. The absence of clearly defined competency requirements [6] that support a pedagogy for learning from incidents for continuing airworthiness staff could also be considered an impediment to effective learning in the domain.

The featured industry sector is regulated by the application and upkeep of numerous requirements in the jurisdictions of operation. In general, a costly regulatory overhead tends to be carried by regulating states and operators to support safe and viable activity. However, a growing tendency to increase regulatory requirements in pursuance of safer activity across the segments may not always offer the same returns as previously realized by states. Brunel [17] (p.45) suggests, " ... it is impossible to make men perfect: the men will always remain the same as they are now and no legislation will make him have more presence of mind ... ". Furniss et al. [18] reviewed the Hollnagel [19] Functional Resonance Analysis Method (FRAM) which explores how functional variability resonates within systems, i.e., how well comprising elements function in a system. They also consider how FRAM can be modified to support complex socio technical system improvements. Perhaps as the paradigm supporting the linearity of regulatory oversight shifts, proactive regulatory inputs will also influence more effective safety outputs as intricacy increases.

1.1. *Systematic Literature Review*

The primary reason for conducting a systematic review was to examine how learning from incidents occurs in aircraft maintenance and continuing airworthiness management. Other sectors and the issues impacting learning in these areas were also considered. The literature review sets out to establish factors that contribute to or potentially constrain learning from incidents in the subject domain. Applying a qualitative research approach is advantageous as it can provide a deeper contextual understanding of the literature and can assist with better research integration. The application of rigor and comprehensiveness can assist with advancing knowledge and identifying research gaps and aspects for further research in this area. Okoli and Schabram [20] suggest "a dedicated methodological approach is necessary in any kind of literature review". A preliminary search of literature highlighted a scarcity of best-practice guidelines for conducting systematic literature reviews in this area.

Qualitative research involves handling considerable volumes of data and a degree of discipline is required so that search results, decisions regarding subject inclusions and exclusions are recorded and references are well managed. Endnote was used in support of the literature review during this research. An electronic database is useful for supporting a search strategy, arranging publications and storing references [21]. The qualitative data analysis software NVivo (NVivo 12, QSR International, Melbourne, Australia) was used to augment the data management, storage and analysis associated with the literature review. NVivo possesses many functions, such as facilitating the synthesis of a review [22]. A systematic search of in excess of 1000 publications was performed in the following databases: Web of Science, Scopus, IEEE Xplore, ProQuest and EBSCO. The following predefined search terms were applied: "learning from incidents", "learning from experience", "aircraft maintenance", "aircraft management" and "safety management systems". A practical screening of title and abstract was applied to each manuscript using predefined terms (e.g., subject, setting, publication, year). This part of the process had to be broad enough to create a sufficient number of applicable publications but also had to be practically manageable. The following criteria were implemented for the practical screening of the source bibliographic details, title and abstract:

- Subject—Related to learning from incidents and past experiences.
- Setting—Any high reliability industry or sector where learning from incidents is critical.
- Publication—Journal or peer-reviewed conference proceedings.
- Date range—published post 1992. The year 1992 was the starting point for the screening process, since at the time of planning the research project, 25 years was considered to be a reasonable timespan to include material pertaining to learning from incidents.

The output of the practical screen step produced a list of publications denoted as the screened set of publications. An Endnote library was then created to store and manage the full text of the retrieved publications. The next step involved filtering the publications into primary and secondary publication subsets using only primary research manuscripts in the next phase. Applying a set of criteria helps to reduce any researcher bias in the screening system. A set of inclusion and exclusion criteria [23] was developed in accordance with the guidelines included in [24,25], listed in Table 1. Two researchers were involved in the screening process.

Table 1. Inclusion and exclusion criteria used for the filtering of the subset of primary publications.

Included	Excluded
Research studies	Literature reviews
Qualitative and mixed methods	Quantitative methods
Perceptions and experiences	Focused on decision-making and legislative requirements
Reference to just culture	Not about "no blame" or a punitive approach
High reliability settings	Non high reliability settings
Published post 1992	
Peer-reviewed publications	
Industry based settings	
Original studies	

The final set of 18 papers was imported into NVivo and the following analysis approach, as defined by Bandara et al. [22], was used for the selection of the codification themes:

- Deductive—themes reported on are predetermined to some extent. In this case, these predetermined themes were the output of a focus group process.
- Inductive—themes reported are derived from analysis of the literature.

NVivo is limited in terms of providing thematic classifications based on the occurrence of key words but can assist with identifying relationships between words and phrases amongst publications. It also provided thematic classifications of data based on the occurrence of key words and phrases. The coding process consisted of selecting relevant passages of text that were captured in one or several of the framework nodes. Maykut and Morehouse [26] defines a propositional statement as "a statement of fact the researcher tentatively proposes, based on the data". Memos were used to draft these summary statements which formed part of the literature review. Central to the idea of learning is how an incident is generally moderated during its useful existence. Section 1.2 documents this approach.

1.2. The Notion of a Generic Incident Lifecycle

Figure 1 illustrates how an incident tends to be managed through its quiddity. This view is one possible way of representing the elements comprising a lifecycle view. Cooke and Rohleder [27] suggest it should also be evident that an incident system will operate most effectively when a safety management system has already been put in place and avoidable risks are addressed. They propose an effective system that addresses: identification and response, reporting, investigation, identifying causal structure, making recommendations, communicating and recalling incident learning, and implementing corrective actions. Drupsteen et al. [28] also consider an incident from a learning perspective in its cycle. Their main constituents are investigating and analyzing incidents, planning interventions, intervening and evaluating (each of these four stages are further sub-divided into eleven sub-components). Continuing airworthiness-related incidents are notified by way of a formal mechanism of reporting. During the data gathering phase of this research, the steps outlined in Figure 1 were found to be dictated by regulatory requirements [6,8]. Once the incident enters its lifecycle, it ideally transverses a process that transforms the information gathered into knowledge. Figure 1 and the contiguous paragraph offer an overview of how the capture and processing of the incident information occurs in practice.

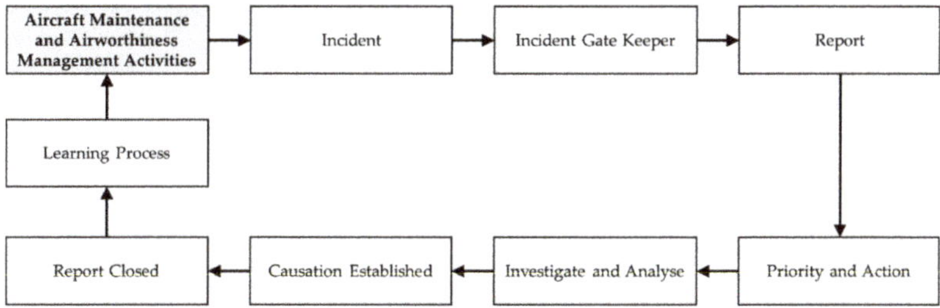

Figure 1. An example of an aviation incident lifecycle within the continuing airworthiness and aircraft maintenance sector.

Continuing airworthiness-related serious incidents are rare but often due to environmental, cognitive and mechanical demands, reportable and unreportable events do occur. All organizations in the industry segment subscribe to a reporting system and reports can be made electronically or in paper form in smaller organizations. The main underpinning regulation in Europe, EU regulation 2018/1139 [29], refers to a management system and mandates an organization to implement and maintain such a system to ensure compliance with these essential requirements. In practice, although a reporter can report events directly to an aviation authority, all organizations are required to have an internal reporting system also. A focal point/gate keeper will process these reports either internally and/or inform third party stakeholders such as aviation authority or aircraft manufacturer as required by procedure. Depending on the event, technical management may determine there are immediate actions required to recover a situation or restore serviceability. While a small number of scenarios will require an event to be investigated fully before an aircraft returns to service, many incidents are investigated post event. As soon as causation is established, if accepted by the relevant technical function, the report is closed. This management system is strongly influenced by regulatory requirements and procedural form and is a pre-eminent influence on how an incident and its actors behave from the time a report is made to the time its impact has been terminated. One of the limitations inherent in this cycle is that lessons tend to be delivered at a later point in time mostly through the medium of recurrent training programs such as continuation and human factors training.

Therefore, there is often a hiatus in the feedback cycle. However, the effectiveness of the process and the perceived contribution to learning are not fully reflected in this view.

1.3. A Potential Learning Cycle Emerges

According to Lindberg et al. [30], in order to prevent accidents, it is essential to learn from previous accidents and incidents. Lukic et al. [31] suggest that in order to increase the effectiveness of learning from incidents, it is necessary to understand who should be included in the learning process. In Figure 2, the incident lifecycle is aligned with the learning process in order to highlight where potential improvements might be made. As the incident is managed and causation is established, there are potential avenues open for learning. The ultimate desired outcome is that adequate measures are put in place to prevent a recurrence of the event. However, the lessons available in a potential learning product are not always used to best effect when considering the Figure 1 process. Drupsteen et al. [28] state that "many incidents occur because organizations fail to learn from past lessons", because the traditional approach often stops short of preventing future incidents. Their research examines: investigating and analyzing incidents, planning and prevention, intervening and evaluating steps in a learning process. Ward et al. [15] found that the resulting relationship between the individuals and the systems have a direct impact upon the system and prevailing environment. Silva et al. [3] examined how organizations use accident information to reduce the occurrence of unwelcome events. Drupsteen and Wybo [10] found that hindsight can determine if an organization did learn from an event but there are no models to assist with gauging the "propensity" of an organization to learn. Drupsteen and Hasle [1] suggest that learning can be improved if limiting factors are addressed.

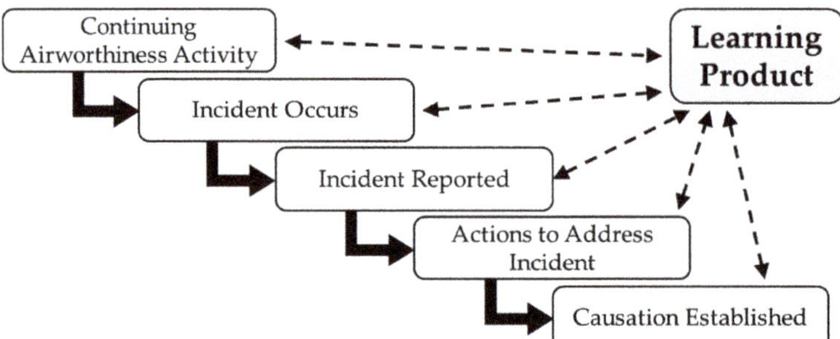

Figure 2. Incident learning product and process (broken line denotes iterative learning feedback).

The proposed enhancement (shown in Figure 2) to the generic lifecycle in the "traditional" approach represents a novel view and brings the learning product into focus. This figure highlights the benefits of ensuring the feedback loop of an incident is centered on the learning product. Treating its development as an iterative process ensures all steps in the cycle are included and where deficiencies are noted, they can be identified and communicated during the iterations. This can assist with delivery of timely and sustainable learning and help prevent an inability to think, talk and see what actions are proper in specific situations [32]. According to Drupsteen et al. [28], it is necessary to gain an insight into the steps of the process to identify factors that hinder learning in order to make improvements. The research suggests an emphasis on accessible, timely and appropriate learning content could provide all stakeholders in the process with better value for their efforts. Perhaps one reason that the customary incident lifecycle and its limitations prevail is related to management theory. While innovators like Taylor [33] are responsible for advances in management, such theories have not always fully considered safety and learning. The early 1900s witnessed a time when it was necessary to inaugurate efficiencies in production by initially decomposing tasks in order to introduce liner efficiencies. The limitations experienced in incident learning processes today may relate to this circumscribed tradition.

2. Materials and Methods

2.1. Philosophical Underpinnings

The fields of science and philosophy consider ontology and epistemology in terms of What is the nature of reality? and How is human knowledge constructed? The ascendant ideologies of positivism and interpretivism can be applied in support of these philosophical differences [34–37]. Hirschheim [38] puts forward the aim of positivism to, "seek to explain and predict what happens in the social world by searching for irregularities and causal relationships between its constituent elements". In contrast to this stance, Schwandt [39] suggests the aim of interpretivism is to gain understanding. Interpretive research seeks to develop a richer understanding of the complex world of lived experience from the point of view of those who live in it. *"This goal is variously spoken of as an abiding concern for the life world, for the emic point of view, for understanding meaning, for grasping the actor's definition of a situation, for verstehen"* [39] (p.118).

The intent of this qualitative study was to understand how various situations impact on learning from incidents in the aircraft maintenance and continuing airworthiness management domain by interacting with the participants on a social plane. Thus, in order to gain an empathetic understanding of the participants and their actions, the pursuit of "verstehen" considers adopting an interpretive paragon as an approach. This approach is not initiated with the aid of a hypothesis intended for testing but rather using a lodestar that guides the researcher to a point of discovery supported with an inductive modus operandi. The study is unwavering in its support for the view that (individual and combined) qualitative and quantitative approaches possess equal value in terms of their investigative potential in this area of focus. In summary, the project employs a qualitative research methodology in an effort to generate "rich" findings in support of gaining a good understanding of the learning environment in the featured domain. According to Maykut and Morehouse [26], the purpose of qualitative research is to discover the inner world of perceptions and meaning-making in order to gain an understanding to describe and explain certain social phenomenon from participants' perspectives. In order to accomplish this, focus group activity was managed concurrently with the literature review. These activities cumulatively generated five themes which were used as the basis for a semi-structured interview template. The project employed a qualitative research methodology in an effort to generate "rich" findings in support of gaining a good understanding of the learning environment in the featured domain. The outcome of a qualitative research initiative was contextual findings as opposed to broad generalizations.

2.2. Focus Group

According to Kitzinger [40], "focus groups are group discussions organized to explore a specific set of issues such as people's views and experiences ... ". The idea of conducting group interviews is not a new one. Bogardus [41] is an early example of a reference to utilizing the group interview. Frey and Fontana [42] suggest that group interviews can be formally structured for a specific purpose or can be performed in a more informal setting where a researcher can "stimulate a group discussion". Specific examples in the literature of focus groups being developed systematically within the area of aircraft maintenance and management are scarce. Frey and Fontana [42] state that although group interviews have implicitly informed research, often they are not formally acknowledged as part of the process. Powell and Single [43] remind us that when recruiting focus group participants, one must be mindful of systemic biases. Averting this was ensured by being careful to enlist the participants from different organizations and different positions of responsibility. Three sessions comprising of three industry professionals within each group were successfully moderated by the researcher. During the three phases of working with the focus group, statements and terms were recorded as dialogue amongst the members and observed. The second meeting of the focus group developed four codes (safety, regulatory compliance, root cause, reporting) that had emerged from the group's earlier outputs. These four codes were further distilled during the focus group activity and were consolidated into two themes (reporting, root cause) that were to eventually form part of the piloted semi-structured interview instrument.

Reporting and root-cause themes were the result of the draft consolidation of the comments and emerging codes. In concert with the focus group activities, a literature review was performed by the researchers and this generated three further themes as reflected in Table 2.

Table 2. Codification themes used in the NVivo analysis of the final set of publications.

Codification Theme	Description	Origin
Root Cause	Reason to establish causation	Focus Group
Reporting	Value of reporting to learning from incidents	Focus Group
Learning from Incidents	Outcomes of learning from incidents	Literature Analysis
Just culture	Impact of just culture on learning from incidents	Literature Analysis
Precursors	Contribution of precursors to learning from incidents	Literature Analysis

The resulting draft semi-structured template (provided in Appendix A) containing the five themes was scrutinized by the focus group. The constituent questions relating to each theme and the running order of the document was subject to many minor changes during the individual piloting of the instrument with the three group members.

2.3. Data Collection

Data were gathered from seven organizations using a semi-structured interview template. The participating organizations were involved in aircraft maintenance and continuing airworthiness activities. Building trust and commitment, as proffered by Chatzi [44] and Chatz et al., [45] was deemed to be a necessary tenet of a successful data collection exercise. Managing the interview process with the support of senior staff complimented visible top-down support for the research and ensured there would be no confusion regarding access to what some organizations often classify as sensitive commercial data. The aim was to explore how learning from incidents occurs and what can constrain learning in the area of focus. The pilot phase ensured the desired outcome of the main data collection phase would be congruent with the aims of the study. The interviews were recorded, transcribed and participants could not be identified from the recordings or transcripts. Full ethical approval for the data gathering was granted by the University.

2.3.1. Instrument

Data were collected using semi-structured interviews, lasting on average sixty minutes. The "aide memoir" was arranged so that participants could offer a flexible response and any emerging themes could be identified. The semi-structured approach facilitated emphasis being placed upon any points that warranted further focus or examination by the researchers. An example of the interview template is included in Appendix A. Interviewees were asked to give an example of a recent incident they were familiar with. The structure of the template, (a) probed process around reporting and (b) elicited the participants perception of learning from incidents within their organizations. Following on from the initial contact on reporting, the participants discussed just culture, learning, root-cause and incident precursors during their individual engagements with the researchers.

2.3.2. Participants

The "key issue in selecting and making decisions about the appropriate unit of analysis is to decide what it is you want to be able to say something about at the end of the study" [46] (p.168). The objective of this study was to investigate individuals' perceptions of how learning from incidents takes place and the obstacles present in the maintenance and continuing airworthiness management domain of the aviation industry segment. There were thirty-four (34) participants in total, as presented in Table 3.

Each of the organizations maintained between 6 and 300 aircraft at the time of the study. While traditional reporting and learning themes were evident outputs from the focus group meetings, it was decided that the data would be collected through one-to-one semi-structured interviews. Semi-structured interviews permitted the researchers to get a deeper understanding of complex organizational and social interactions and at the same time follow a construct. The participating organizations were selected based upon them being accredited to perform aircraft maintenance and continuing airworthiness activities since the inception of EU regulation 1321/2014 [6]. Within this domain, there are categories of staff that are required to be aware of incident reporting and make a report as necessary (e.g., technical managers, certifying staff, quality assurance staff, stores personnel, technical services). Each organization is required under legislation to employ a satisfactory level of staff regardless of their aviation activities. As a minimum, at least one of each of these roles was represented in the study. It was ensured that at least one staff member from each discipline was included in the study and had made a report in the previous twelve months. As certifying staff, technical managers and quality assurance staff are by virtue of their position active reporters (due to their exposure to active operations), staff in these disciplines were well represented in the study's cohort. Participation in the study was on a voluntary basis and all who participated were acquainted with the project prior to performing the interviews. All participants signed consent forms.

Table 3. Participants in the study (n = 34).

Participant Roles	Number
Category B1 Engineer	4
Supervisor	3
Category A Mechanic	3
Quality Assurance Engineer	3
Category B2 Engineer	2
Shift Controller	2
Contract Composite Inspector	1
Inspector	1
Aeronautical Engineer	1
Category B1/B2 Engineer	1
Maintenance Manager	1
Technical Safety Manager	1
Technical Services Manager	1
Line Maintenance Manager	1
Deputy Quality Manager	1
Maintenance Control Manager	1
Maintenance Planner	1
Maintenance Safety Officer	1
Apprentice Technician	1

2.4. Data Analysis

Thematic analysis was the method chosen to support the analysis of the study's data. The Braun and Clarke [47] six-step proposition, which consists of eight discreet cycles, in conjunction with the QDA Training [48] material, formed the basis of the analysis technique. A practical iterative approach was adopted throughout the analysis where the data were formally arranged into discrete phases. The eight individual stages of analysis distributed over the six phases were designed to support a robust and rigorous analysis of the data. Table 4 below illustrates the stages and processes outlined and performed in NVivo and links this to the practical guidelines set out in Braun and Clarke [47]. Their six-step approach that supports the application of thematic analysis is shown in column one and the corresponding application in NVivo is shown in column two. The third column features the strategic elements of coding as the researcher moved from the initial participant-led descriptive coding, to the secondary coding which was more interpretative in nature indicating this phase of coding was both researcher- and participant-led. The final abstraction to themes was researcher informed only. This phase was designed to allow the researchers to engage the participant in direct dialogue with a wider arena such as literature and policy or strategy for example. The fourth and final column illustrates the more iterative nature of the coding, analysis and reporting of proceedings that terminate in a conclusion.

Phase 1 activity involves familiarizing oneself with the transcribed data. In this first phase, the data were loaded into NVivo. It was checked and re-read several times to ensure accuracy of the uploaded transcripts. At the end of the phase activity, initial codes were noted down and retained.

Generating initial codes (open coding: phase 2)—According to Lincoln and Guba [49] (p. 345), a data unit can be defined as the "smallest piece of information about something that can stand by itself, that is, it must be interpretable in the absence of any additional information other than a broad understanding of the context in which the inquiry is carried out". The open coding is intended to systemically organize the data and uncover the essential ideas found in the data [50]. Each discrete unit of data is labelled in line with the phenomenon it represents. The second phase required broad participant-driven open coding of the interview transcripts recorded during the data gathering step of the research study. Features of interest were coded in a systematic way across the complete dataset where data relevant to each code were collected. Clear labels were allocated to these codes and definitions to serve as rules for inclusion [26].

Table 4. Stages and Process Involved in Qualitative Analysis. Adapted from Braun and Clarke [47] and QDATRAINING Training [48] material.

Analytical Process (Braun and Clarke, 2006) [47]	**Practical application of Braun and Clarke in Conjunction with NVivo**	**Strategic Objective**	**Iterative Process Throughout Analysis**
1.Familiarizing yourself with the data	Transcribing data (if necessary), reading and re-reading the data, noting down initial ideas. Import data into the NVivo data management tool	Data Management *(Open and hierarchal coding through NVIVO)*	Assigning data to refined concepts to portray meaning
2. Generating initial codes	Phase 2. Open Coding:Coding interesting features of the data in a systematic fashion across the entire data set, collecting data relevant to each code	Descriptive Accounts *(Reordering, 'coding on' and annotating through NVIVO)*	Refining and distilling more abstract concepts
3. Searching for themes	Phase 3. Categorization of Codes:Collating codes into potential themes, gathering all data relevant to each potential theme		Assigning data to themes/concepts to portray meaning
4. Reviewing themes	Phase 4.Coding on:Checking if the themes work in relation to the coded extracts (level 1) and the entire data set (level 2), generating a thematic "map" of the analysis	Explanatory Accounts *(Extrapolating deeper meaning, drafting summary statements and analytical memos through NVIVO)*	Assigning meaning
5. Defining and naming themes	Phase 5. Data Reduction:On-going analysis to refine the specifics of each theme, and the overall story (storylines) the analysis tells, generating clear definitions and names for each theme		Generating themes and concepts
6. Producing the report	Phase 6. Generating Analytical Memos. Phase 7. —Testing and Validating. Phase 8. Synthesizing Analytical Memos. The final opportunity for analysis. Selection of vivid, compelling extract examples, final analysis of selected extracts, relating back of the analysis to the research question and literature, producing a scholarly report of the analysis [47,48]		

A set of provisional categories was generated for the segmented data to be coded to. These categories were descriptions of concepts and themes in broad terms. They took two forms: researcher-driven and participant-driven. The former was derived from a theoretical framework underpinning the study and the latter from the knowledge gained of the participants' language and customs. Hammersley and Atkinson [51] (p.153) consider the importance of participant-driven categories: "the actual words people use can be of considerable analytic importance as the 'situated vocabularies' employed provide valuable information about the way in which members of a particular culture organize their perceptions of the world, and so engage in the social construction of reality".

Searching for themes—In phase 3, codes from phase 2 were collated into categories of codes by structuring all the data relevant to each potential category into a framework that could be used in support of further analysis. This phase also included distilling, re-labelling and merging common codes that were generated in phase 3 to ensure the labels and definitions for inclusion were an accurate reflection of the coded content. These first-round categories are best described as broad descriptions of concepts and themes. During the analytical process they underwent content and definition change and the existence of the two forms of category provides an important means of traversing between "natural" and "theoretical" discourses. Araujo [52] (p.68) suggests that "codes should be viewed in two ways: as part of the analyst's wider theoretical framework and as grounded in the data.; the process of coding data should be regarded as an important intermediary step in translating social actors' frames of meaning into the frame of theoretical discourse; coding frames therefore, mediate between the 'natural' everyday discourse and the theoretical discourses in social science".

Reviewing themes (coding on) in phase 4 required further decomposition of the study units of data identified in phase 1. This activity was intended to support a greater understanding of the highly qualitative elements and gain a deeper appreciation of the meanings contained within. It should be noted that not every task could be further broken down and this meant that the activity was performed only as required. Restructured codes were broken down into further sub-codes in order to augment a greater understanding of the meanings embedded within them. These distinctive aspects included communication with management, discovering latent issues, just culture, learning lessons, reporting, root causes and story of an incident.

Defining and naming themes in phase 5 of the data analysis was concerned with analyzing the tentative categories identified in phase 2 for their properties and characteristics. This is a pre-cursor to drafting a propositional statement for each category. Developing analytical memos moves the process beyond identification and description of broad categories to a position of analyzing and fusing meanings in the data under each category. This progressed to drafting a statement that aspires to illustrate the concerted meaning of the segments of data coded to each category. Maykut and Morehouse [26] (p.140) defines a propositional statement as, "*a statement of fact the researcher tentatively proposes, based on data*". This phase in addition to further data analysis to refine the specifics for each theme, generated clear definitions and a name for each theme. It also involved data reduction by consolidating categories from all three cycles into a more abstract, philosophical and literature-based thematic framework and conceptually mapping and exploring their relationships with one another for reporting purposes.

Producing the report in phase 6 required analytical memos to be written against the higher-level themes to present an accurate summary of the content of each category and its codes and to also propose findings. The tasks associated with phase 6 included (i) generating analytical memos, (ii) testing and validating and (iii) synthesizing the memos coherently and cohesively, and were performed simultaneously. Writing the analytical memos against the higher-level codes (i.e., learning from incidents, learning process and learning product) required an accurate summary of each category and its codes and findings against categories. These memos considered a few key areas:

1. The content of the cluster of codes which were being reported on.
2. Patterns where relevant.
3. Considering background information noted against participants and examining any patterns relating to participants' profiles.
4. Considering any relationship between codes and their importance in relation to the research questions.
5. Noting any primary sources relating to the context of the relationship with the literature in addition to highlighting any gaps in the literature.

Testing, validating and revising analytical memos was performed in phase 7. The purpose of this was to provide a self-audit of the proposed findings by soliciting evidence in the data beyond just textual quotes in support of the recorded findings and to also expand on deeper meanings within the data. This required the data to be interrogated, not only relying on relationships across and between categories, but also a degree of cross tabulation with demographics, observations and the literature. The outcome of this phase was evidence-based findings as each proposed finding was validated by being rooted in the data themselves and was reliant on the creation of reports in support of substantiated findings.

The discipline of writing analytical memos was used during the data analysis process. Birks et al. [53] believe "memoing serves to assist the researcher in making conceptual leaps from raw data to those abstractions that explain research phenomena in the context of which it is examined". In general, memos were employed at the "ideation" stage when the researcher was developing thought processes and early in the data capture phase. As decisions were made, the early processes and rationale for final analysis iterations were recorded using this medium. Memos were further employed to preserve an objective closeness to the harvested data and to maintain the context of each semi-structured interview at the participating individuals' level. Developing ideas, reasons for considering possible category relationships and connections was also possible through the application of the analytical memo process. The rigorous support memoing offered served to guide the analysis of the data through different levels of abstraction [54]. The rule of this activity served to ensure a high degree of continuity between the outputs of ideation and the evolving interpretation that were honed through the researchers' articulation, exploration and their iterations of the data. Overall, this drew out the meanings in the data through the increased sensitivity the researchers were offered by applying the memoing process [53].

In phase 8, the analytical memos were synthesized into a coherent and cohesive report with the findings well supported. The final phase involved the assembly of the narrative with the data extracts while appreciating the product of this amalgam in the context of the related literature. The example features the finding, clear links to the interview data and literature and an explanatory narrative in the form of a memo. This finally resulted in the compilation of the report which contained the results and discussion elements of the body of work.

In summary, this study adopted an interpretative approach pivoting on the fact that it was of an exploratory nature. The study performed thirty-four interviews in eight aircraft maintenance and management organizations based in Ireland. An analysis of various potential research methods and means of data collection resulted in the following research design being implemented. A thematic analysis approach was employed as a research methodology:

- Unit of analysis is an individual;
- Semi-structured interview guide was constructed following a systematic analysis of literature and the use of a focus group;
- Data were collected through qualitative interviews;
- Thirty-four interviews were collected in locations endorsed by eight organizations;

- Qualitative analysis based on the guidelines from Braun and Clarke [47] (thematic analysis) employing a six-phase approach was used in the study.

3. Results and Discussion

3.1. Framework

Figure 3 presents a framework that offers an insight into how the present study applied the research inputs and produced the results.

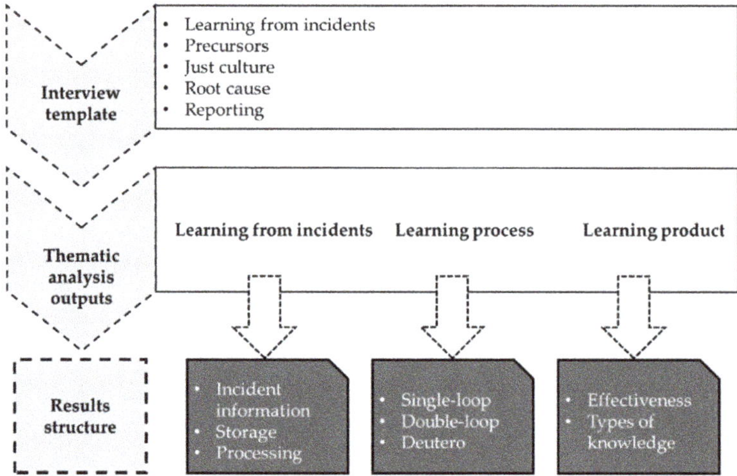

Figure 3. Research study framework.

The top layer reflects the five themes that formed the basis for the data gathering template. These themes were developed through an iterative process of conducting focus group sessions with two themes emerging, i.e., root cause and reporting. Concurrently, a systematic literature review was performed using NVivo software to assist the researchers manage over 1000 screened publications. Following a thematic analysis of the data, three main themes (Appendix B) emerged from a final cache of 18 publications, i.e., learning from incidents, precursors and just culture. The five themes informed the structure of a data gathering instrument that supported 34 semi-structured interviews in the continuing airworthiness segment of the industry. Following transcription, the data were uploaded to NVivo where they were thematically analyzed using the Braun and Clarke [47] framework. The outputs from the thematic analysis distilled the interview analysis into three main outputs, i.e., learning from incidents, learning process and learning product. The lower tier represents the elements the themes were comprised of and the findings are presented under these headings (Table 5).

Table 5. Summary of results. [1] Learning from incidents (LFI) is a safety management activity with a desired outcome of preventing unwelcome event recurrence.[2] A learning process facilitates a change in knowledge and behavior intended to support LFI.[3] Safety related information arising from the LFI process.

Learning from Incidents [1]	Learning Process [2]	Learning Product [3]
The decision to report an incident can be impacted by the perceived commercial pressure and the potential for embarrassment associated with making a mistake, amongst front line maintenance staff.	The release of a safe aviation product is the primary goal all operational maintenance and management staff espouse to.	In the organizations supporting the study, it was apparent that incidents are managed with the support of a consistent life-cycle methodology.
Identifying and understanding organizational behavioral and human factors are important elements affecting decisions to report.	Single-loop learning is a level of learning that can exist in a dynamic operational environment where a "find and fix" ethos exists.	Learning products that arise from the managed lifecycle of an incident are intended to impart sufficient learning to prevent recurrence or occurrence of same or similar events.
Inadequately resourced investigation and follow up of incidents does not support the determination of accurate event causation and measures to prevent similar incidents reoccurring.	The mandatory human factors continuation training program is considered by study participants to be an effective enabler of double-loop learning.	While aircraft manufacturers generally provide feedback on notified incidents, component manufacturers provide less feedback with little or no feedback arising from aviation authorities on submitted reports in the jurisdiction of the study.
The recognition of the extended impact of under-reporting on "levels of learning" is not always a priority in some organizations.	Evidence amongst study participants where a review of single and double-loop learning within organizations was not available during the study.	The cost of classroom delivered continuation training is a primary consideration for most organizations.
The absence of a potential learning product that results from effective reporting is an impediment when attempting to gauge the effectiveness of learning.	No formal requirement for competence in the areas of learning for managers and accountable persons exists in EU regulation 1321/2014.	Computer based training is an option that is under trial by some organizations but there are concerns amongst operational staff regarding its overall effectiveness in its current form.
Pressure to prematurely close incident reports does not promote thorough event causation and measures to prevent similar incidents reoccurring.	No competence requirements for staff involved in the development or delivery of formal human factors continuation training programs.	Just culture has a positive impact on reporting rates.
		Feedback to staff on incident causation factors from an information and learning perspective is important.
		Poorly designed continuation training syllabi do not support effective learning.
		Timely follow up to incident reports supports more effective learning outputs from the reporting process.

3.1.1. Learning form incidents—Acquiring, Processing and Storing Data

Incident reporting is accepted as a worthwhile activity amongst those participating in the study. This is based on the collective notion that the initiative raises awareness of incidents and potential hazards and can therefore help prevent event recurrence. The authors recognize that awareness is an important component of learning from incidents. Situations do arise where due to lack of report data, it is questionable if all the necessary reports are being submitted as required. Amongst the constraints to making a report are perceived production pressures and the potential embarrassment that could arise from making a mistake and highlighting it [5]. There are just culture concerns amongst some staff because they do not always know what the impact for them personally will be if they submit an incident report [44].

A dedicated focal point in organizations is essential for the systematic management of reported incidents. Where this discipline is applied, the process owner is responsible for highlighting reported issues and raising the necessary awareness amongst operational staff. Once an incident is acquired through the efforts of a reporting system, some form of processing and analysis is necessary. The availability of adequate resources for determining causation and implementing measures to prevent recurrence was identified as a primary point of concern. Perceived premature closure of reports was also highlighted amongst participants. There was a call for improved accountability and transparency on decisions relating to some closure actions. Respondents associate the practice of applying commercial key performance indicators to safety management as shallow efforts are sometimes made by organizations to expeditiously and prematurely close reports on occasion. Incident reporting and safety management initiatives have been in existence for some time. Large repositories of associated safety data are stored in many organizations. Although entities are mandated to inform key stakeholders, there is a strong opinion amongst some participants that the data repositories could be aggregated and put to better use in support of learning amongst all operators.

3.1.2. Learning Process—Single-Loop, Double-Loop and Deutero Learning

The interview data confirms that safety is a primary underpinning value in the organizations that participated in the study. The release of a safe product, i.e., an aircraft or component, is a formative pursuit and measure of learning. In organizational environments where a "find and fix" ethos may prevail, single-loop learning [11] is evident in the examples presented.

A desired outcome of double-loop learning [11] is often witnessed for example through the adjustment of environmental, behavioral and procedural norms. Instances of double-loop learning can be evident following unsuccessful attempts through single-loop learning where causation is then adequately understood and actioned. Continuation (mandatory in-service) training was considered by study participants as an effective mechanism that enables double-loop learning. During the study, it was apparent that single and double learning loops are recognized amongst many participants as having differing capabilities in terms of delivering an effective learning product. However, there was no evidence of formal reviews of single and double-loop learning being performed within the participants' organizations. Although deutero-learning [11,55] may be considered as a natural extension of other levels of learning, the concept did not feature strongly amongst the participants. A review of the EU1321/2014 [6] implementing requirements confirms an absence of any mandatory requirement to review learning processes.

3.1.3. Learning Product—Effectiveness and Types of Knowledge

Continuation training is a mandated European requirement [6] for all aircraft maintenance and continuing airworthiness management organizations. It is a product as well as a medium for imparting learning from incidents and safety related hazards. It was identified during the study that the learning product is shared amongst staff through three primary means of distribution: formally delivered continuation training, tool-box talks and safety briefings and electronic, paper, notice board and "read and sign" safety publications. The study suggests a learning product can arise as a result of an output from an incident lifecycle. Feedback from submitted occurrences to stakeholders varies from very good to poor. Cost is seen as a major consideration in some of the participating organizations when planning continuation training delivery. Although computer-based training is being considered in some companies as a viable option to class-room delivery, concerns are evident in respect of effectiveness of this medium in its current form. Bedwell and Salas [56] suggest computer-based training (CBT) can be used as a methodology for providing, "*systematic, structured learning; a useful tool when properly designed*".

The perceived overburdening of operational staff with complex learning products and excessive cognitive loads was recorded as an impediment to learning during the study. Participants suggested this can arise from poorly designed training syllabi delivered during periods of high operational activity.

Four knowledge types were identified and relate to: conceptual, dispositional, procedural and locative knowledge forms [57]. One of the key objectives of learning from incidents is to identify the type of knowledge needed to prevent an issue recurring. When a reportable issue, for example, is discovered, the submitted report will identify "what" happened. Subsequent follow up will set out to determine "why" the issue occurred. The guiding principles of "how" to perform the task or operation are often contained in procedures or data particular to the task. The information contained in procedures will enable a person to utilize other forms of knowledge. Prevailing safety culture within an organization will have an impact on learning from incidents. If a strong commercial/production culture exists, this may have an impact on, for example, the depth and breadth of learning from incidents within the company. Induction and initial training are important when accessing information for new staff. Accident data repositories contain well-documented human factor-related examples often relating to access to approved data and consequently resulting in potentially preventable incidents. Examining the limitations of each type of knowledge when continuation training programs are being developed was flagged as important by some participants. During the study, no discernible differences were recorded in how the types of knowledge were differentiated in participant organizations. A review of the EU 1321/2014 [6] human factors syllabus requirements did not highlight a need to appreciate or account for these human centered limitations when designing and delivering training lessons. Improved regulatory guidance on the design of effective human factor related material should therefore be developed. Information on how training should be structured in order to appreciate types of knowledge and capitalize on it as a minimum are required to ensure the most efficacious outcome from incident-related training.

4. Conclusions

An ameliorating feature of learning from incidents is the potential to effect sustainable improvements in aviation safety. A review of safety from the perspective of maintenance and continuing airworthiness staff is key to understanding the relationship between safety and the concept of learning from incidents [31]. From the study's qualitative data, we were able to identify how learning occurs in the airworthiness segment, and issues that support and constrain learning. Recurrent mandated training initiatives such as continuation training were found to be pivotal in enabling learning. Aspects such as prevailing culture and poor event causation were noted to have a negative impact on learning. Our proposed incident learning process (Figure 2) offers a panoramic of where potential learning opportunities and procedural improvements can arise within the lifecycle of an incident. This perspective could be applied in support of developing regulatory working group specifications and validating continuation training initiatives. In addition, it could also be used to develop a holistic review approach to learning from incidents within other organizations both in the aviation industry and outside. Two notable limitations to our research arise. First, the scarcity of prior studies capable of supporting the basis for the research was pronounced.

However, prior studies in parallel domains were successfully leveraged in support of the literature review. Second, the study's population (n = 34) was relatively small. As the study participants were representative of all affected domain functions and a point of saturation was reached, it was deemed adequate.

This research is capable of supporting other papers on additional benefits associated with learning from incidents (LFI). Notably, with the imminent implementation of a safety management (SMS) requirement for continuing airworthiness organizations, potential improvements to hazard identification arising from learning from incidents (LFI) could be highlighted.

Author Contributions: Conceptualization, J.C. and K.I.K.; methodology, J.C.; formal analysis, J.C.; investigation, J.C.; validation, J.C and K.I.K.; data curation, J.C.; writing—original draft preparation, J.C. and K.I.K.; writing—review and editing, J.C. and K.I.K. All authors have read and agreed to the published version of the manuscript.

Funding: This research received no external funding.

Institutional Review Board Statement: The study was conducted according to the Institutional guidelines and approved by the Institutional Ethics Committee of University of Limerick (Research Ethics Approval Number: 2015_12_06) dated 6 December 2015.

Informed Consent Statement: Informed consent was obtained from all subjects involved in the study.

Data Availability Statement: The data presented in this study are available on request from the corresponding author.

Conflicts of Interest: The authors declare no conflict of interest.

Appendix A Semi-Structured Interview Template

Table A1. The Semi-Structured Interview Template used in this study

Code 1	Code 2	Previous Positions	Years in Previous Positions
Position	Years in position	Qualification	Type of organization

a. Reporting

- Could you describe an occurrence/incident that happened recently?
- How is a report made?
- Who decides what events to report?
- Where does the requirement to report come from?
- How is the importance of reporting highlighted in the organization?
- What do you think the aim of reporting is?
- Have you received feedback from reports you have submitted?

b. Just culture

- Do you think there is a good safety culture in the organization?
- Why is this?
- Is it easy to communicate with management on safety issues?
- Do you feel a just culture exists in the company? (Why is that?)
- How does just culture impact on reporting?

c. Learning

- How are lessons that arise from occurrence/incident reporting delivered to staff in your area?

- How is learning achieved? (What is the process?)
- What obstacles to learning from incidents have you experienced in your position?
- In your opinion, what conditions or developments could improve learning from incidents/occurrences in your organization?

d. Root Cause

- What is your opinion on efforts to establish a single root cause when an incident/occurrence is investigated?
- Is this approach always effective?
- What situations have you experienced where incident causes can be numerous and complex?

e. Occurrence/Incident Pre-cursors

- How important is it to identify and report events not required by the mandatory occurrence reporting (MOR) schemes? (Why is this?)
- Is the organization's occurrence/ incident reporting system capable of managing reports other than MOR's?
- Is there a better way of gathering and using the potential information from non-mandatory events? (What would you suggest?)

Appendix B Defining and Naming Themes

Table A2. Taxonomy used in defining and naming the themes used in this study.

Phase 5—Categories Conceptually Mapped and Collapsed into 3 Major Themes with 8 Sub-Themes	Code Definitions for Coding Consistency	Interviews Coded	Units of Meaning Coded
LEARNING PROCESS	This relates to the three levels of learning suggested by Bateson (1972) and applied by Argyris and Schon (1996)	34	359
Deutero-learning	This relates to when members of an organization reflect on previous learning and thereby setting about to improve its learning process.	26	65
Double-loop Learning	This relates to learning that takes place and organizational norms and theory in use are changed.	26	63
Single-loop Learning	This relates to when an organizations' members detect and correct errors but still maintain the organizations theory in use.	26	63
LEARNING PRODUCT	This relates to what the learning process delivers	33	235
Effectiveness	This relates to measuring effectiveness of learning	31	155
Types of knowledge	This relates to conceptual, procedural, dispositional and locative knowledge	23	74
LEARNING FROM INCIDENTS	This relates to the inputs necessary to enable the assembly of a learning material in support of learning from incidents	17	213
Processing	This relates to how learning information is processed	17	82
Acquiring	This relates to the sources of information that support learning and how there are gathered	16	55
Storing	This relates to how learning information is stored	12	27

References

1. Drupsteen, L.; Hasle, P. Why do organizations not learn from incidents? Bottlenecks, causes and conditions for a failure to effectively learn. *Accid. Anal. Prev.* **2014**, *72*, 351–358. [CrossRef] [PubMed]
2. Chang, Y.-H.; Wang, Y.-C. Significant human risk factors in aircraft maintenance technicians. *Saf. Sci.* **2010**, *48*, 54–62. [CrossRef]
3. Silva, S.A.; Carvalho, H.; Oliveira, M.J.; Fialho, T.; Soares, C.G.; Jacinto, C. Organizational practices for learning with work accidents throughout their information cycle. *Saf. Sci.* **2017**, *99*, 102–114. [CrossRef]
4. Akselsson, R.; Jacobsson, A.; Bötjesson, M.; Ek, Å.; Enander, A. Efficient and effective learning for safety from incidents. *Work* **2012**, *41*, 3216–3222. [CrossRef]
5. Hobbs, A.N. Human Errors in Context: A Study of Unsafe Acts in Aircraft Maintenance. Ph.D. Thesis, University of New South Wales, Kensington, Australia, 2003.
6. European Commission (EU). *Commission Regulation (EU) No 1321/2014 of 26 November 2014 on the Continuing Airworthiness of Aircraft and Aeronautical Products, Parts and Appliances, and on the Approval of Organisations and Personnel Involved in These Tasks*; European Commission (EU): Brussels, Belgium, 2014.

7. Harvey, C.; Stanton, N.A. Safety in system-of-systems: Ten key challenges. *Saf. Sci.* **2014**, *70*, 358–366. [CrossRef]
8. European Commission (EU). *Commission Regulation (EU) No 376/2014 OF THE EUROPEAN PARLIAMENT AND OF THE COUNCIL of 3 April 2014 on the reporting, analysis and follow-up of occurrences in civil aviation, amending Regulation (EU) No 996/2010 of the European Parliament and of the Council and repealing Directive 2003/42/EC of the European Parliament and of the Council and Commission Regulations (EC) No 1321/2007 and (EC) No 1330/2007*; European Commission (EU): Brussels, Belgium, 2014.
9. Gerede, E. A study of challenges to the success of the safety management system in aircraft maintenance organizations in Turkey. *Saf. Sci.* **2015**, *73*, 106–116. [CrossRef]
10. Drupsteen, L.; Wybo, J.-L. Assessing propensity to learn from safety-related events. *Saf. Sci.* **2015**, *71*, 28–38. [CrossRef]
11. Argyris, C.; Schön, D.A. *Organizational Learning II: Theory, Method, and Practice*; Addison-Wesley Publishing Company: Boston, MA, USA, 1996.
12. Drupsteen, L.; Guldenmund, F.W. What is learning? A review of the safety literature to define learning from incidents, accidents and disasters. *J. Contingencies Crisis Manag.* **2014**, *22*, 81–96. [CrossRef]
13. Fogarty, G.J.; Saunders, R.; Collyer, R. Developing a model to predict aircraft maintenance performance. In Proceedings of the 10th International Symposium on Aviation Psychology, Ohio State University, Columbus, OH, USA, 3–6 May 1999.
14. Sms, I. Safety Management Manual (SMM). In *Doc 9859*; ICAO: Montréal, QC, Canada, 2012.
15. Ward, M.; McDonald, N.; Morrison, R.; Gaynor, D.; Nugent, T. A performance improvement case study in aircraft maintenance and its implications for hazard identification. *Ergonomics* **2010**, *53*, 247–267. [CrossRef]
16. Jacobsson, A.; Ek, Å.; Akselsson, R. Learning from incidents–A method for assessing the effectiveness of the learning cycle. *J. Loss Prev. Process Ind.* **2012**, *25*, 561–570. [CrossRef]
17. Brunel, I. *Sessional Papers Printed by Order of the House of Lords, or Presented by Royal Command*; Government of Great Britain: London, UK, 1841.
18. Furniss, D.; Curzon, P.; Blandford, A. Using FRAM beyond safety: A case study to explore how sociotechnical systems can flourish or stall. *Theor. Issues Ergon. Sci.* **2016**, *17*, 507–532. [CrossRef]
19. Hollnagel, E. *Barriers and Accident Prevention*; Ashgate Publishing: Farnham, UK, 2004.
20. Okoli, C.; Schabram, K. *A Guide to Conducting a Systematic Literature Review of Information Systems Research*; Sprouts Farmers Market: Chandler, AZ, USA, 2010.
21. Houghton, C.; Murphy, K.; Meehan, B.; Thomas, J.; Brooker, D.; Casey, D. From screening to synthesis: Using nvivo to enhance transparency in qualitative evidence synthesis. *J. Clin. Nurs.* **2017**, *26*, 873–881. [CrossRef] [PubMed]
22. Bandara, W.; Furtmueller, E.; Gorbacheva, E.; Miskon, S.; Beekhuyzen, J. Achieving rigor in literature reviews: Insights from qualitative data analysis and tool-support. *Commun. Assoc. Inf. Syst.* **2015**, *37*, 154–204. [CrossRef]
23. Gough, D.; Oliver, S.; Thomas, J. *An Introduction to Systematic Reviews*, 2nd ed.; Sage: London, UK, 2017; pp. 83–106.
24. Meline, T. Selecting studies for systematic review: Inclusion and exclusion criteria. *Contemp. Issues Commun. Sci. Disord. ASHA* **2006**, *33*, 21–27. [CrossRef]
25. Wienen, H.C.A.; Bukhsh, F.A.; Vriezekolk, E.; Wieringa, R.J. *Accident Analysis Methods and Models—A Systematic Literature Review*; University of Twente, Centre for Telematics and Information Technology (CTIT): Enschede, The Netherlands, 2017.
26. Maykut, P.S.; Morehouse, R. *Beginning Qualitative Research: A Philosophic and Practical Guide*; Falmer Press: London, UK, 1994.
27. Cooke, D.L.; Rohleder, T.R. Learning from incidents: From normal accidents to high reliability. *Syst. Dyn. Rev.* **2006**, *22*, 213–239. [CrossRef]
28. Drupsteen, L.; Groeneweg, J.; Zwetsloot, G.I. Critical steps in learning from incidents: Using learning potential in the process from reporting an incident to accident prevention. *Int. J. Occup. Saf. Ergon.* **2013**, *19*, 63–77. [CrossRef] [PubMed]

29. European Commission (EU). *(EU) 2018/1139 of the European Parliment and the Council of 4 July 2018 on common rules in trhe field of civil aviation and establishing a European Union Aviation Safety Agency, and amending Regulations (EC) No 2111/2005, (EC) No 1008/2008, (EU) No 996/2010, EU376/2014 and Directives 2014/30/EU of the European Parliment and the Council, and repealing Regulation (EC) No 216/2008 of the European Parliment and of the Council and Council Regulation (EEC) No 3922/91*; European Commission (EU): Brussels, Belgium, 2018.
30. Lindberg, A.-K.; Hansson, S.O.; Rollenhagen, C. Learning from accidents–What more do we need to know? *Saf. Sci.* **2010**, *48*, 714–721. [CrossRef]
31. Lukic, D.; Littlejohn, A.; Margaryan, A. A framework for learning from incidents in the workplace. *Saf. Sci.* **2012**, *50*, 950–957. [CrossRef]
32. Steiner, L. Organizational dilemmas as barriers to learning. *Learn. Organ.* **1998**, *5*, 193–201. [CrossRef]
33. Taylor, F.W. *The Principles of Scientific Management*; Harper & Brothers: New York, NY, USA, 1911.
34. Weber, R. Editor's Comments: The rhetoric of positivism versus interpretivism: A personal view. *MIS Q.* **2004**, *28*. [CrossRef]
35. Guba, E.G.; Lincoln, Y.S. *Competing Paradigms in Qualitative Research*; Sage Publications: Thousand Oaks, CA, USA, 1994.
36. Walsham, G. Interpretive case studies in IS research: Nature and method. *Eur. J. Inf. Syst.* **1995**, *4*, 74. [CrossRef]
37. Oates, B.J. *Researching Information Systems and Computing*; Sage Publications: Thousand Oaks, CA, USA, 2006.
38. Hirschheim, R. Information systems epistemology: An historical perspective. *Res. Methods Inf. Syst.* **1985**, *9*, 13–35.
39. Schwandt, T. Constructivist, interpretivist approaches to human inquiry. *Handb. Qual. Res.* **1994**, *1*, 118–137.
40. Kitzinger, J. The methodology of focus groups: The importance of interaction between research participants. *Sociol. Health Illn.* **1994**, *16*, 103–121. [CrossRef]
41. Bogardus, E.S. Social distance in the city. *Proc. Publ. Am. Sociol. Soc.* **1926**, *20*, 40–46.
42. Frey, J.H.; Fontana, A. The group interview in social research. *Soc. Sci. J.* **1991**, *28*, 175–187. [CrossRef]
43. Powell, R.A.; Single, H.M. Focus groups. *Int. J. Qual. Health Care* **1996**, *8*, 499–504. [CrossRef]
44. Chatzi, A.V. The diagnosis of communication and trust in aviation maintenance (DiCTAM) model. *Aerospace* **2019**, *6*, 120. [CrossRef]
45. Chatzi, A.V.; Martin, W.; Bates, P.; Murray, P. The unexplored link between communication and trust in aviation maintenance practice. *Aerospace* **2019**, *6*, 66. [CrossRef]
46. Patton, M.Q. *Qualitative Evaluation and Research Methods*, 2nd ed.; Patton, M.Q., Ed.; Sage Publications: Thousand Oaks, CA, USA, 1990.
47. Braun, V.; Clarke, V. Using thematic analysis in psychology. *Qual. Res. Psychol.* **2006**, *3*, 77–101. [CrossRef]
48. QDA Training. *Working with NVivo*; QDA Training: Dublin, Ireland, 2013.
49. Lincoln, Y.S.; Guba, E.G. *Naturalistic Inquiry*; Sage Publications: Thousand Oaks, CA, USA, 1985.
50. Baskerville, R.; Pries, J.H. Short cycle time systems development. *Inf. Syst. J.* **2004**, *14*, 237–264. [CrossRef]
51. Hammersley, M.; Atkinson, P. *Ethnography: Principles in Practice*, 3rd ed.; Routledge: London, UK, 2007.
52. Araujo, L. Designing and Refining Hierarchical Coding Frames. In *Computer-Aided Qualitative Data Analysis: Theory, Methods and Practice*; Udo, K., Ed.; Sage Publications: London, UK, 1995; pp. 96–104.
53. Birks, M.; Chapman, Y.; Francis, K. Memoing in qualitative research: Probing data and processes. *J. Res. Nurs.* **2008**, *13*, 68–75. [CrossRef]
54. Miles, M.B.; Huberman, A.M. *Qualitative Data Analysis: An Expanded Sourcebook*; Sage Publications: Thousand Oaks, CA, USA, 1994.
55. Bateson, G. The Logical Categories of Learning and Communication. In *Steps to an Ecology of Mind*; The University of Chicago Press: Chicago, IL, USA, 1972; pp. 279–308.
56. Bedwell, W.L.; Salas, E. Computer-based training: Capitalizing on lessons learned. *Int. J. Train. Dev.* **2010**, *14*, 239–249. [CrossRef]
57. Thorndike, E.L. Fundamental theorems in judging men. *J. Appl. Psychol.* **1918**, *2*, 67. [CrossRef]

Publisher's Note: MDPI stays neutral with regard to jurisdictional claims in published maps and institutional affiliations.

Article

Winging It: Key Issues and Perceptions around Regulation and Practice of Aircraft Maintenance in Australian General Aviation

Anjum Naweed [1,*] and Kyriakos I. Kourousis [2]

[1] School of Health, Medical & Applied Sciences, Appleton Institute for Behavioural Science, Central Queensland University, Wayville, SA 5034, Australia

[2] School of Engineering, University of Limerick, V94 T9PX Limerick, Ireland; kyriakos.kourousis@ul.ie

* Correspondence: anjum.naweed@cqu.edu.au; Tel.: +61-(0)-8-8378-4520

Received: 8 June 2020; Accepted: 19 June 2020; Published: 26 June 2020

Abstract: The very diverse character of General Aviation (GA) within Australia poses challenges for its effective management of risk and safety in the sector. Improvements for human performance and perceptions of safety within the maintenance environment are among the areas which regulators have targeted for continuous improvement. This paper provides a timely empirical exploration of maintenance engineer perspectives around: (1) Changes in the role of the regulator/regulation that have impacted the sector and diminished safe operations; and (2) specific practical and operational challenges that the GA industry must deal with to sustain safe operations going forward. A thematic analysis of transcribed qualitative data revealed five key themes and identified a number of key issues from sector changes including a decline in training and education, drift in working practices, and wider power-distance gap. Issues with auditing and bureaucratization, negative safety climate, and underlying values and philosophies were also found. Practical and operational challenges going forward included an array of concerns associated with safety, the mismatch between GA and commercial aviation, workforce development and the financial burden in the sector. The results draw attention to the interconnectedness between various components of the GA system, and carry timely implications for regulation in the GA sector. Future research directions are discussed.

Keywords: General Aviation; human factors; engineering changes; regulation; safety; management changes; airworthiness; aviation; industry change; maintenance engineering culture

1. Introduction

Aviation is a heavily regulated industry for the purpose of safe operations. All aircraft maintenance, including those in the General Aviation (GA) sector, falls under the scope of these safety regulations, since maintenance is an essential activity for sustaining airworthiness (denoted as continuing airworthiness). The GA aircraft operation, as defined by the International Civil Aviation Organisation (ICAO), is an aircraft operation other than commercial transport operation or aerial work operation [1]. The nature and the needs of the GA sector are invariably different to those of commercial air transport conducted with the use of a large complex aircraft. The GA sector has a lower risk profile (as measured by the impact of an accident). In particular, if one considers the systems safety view of risk, whereby risk is a combination of the likelihood of the hazard and the consequences of any ensuing accident [2], then, although the probability of having an accident in GA is higher than in commercial aviation [3], the lower severity of such accidents results in a lower risk profile. Moreover, GA is characterized by highly diverse operations, operators and (mostly ageing) aircraft fleets, in conjunction with scarce/unavailable financial, infrastructure and human resources (mainly attributed to the aircraft owner's limited financial capacity). This unique mix has an effect both on the continuing airworthiness rules for GA aircraft and also on how maintenance is actually practiced within this sector of the aviation industry. Thus, we often see that in the drafting of regulations and regulatory oversight in GA, a "one size fits all" approach is not the norm at a worldwide level.

In an attempt to regulate GA operations in a proportionate (and cost effective) manner, continuing airworthiness rules and oversight are typically less stringent than in commercial air transport [4]. Therefore it is interesting to examine the safety performance of the GA sector, namely the safety record in relation to maintenance-related accidents. The GA aircraft maintenance has been identified as an accident-precipitating factor [5], with 30% of accidents in Australia attributed to this [6]. For the Australian GA sector, this observation is alarming when taking into consideration the size of the fleet (nearly 90% of all aircrafts), its flight activity (40% of total flight hours are produced by GA aircraft) [7], and the maintenance burden cost (close to 20% of the total operational expenses for GA aircraft) [8]. In practical terms, retaining the same safety record would increase the absolute number of maintenance-related accidents if there is a continued growth of the activity.

The Australian civil aviation regulator, CASA (Civil Aviation Safety Authority), has recognised the need for safety improvements, via a regulatory reform and modernization/alignment against best practices at the international level. Special attention has also been placed on ageing aircraft issues (structural fatigue, corrosion, etc.), as the Australian fleet has an increasing average age [8]. It is of note that CASA approached this matter with a change management mindset, attempting to achieve a consensus from the regulated parties. The management of change, as a systematic approach/methodology to enable change at an individual and organisation level [9], has also been promoted by CASA for safety management purposes in the wider aviation industry [10]. Past CASA attempts to impose regulations or dictate mandatory actions did not have the expected result, as compliance was found to be problematic. An extensive industry consultation process was employed for this purpose during 2018–2019 [11]. Tailoring of maintenance regulations (framework, oversight, and practice) for the GA aircraft was a topic on the industry's "wish list" from this consultation. This finding has also confirmed the feedback collected by CASA from their previous interaction with the GA sector.

Eventually, CASA decided to introduce a new set of maintenance regulations that mirrored the United States (US) Federal Aviation Administration (FAA) rules for GA aircraft, with these outlined in CASA's consultation accompanying documents [11]. The FAA rules' approach was by far the preferred industry choice, recording a 78% acceptance rate [11]. The decision to follow the FAA regulations for GA aircraft is not consistent with the CASA's overall regulatory structure and philosophy, as the Australian Civil Aviation Safety Regulations (CASRs) largely follow/mirror the European Aviation Safety Agency (EASA) rules. This triggers questions on the underlying reasons that provoke this type of change as well as what challenges may manifest as part of transitional effects.

Given the alarming statistics in the Australian GA sector, research by Naweed and Kingshott [12,13] undertaken in the interstices of work and safe operations in the Australian GA sector, has examined how affect influences decision making and action tendencies in maintenance engineering scenarios. This body of research has identified numerous contextual factors that feature in ways of working and give rise to system behaviours that can have a direct impact on safe operations. Identified contextual factors included distraction under pressure, incorrect manuals, maintenance costs, perceived customer disloyalty, managerial interference, and negative rumination. A number of the findings also echo normalisation of deviant behaviour, a theoretical stance where maladaptive practices come to be accepted as the norm [14,15], and which reflect a complex array of interacting factors at play where Rumsfeldian "unknown unknowns" can often loom large [16].

Empirically exploring the industry's perception on the role of regulations and the regulator, as well as other factors that may directly impact the practice of maintenance on GA aircraft, is necessary to obtain a better understanding of what is happening at an end-user level. The technical staff directly involved in conducting and certifying maintenance, as well as the aircraft owners, are both key stakeholders in the airworthiness sustainment business.

Aims and Objectives

The very diverse character of GA poses challenges to effective risk and safety management within the aviation maintenance environment. Improving human performance and perception around safety are among the areas that have been targeted by regulators for continuous improvement. The present study set out to build on previous substantive research [12,13] and examine this in a comprehensive way through semi-structured interview data elicited from Australian aviation maintenance engineers in GA using the following research questions:

1. What perceived changes in regulation and/or in the role of the regulator have impacted the GA sector in Australia, and to what extent are these changes perceived to have diminished safe operations?
2. What specific practical and operational challenges does the GA industry have to deal with to sustain safe operations going forward?

2. Materials and Methods

2.1. Study design

The research questions driving this study were applied as part of a focused analysis on an existing dataset, collected in late 2016 using a qualitative research design to gain insight into specific participant perceptions and experiences. Data collection was facilitated by semi-structured interviews and application of the Scenario Invention Task Technique (SITT) [17,18]. The SITT is a pen-paper task that involves scenario-creation and combines principles of the Critical Decision Method [19] and Rich Picture Data method [20] to scaffold interviewing. Participants are asked to invent a challenging scenario specific to their work using illustrations to assist with verbalization and articulation in ways that encourage transitions from analytical and creative thinking, through to systems thinking [21] processes (i.e., holistic perceptions, consideration of changes over time) when answering their questions. The SITT has been used to elicit subject matter expertise and identify critical themes associated with safety, risk, training, and/or ways of working in a number of related complex domains [12,22–26].

2.2. Participants and Recruitment

A total of 10 GA maintenance engineers took part in the study. Participants were recruited using a purposeful sampling approach with ages ranging from 21 to 60 ($M_{\text{age}} = 41$, $SD = 11.37$). Nine males and one female took part, broadly representing the gender ratio in GA maintenance engineering. Experience in GA maintenance engineering ranged from 1 to 42 years ($M_{\text{exp}} = 17.8$, $SD = 10.5$).

2.3. Procedure

Table 1 provides an overview of the interview protocol. Each interview session took ~60 min to complete. The first part developed rapport with participants and elicited views of any substantive changes that had occurred across the industry. The second part applied the SITT and required the creation of a challenging workplace scenario (real or hypothetical). An A3-sized paper and felt-markers were provided to develop a pictographic scenario representation (see Figure 1 for example). With the aid of the illustration, the scenario was used to identify decisions, feelings and perceptions, and probe the role of the safety, regulation, impact of training, and the influence of someone with more/less experience. Pragmatic validity was ascertained through follow up checks of understanding with scenarios serving as concrete examples for broader views of safety and industry impacts. The study was approved by the Central Queensland University ethics committee (Approval no. H16/05-146).

Table 1. Overview of the interview protocol.

Topic	Example Content	Example Questions
General experience in aviation	Background, entry into industry, rapport building questions	"What is your background in aviation?" "Do you remember the first aircraft you fixed?"
Current processes and changes in the industry	Training delivery, industry status past and present, challenges	"How has the aviation industry changed over the years?" "What aspects of the industry do you find most challenging?"
Safety in aviation maintenance	Changes in safety, positive and negative influences	"How has the safety of aviation maintenance changed over the life of your work?" "What has a negative impact on safety?"
Factors impacting a maintenance task	Distractions, pressures in day-to-day maintenance work	"What pressures do you feel when completing a job?" "How do you limit distraction during work?"
Scenario Invention Task Technique Activity	Create scenario	"Imagine you are at work and completing a difficult maintenance engineering task. Use the pen-paper to describe this task, using any drawing convention you wish."
	Recall and retell	"What else would be going on around you?"

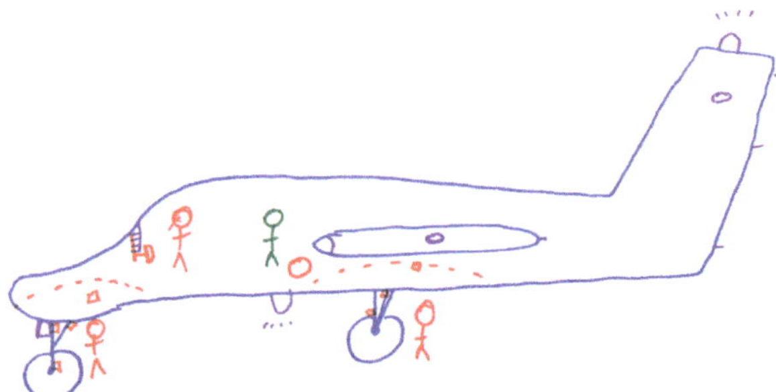

Figure 1. Example pen-paper scenario drawing of a re-rigging landing gear task created by a participant in the study. The green figure (middle-right) depicts the avionics technician and the three red figures depict airframe engineers. Here, the Scenario Invention Task Technique (SITT) is being used to illustrate a coordinated effort of teamwork required between multiple airframe engineers and an avionics technician, highlighting the complexity and dynamism in the aircraft maintenance workplace, but also the perceived threat arising from interruptions, excessive bureaucracy, and issues related to education and training factors, and attitudes on ways of working.

2.4. Data Analysis

All data was analyzed thematically with the aid of CMapTools (v. 6.01.01), a visual concept mapping tool under an inductive approach (i.e., without developing categories *a priori*) [27] to identify perceived changes across the Australian GA industry in recent years, and to draw insight into meaningful relationships between: The role of the regulator/regulation, safety in operations, and links with ways of working. Figure 2 shows an overview of the data analysis process; in short, statements were arranged into clusters (i.e., "meaning units") through gradual and systematic application of open, axial and selection coding via concept mapping to identify ideas and draw connections. Figure 3 shows how the data looked like in CmapTools at different stages of the analysis. Through this process, data were thus coded semantically from description to interpretation, and grouped into overarching themes, and each were individually analysed to identify patterns between themes.

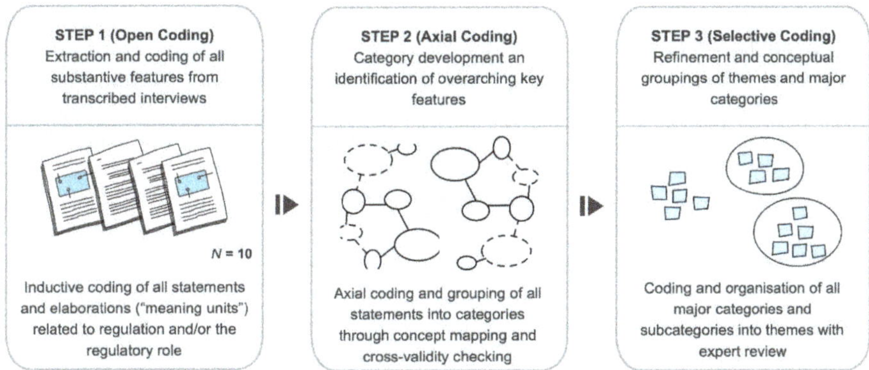

Figure 2. Overview of the data analysis process showing the three main steps of open, axial, and selective coding undertaken. Through this systematic process, data were coded semantically from description to interpretation.

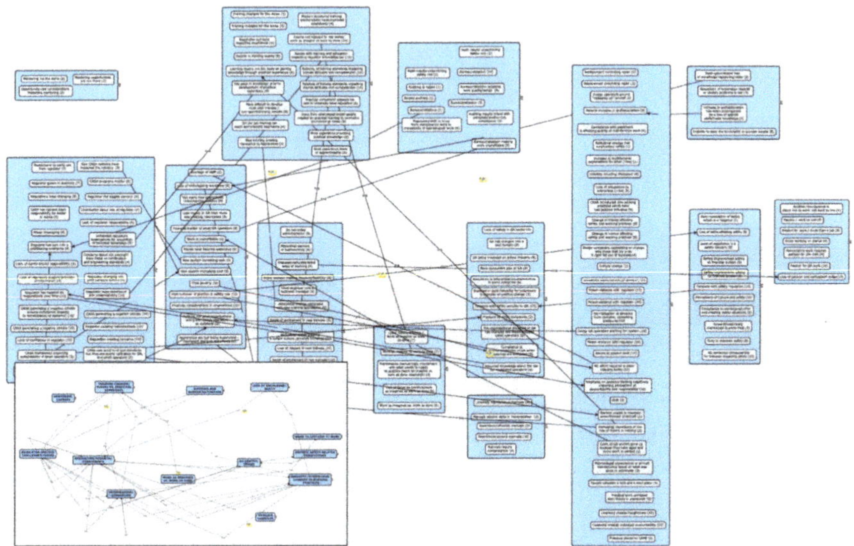

Figure 3. Illustration of concepting mapping of coded data in CmapTools. The large screenshot depicts arrangement of data midway through Steps 2 and 3 (axial and selective coding) and the inset (bottom-left) highlights connections between meaning units across thematic groupings in a collapsed form. Note: writing in picture is designed to provide an overview and is intentionally illegible.

Data collection was performed by a professional working within the GA sector in the capacity of a junior researcher in the service of a psychology Honours project; however, in an effort to maintain internal consistency, reduce chances of personal bias, and address questions of reliability and trustworthiness of the findings, all data analysis for this study was subsequently performed by a senior academic who was well-versed in the research design, but who also worked outside the GA industry (A.N.). Review and checking of findings and final codes was undertaken by an academic with intimate knowledge of the aviation industry (K.I.K).

3. Results and Discussion

Table 2 presents a summary of the thematic findings from the analysis. A total of five superordinate themes were identified. They were: (1) Changes to Industry and Working Practices; (2) Role of the Regulator; (3) (Re)calibration of Underlying Values and Philosophies; (4) Work as Imagined vs. Work as Done; and (5) Practical and Operational Challenges for GA. The first and fifth theme were the most represented within the data, each attracting more than a quarter of the overall coding of statements. Each of these themes also featured the greatest number of major (i.e., subordinate thematic) categories (4). The next sections present the findings associated with each theme. Supporting transcript excerpts are given where necessary via an anonymous ID-tag in parentheses where "(P_x)" indicates "(Participant_number)".

Table 2. Summary of thematic analysis with indication of frequency of statements and theme totals.

Themes	Major Categories	*N* [1]	Frequency of Statements	Theme Totals
Changes to Industry and Working Practices	General decline in training and education	7	16 (10%)	47 (29%)
	Drift in working practices	6	12 (7%)	
	Emphasis and growth of safety culture	5	11 (7%)	
	Wider power-distance gap	3	8 (5%)	
Role of the Regulator	Auditing and bureaucratization	5	10 (6%)	29 (18%)
	Lack of clarity and support	3	7 (4%)	
	Negative safety climate	2	12 (7%)	
(Re)calibration of Underlying Values and Philosophies	Working to live, not living to work	6	15 (9%)	27 (16%)
	Attitudinal stability	5	12 (7%)	
Work as Imagined vs. Work as Done	Theory vs. practical experience	7	10 (6%)	16 (10%)
	Restrictive maintenance manuals	3	6 (4%)	
Practical and Operational Challenges for GA	Mismatch between GA and commercial aviation sector	6	13 (8%	45 (27%)
	Safety-risk and safety concerns	5	11 (7%)	
	Workforce development	4	10 (6%)	
	Financial burden	4	11 (7%)	

[1] *N* indicates the number of individual participants who mentioned the statement (total = 10).

3.1. Changes to Industry and Working Practices

Table 3 decomposes the findings associated with the theme of *Changes to Industry and Working Practices* to show the minor categories in addition to the major ones. The highest represented of all major categories in this theme was *general decline in quality of training and education*, within which the *training changed for the worse* minor category was the most prevalent, both within the data and across participants. As reflected by the title, the consensus was that levels of training and quality had declined: *"the training has changed and I don't think it's been to the benefit of the industry"* (P_8); *"I think these days [the training is] no good"* (P_5), such that there was no longer any real effort required to enter the industry, *"I've seen it myself, [trainees will] do the exam three or four times, and the instructor will go, 'you're just not getting it. Fifty-one percent, okay, next'"* (P_10). This was attributed to a variety of systems factors, such as restrictive curricula, *"[trainees] don't have the … as broad of experience s previous times […] these days [the training curriculum is] a lot more restrictive"* (P_3), and key gaps in knowledge prior to the development of practical experience:

> *I think they should learn an air law before they come out on the shop floor for their apprenticeship. In that way, that's just ridiculous to not have air law knowledge before you actually go and work in the industry. It's not a part of the setup which should be.* (P_9)

Table 3. Summary of findings for the Changes to Industry and Working Practices theme.

Theme	Major Categories	Minor Categories
Changes to Industry and Working Practices	General decline in quality of training and education	Training changed for worse Mentoring Regulator knowledge
	Drift in working practices	Unscrupulous operators Aircraft owners Maintenance work
	Emphasis and growth of safety culture Wider power-distance gap	

A perceived relaxing of training standards was also felt to impact trainee attitudes and competencies:

> *[Trainees today] never really learn and they never become accountable for actually [becoming competent]. Since they know that there is—"hang on a minute, if I just cock this up and I don't put any effort in, but I keep turning up, I'm going to get through anyway".* (P_10)

In many cases, these perspectives explored the tensions between practical work and theory, with views that practical components appeared too late in tertiary education and vocational training, *"there's a lot of university degrees where it's not until the third or fourth year where any practical's done"* (P_3), practical work coincided with theory in yesteryear, and input from older and more experienced people was needed on practical training to overcome shortcomings:

> *I did one year full-time of the Cert IV. That got me, basically, two-thirds of my theory out of the way. During that time, I was doing work experience with four different operators.* (P_2)

> *Younger people coming through with these training organizations need to be trained practically on the job, and they just need the input from the older people.* (P_8)

Another key aspect to a perceived general decline in training and education was issues with *mentoring*, where opportunities for this were no longer there or no longer the same:

> *[Apprentices are] coming out very insecure and very poorly trained, I suppose, because they just haven't had that time being mentored by the good engineers [...] there's just not that same level of mentorship, I feel, anymore.* (P_2)

Even when opportunities were there, prospective mentors were no longer taking on apprentices because of resistance to new approaches:

> *A lot of employers these days don't have the money or the time to baby in a practice. They want them to come in as a second-year apprentice, basically. They are new one year, come in as a second year, and be able to let them go into the work. You can't do that. That happens on a daily basis. That is just how it happens. That's why I don't have apprentices anymore and I will not, I will not, I don't have the time.* (P_5)

Issues with learning and education were also considered to have impacted regulator knowledge, *"They've all got Bachelors of Aeronautical Engineering, but not one of them actually knew their own rules, which undermined the system"* (P_10).

A gradual *drift in working practices* within the industry was attributed to changes to various roles. *Unscrupulous* and *"dodgy" operators* and companies was an important perception here, and a loss of moral principles and honesty was being seen, both with respect to how aircrafts were being maintained and how the system was being *"worked."* For example, *"you still find things in aircraft that have been dodgied up and stuff"* (P_3);

People in General Aviation either can't afford [to maintain their aircraft] or don't want to pay. They will shop around and find the cheapest place, which means that there are unscrupulous organizations out there that will cut corners, that will do it cheaper; (P_4)

And

I've seen airplanes go into a hangar and then an hour later, they roll out and the annual inspection has been done on them. You can't do that in GA. There are some shops around that do it, but they tend to get found out. The industry, as you said before, is quite a small industry. There's enough scuttlebutt around that you hear, that you don't take your pioneer. They're dodgy. Or, if you want a hundred hourly done, fly over the airfield and it'll turn up in the mail a week later; (P_10)

Note: "Scuttlebutt" here is used a slang reference for a rumour, or "word on the street."

In terms of *maintenance work* and practice, replacement was perceived to be overriding repair:

"[You fix faults] in accordance with the maintenance manual and that's it [. . .] [The new Maintenance Engineer's] idea of fault finding is, 'oh that's loose, or there's the fault.' Fix it by replacement instead of repair quite often. (P_3)

Compliance with regulations/rules was thought to be affecting the quality of paperwork:

[The regulators are] bogging down the engineer with paperwork and not letting the engineer do what engineers do best, which is work on aircraft, or supervise people that are working on aircraft. (P_4)

Perceived changes in the practices of *aircraft owners* were contextualized by changes in the financial landscape which affected safety and working practices. These were considered to provoke unrealistic and mismanaged expectations based on what was done in previous years, greater pressure on (licensed) maintenance engineers, and the burgeoning reality that some owners were simply no longer able to maintain the airworthiness of their aircraft:

We're doing a job at the moment where we're being told to leave things to another year to reduce the cost and then spread it across. More people saying, "Oh, we're going to be changing that in a year and a half. Can you just sign it off?"; (P_9)

And

Some owners simply don't have the funds to be able to sustainably maintain their aircraft in an airworthiness state [. . .] the LAME has to be the one, I suppose, who makes the decision to go, "I'm just not going to sign this airplane out again until we do $30,000 worth of work"; (P_2)

Note: "LAME" is an acronym for Licensed Aircraft Maintenance Engineer.

By way of extending on the foregoing points, a perceived *wider power-distance gap* between the maintenance engineer and the regulator and customer was seen as a significant change. Power distance has been widely researched as a key dimension of organisational cultures, broadly defined as the extent to which less powerful members of organizations and institutions accept and expect that power is distributed unequally [28]. In line with the findings, the perceptions that power-distance distributions had changed was viewed to affect working relationships, for example if an unhappy aircraft owner contacted the regulator to report perceived issues with an aircraft service *"it seems to be enough that you've only got to say that to the regulator and you're guilty until proved innocent"* (P_10). These perceptions were multifaceted in that they were also associated with normalization of deviance born from complex and competing pressures:

You may be put under a position where it's Friday night and an owner is screaming, to go away kids and family. You go, "Bring it in. We'll do an oil change. I had a good look at it this last time." You know the member. You know the aircraft. You're intimate with it. Maybe you won't do every checklist in there, because last time, you did the whole thing. You sign it off, and you might, six weeks later, have a disgruntled person that leaves there. They then ring the regulator and go, "He didn't do this proper annual". (P_10)

The power that customers had over maintenance engineers were reflected in working practices where aircraft were always serviced with the *"bare minimum"* because engineers were encouraged to *"keep doing these 3 or 4-day hundred hourly's"* (P_2) through lack of provision for major refurbishment.

Power-distance with the regulator was also reported in the context of examination for licensure which impacted working relationships. For example:

> *There is no right of reproach. You cannot go to anyone within [the regulator] and argue it. Recently I did my [aircraft] exam for my license […] Doing the exam, I noticed there were some discrepancies in the wording […] When I rang [the regulator], they effectively said "stick your head up your ass, we don't care, we are the regulator and we will do what we want to do." They are not engaging with it.* (P_10)

In terms of general industry changes, the consensus was that the regulator had grown in austerity over the years and become *"stricter"* (P_7), with an added perception that in an industry-wide regulation satisfaction survey, *"they ended up with a 99 percent disapproval"* rating (P_10).

Despite negative participant perceptions around training and education, drift, and ostensible widening of power-distance, an *emphasis and* indeed even *growth of safety culture* was a prominent perceived change in GA and its working practices. More frequent inspections, helpful programs, increase in multifactorial explanations for error, introduction of new working practices with positive influence, increases in work flexibility for volunteers, and general increases in professionalism and attitude were all attributed to a change and maturing in safety culture, for example: *"I think the professionalism [in the General Aviation industry] has increased overall"* (P_3); *"the authorities come and keep us on guard as well, coming in for inspections here, there and everywhere, which is fair"* (P_9); *"that just culture of no blame and trying to find the cause behind errors is really a big, it is a big thing now"* (P_1).

The focus on compliance was perceived to be *"flushing out"* (P_2) dishonest or unreliable operators, and while some perceptions showed ambivalence, there was consensus that new regulator systems had had an impact on the industry:

> *[The regulator is] trying to help more than they have done, and it is happening. Some of the things I've learned over the past and put into the company like doing safety meetings, things like that … Yeah, that's all stuff that we never used to do in the past, but it's a positive;* (P_9)

> *The small operators are being pushed out now. There's an argument that those smaller dodgy operators get filtered out of the industry. Perhaps that's one of the goals of this increased compliance;* (P_2)

> *[The Regulator] has devolved responsibility in a number of different areas, some good, some bad;* (P_4)

And

> *[The Regulator] bought out a SID's program, which is causing a few dramas and ripples around the industry but in some ways, is actually a good thing because we're finding … like we recently found some tail plane cracks on a Cessna because of … we would have found them anyway, but it was a part of the SID's program;* (P_9)

Note: Supplemental Inspection Documents – a regime that aims to maintain the structural integrity of the airframe of CESSNA 100 series aircraft as they age, due to growing concern of the safety of the ageing fleet [29].

3.2. Role of the Regulator

By way of extending on the foregoing theme, Table 4 decomposes the findings associated with the perceived *role of the regulator*, which invited much critique from participants. Most in this theme commented on *auditing and bureaucratization* where the level of auditing was considered to be *excessive*, *"rigged,"* and with results that *undermined* the safety role:

... we get audited constantly, and you know just picking up little things like where it tool—we used a tool tag system for tool control, one of the numbers was a bit worn off on a tool tag, it then becomes an audit finding [...] at the end of the day that bit of paint worn off a tool tag is not going to make an aircraft come down [...] I've never seen an audit come through with a clean slate. (P_1)

Table 4. Summary of findings for the Role of the Regulator theme.

Theme	Major Categories	Minor Categories
Role of the Regulator	Auditing and bureaucratization	Auditing excessive, rigged and undermining safety Excessive bureaucracy
	Lack of clarity and support Negative safety climate	Regulation turned into a profiteering exercise Negation of responsibility No confidence

The regulator was also perceived to be too bureaucratized, where this excessive bureaucracy was viewed to eclipse work quality/safety and made work unprofitable for many in the sector:

Paper work's always good. You have to have it but, jeez, it's gotten ridiculous now. Back in the day, you'd change a light bulb and go to the pilot, "Yep, you're good to go." I change a light bulb and now, is still takes me ten minutes to change it, but it is an hour of paperwork. (P_5)

The regulatory body is less concerned about the quality of the work that the engineers are doing and more concerned about the paperwork that they're producing. (P_4)

In many ways, these perspectives resonate with notions of a general bureaucratization of safety and bureaucratic accountability, which have gained force in recent years, and refer to the types of activities that are expected of organization members which account for the safety performance of those they are responsible for [30].

A lack of clarity and support from the regulator were also shared, with some participants highlighting perceived deficiencies around the regulator's responsibility, level of experience and understanding, and its provision of information:

... recently, we had three airworthiness inspectors [from the regulator] turn out at a local airfield. They spent three days going over a flying school. They called up the LAME, and they were standing there with a fistful of papers saying, "Blah blah blah blah blah blah blah. You've done wrong. You've done wrong." The [LAME] turned around and said, "Well, actually no, what you're citing as a reference isn't applicable, because that's for an airline. These aircraft are a lesser weight". (P_10)

It's a lack of education. It's like when I registered the aircraft in my name, or put my name down as a registered operator, did I get a leaflet from [the regulator] saying, "as the registered operator you are required to bang, bang, bang?" I think there's a misunderstanding in the industry, particularly in the private aircraft sector of who holds responsibility and what those people are required to do [...] It's law you should know it. I personally think there should be some onus on CASA too, you're the registered operator of this aircraft. (P_4)

There was some uncertainty about the regulator's overall role, and feelings that regulation kept changing, *"A few of the dramas we have is that it's just the changing regulations. It's hard to keep up with all of that"* (P_9). Concerns were also shared about little to no oversight from the regulator on certification and engineering standards:

I have a lot of concerns, not only about the engine. I have a lot of concerns about, first of all, the certification of the aircraft, I have a lot of concerns about the engineering of the aircraft, the standards of the engineers. How they check the engineering standards, there was no oversight happening from [the regulator] at all. (P_4)

In a stark contrast to the overall growth in safety culture construed by the previous theme, the regulator itself was suggested to be contributing to the generation of a *negative safety climate*, particularly around compliance on issues that were viewed to not be safety-related:

> *[The regulator] impacts the industry because [maintenance engineers] will then do things that they shouldn't do. They know [what to do] they've done it that way for years. They know that it isn't a safety case, it's more a compliance issue, but they just do it because they've done it that way for years. Then, the regulator comes in and smacks them down, so then they don't do it. That increases the cost. It increases the time. People don't want to own aircraft. They sell aircraft, so we see these gradually diminishing infrastructure and system that shouldn't need to be that way [...] we have a regulator who sees it's better to regulate than to educate and to assist.* (P_10)

Central to perceptions of mixed-messaging and excessive regulatory oversight were notions that *regulation* had *turned into a profiteering exercise* and paid advertising service:

> *It's this regulatory oversight that's happening on everything all the time. Every time I pick up the phone and talk to [the regulator] they want to send a bill, so the government funded safety authority, that is also funded out of the levy that is put on fuel. Yet every time I do something, they want to charge me for it. I'm not sure how all of that works, are they double dipping? Are they now a profit center for the government? Yet they [are] supposed to be the safety authority [...] at the moment it's as the authority and the regulator aren't telling you "you need to do this", [they are saying that] to do that you need to come to us and when you come to us we're going to charge you.* (P_4)

This was accompanied by views that the regulator had negated its responsibility, created concerns, and ultimately, feelings of *no confidence* in the regulator for many:

> *I've worked on the spinners with guys that were too lazy to do their LAME license. They were more than happy for me to sign for their work. Those people are now [working in the regulator office], in regulatory roles, enforcing rules that they really don't have the privilege to [enforce].* (P_10)

3.3. (Re)calibration of Underlying Values and Philosophies

Table 5 decomposes the findings associated with perceived *(re)calibration of underlying values and philosophies*. A large representation of this theme was within a subtheme highlighting that staff were now *working to live, not living to work* (i.e., working to survive rather than feeling highly motivated or enthusiastic to work). *Engineers* had purportedly *now become business managers*, *"There is no getting away, if you take on the role of chief engineer you're taking on a life of paperwork"* (P_4)—a perspective that was ascribed various feeling and behaviours, including: no joy around administration, *"I don't particularly enjoy the days I'm sitting in an office all day and trying to deal with issues with customers and trying to look after all the staff on the shop floor"* (P_9); pre-occupation with non-skill-related activities and office work, *"those with the most skills and abilities and experience, rightly or wrongly, are ending up driving the office and shuffling the paperwork"* (P_2), and dispassionate and distracted ways of work.

Table 5. Summary of findings for Calibration of Underlying Values and Philosophies theme.

Theme	Major Categories	Minor Categories
(Re)calibration of Underlying Values and Philosophies	Working to live, not living to work Attitudinal stability	Engineers are now Business Managers GA requires passion Licensing culture Teamworking Trainees and their sense of entitlement

These feelings were strongly advocated on the basis that maintenance work in *GA requires passion* and is more than just a job but something which should provide enjoyment, decision-making freedom and flexibility:

Fault finding is what I love doing. I like having a customer come in and say I have a problem with this, and I go search the memory bank and you try and find a solution to it. That, to me, is the best part […] I enjoy being the chief engineer because I can make the decisions. (P_8)

In short, the sense of duty and responsibility to maintain the level of safety that behoves maintenance work was a by-product of real passion and enthusiasm:

Engineers are either passionate and believe in what they're doing and want to do it the right way […] or they're tainted by the industry, they have no respect for the industry, they have no respect for what the industry is trying to achieve. Consequently, they'd rather spend more time on their telephone organizing the building of their new house, while they're charging the client. (P_4)

Attitudinal stability was a key emerging subtheme in the *(re)calibration of underlying values and philosophies*. For instance, attitudinal issues created a certain *licensing culture*; while on the one hand, it gave rise to individual accountability, on the other, it was also felt to produce arrogance:

I've never come across such a toxic, mind state as aviation. Everyone's an expert. The only people they should listen to LAMES because they've earned their right, but anyone feels that, well, hey, I can work on a mower, I can work on an airplane. Why would I need to pay you 120 dollars, 150 dollars an hour to do something I can do myself? Then they will use commercial-grade hardware. We refer to it as "Bunnings Aerospace"; (P_10)

Note: "Bunnings Warehouse" is an Australian household hardware chain; and

You, effectively, really only start to learn when you become licensed. From that day, when you're accountable for you own action is when you actually start really learning. (P_10)

Attitudinal barriers were also encountered in *teamworking* and linked to attitudinal change associated with new training approaches, mismanaged expectations, and a broader "breeding" of disrespect:

I've worked with some engineers that I've basically said, "I don't want those engineers working on my aircraft again […] I can see what they're doing, I don't like what they're doing, I don't like their attitude"; (P_4)

[Trainees] have all sorts of funny preconceived ideas about what their first job's going to look like and how the industry actually is or is not; (P_2)

And

The guys that come in for work experience, they've already done their one-year course and they don't want to be there. They already believe that they know more than you. (P_5)

Much of this was attributed to an entrenched *sense of entitlement in trainees*; this was based on a lack of respect and accountability which also made it more difficult to trust, and therefore, provoked distrust in non-apprenticeship models of learning:

[Trainees] have done their year course, they've paid all the money, they're ready to work on planes—and they're not; (P_5)

I believe that there is no respect anymore. Kids these days, we see them coming through industry. They want to sit there with their thumbs on their phone […] There is certainly a lack of accountability amongst younger people these days. They want everything for nothing. They don't want any hard work to do it; (P_10)

and

You can't put full trust in somebody straightaway as they walk in the door, but you need to trust that they can use a screwdriver or spanner properly and things like that. They don't seem to have that sort of thing when they come into the shop as an apprentice. (P_9)

3.4. *Work as Imagined vs. Work as Done*

Many of the issues observed within the GA industry were thematically consigned as problems of a *Work as Imagined* vs. *Work as Done* mismatch (see Table 6), a category named after recent theoretical conceptualizations where explicit or implicit assumptions of how work should be done is different to how something is actually done [31], [32]. A large category here was *theory* vs. *practical experience*, and in the true essence of this theme, illustrated some very varied perspectives of the *relevance of classroom teaching to the real-world setting*, as well as the value of *work experience*. The GA work as imagined was thought to bear little to no resemblance to work on the "shop floor". In this way, the classroom was thought to provide lower fidelity training, a lack of focus on practical work, and exams were seen to have little relevance:

> *People coming from the college were a bit off with experience, and they've been working on an engine or aircraft that's in their facility, that's been pulled apart last month and put back together, whereas they come here and it's a plane that hasn't been pulled apart for ten years and everything's rusted and corroded;* (P_7)

> *Everyone that I've come across has come out without any knowledge of what we're actually doing on the floor;* (P_8)

> *When you get trained it's always this perfect picture, and then the real life. It's not so perfect. Definitely certain places have different methods to other places. Some I agree with, some I don't;* (P_6)

And

> *Because [the exam is] written by bureaucrats, there's no real relevance to what you actually do in the field to what you learn in the exam.* (P_10)

Table 6. Summary of findings for Work as Imagined vs. Work as Done theme.

Theme	Major Categories	Minor Categories
Work as Imagined vs. Work as Done	Theory vs. practical experience	Relevance of classroom to real-world setting Work experience
	Restrictive maintenance manuals	Inflexible Require interpretation

Alternatively, some views ascribed increasing importance to the role of theory in training, "nowadays, employers are wanting to see the apprentices actually complete all their theory before they'll want to take them on" (P_2). Similarly, work experience was heralded as a mechanism that provided practical knowledge, and afforded knowledge acquisition in a way that learning of theory could not:

> *[Work experience] giving the apprentices the appreciation as to what it is actually like in the working environment, I think there's a lot of benefit in that. Yeah and in my case, it led into an apprenticeship at the end of it. I highly recommend that apprentices and people in the airline industry do that.* (P_2)

However, work experience in the form of on-the-job training was also felt to be a platform that could easily teach and transmit bad habits:

> *I think, go back many years it was all on the job training, and so the bad habits that an engineer taught to another engineer, taught to another engineer, taught to another. Just got passed down and then got even made worse to the point where you have an apprentice who's being taught by someone whose got a license, yet the work's crap.* (P_4)

Restrictive maintenance manuals were given as a specific example of how work as imagined seldom translated into how work was actually done. Maintenance manual logic and perceptions of compliance were considered to be incongruent with what was required in practice, therefore considered restrictive, *inflexible*, demand certain skills, and *require interpretation* in ways that ultimately made them very unwieldy.

> *You read the maintenance manual and it just says, "Fix it." There's no interpretation, you know? […] Quite often the manual will give you all of the information you need to know on how to build it from the nut up, but you're not doing that. You need to dive into that particular part that you're involved with, you know? Then use the manual to repair it from there, not from go to whoa. Understanding the manual and interpreting it is a real challenge.* (P_3)

> … *having this mindset that the regulator tells you that you are not to refer to anything mentally. You are to refer to a manual to do any servicing and maintenance.* (P_10)

3.5. Practical and Operational Challenges for GA

As shown in Table 7 the last major theme in the study (representing 27% of all coded data) was centered around the *Practical and Operational Challenges for GA*. Following on from notions given within the forgoing theme, a key element here was the perceived *mismatch between GA and the commercial aviation sector*. Due to this, *regulation and compliance was seen to have a poor fit* and calibration with GA and small operators:

> *[In] Australia, unfortunately, you haven't really learnt from the mistakes made in the UK, so they're trying to make the General Aviation industry the same as the [commercial] airline industry, which doesn't work […] [they] are very different ballgames.* (P_9)

Table 7. Summary of findings for Practical and Operational Challenges for the GA theme.

Theme	Major Categories	Minor Categories
Practical and Operational Challenges for GA	Mismatch between GA and commercial aviation sector	Regulation and compliance have poor fit with GA needs Implications for sustainability
	Safety-risk and safety concerns	From over-modulation of safety From loss of skills
	Workforce development	Loss of knowledge and skills Staff turnover Mentorship
	Financial burden	Safety improvement increasing cost Age and cost of aircraft

By modelling GA on the commercial airline sector, regulation and compliance was felt to be impractical, counterintuitive, and cause restrictiveness in some scenarios, for example when resources were stretched:

> *"One of the problems is we do a lot of the maintenance on vintage aircraft, and some of the parts just, you can't get a new part with a release note and then there's perfectly good serviceable parts that are available from the parts bin. You're not allowed to use it theoretically;* (P_7)

> And

> *The levels that [the regulator] are wanting people to come up to, and the standards that they're wanting us to comply with, are good. But they need to be careful, though, particularly with small operators, that the effort and the cost and the time required with compliance doesn't detract from actually getting the job done properly and getting the job done safely. It can start to be counterproductive, particularly in small organizations where resources are having to be divided to the compliance side of thing.* (P_2)

The mismatch was also thought to *create implications for sustainability*. By placing undue pressure on small companies, they were reportedly being caught 'between a rock and a hard place', and the requirement for more concurrent jobs coupled with time poverty was having other effects, for instance, *"flushing out good organisations along with dodgy operators"* (P_2);

> *The most difficult point is when [your company] grows to the size that you can't afford the overhead of having someone like a business manager, but you can't afford not to because you physically can't do it and it's not your forte;* (P_4)

> *How do we still make profit trying to do all this paperwork, trying to help apprentices? It goes back to that again—you don't get time;* (P_5)

> *[The implication of the regulator's] high standards that they're trying to promote throughout the industry is there's a real ... Small operators are real disadvantaged [...] It's basically only able to be done sustainably with bigger organizations. Middle level and larger organizations seem to be the only organizations that can continue to be able to be sustainable;* (P_2)

> And

> *Most of the maintenance goes abroad these days for the heavy stuff. To the lower maintenance, you still have 2 or 3 engineers that are available to work on their airplanes, where in General Aviation, we generally have maybe 1 to 2 for the entire airplane. We have 25 multiple jobs happening at (any) one time.* (P_9)

Going forward, there were also practical and operational challenges around *safety-risk and safety concerns*. An *over-modulation of safety* was attributed to tensions with the regulator and were peculiar to the GA industry, *"[the regulator has] potentially, maybe [gone] too far in some respects they've gone over what (level of safety) it has to be"* (P_1), were perceived to leading to the development of a tauter system, where airworthiness was piece-meal and the compliance, counterproductive:

> *[Aircraft owners] simply just can't afford to repair their aircraft, so there's compromises need to be made in terms of, "Okay, we'll do these jobs this time. We'll stretch these other jobs out for another hundred hours for the next inspection," or what have you.* (P_2)

Paradoxically, the product of this over-modulation was therefore the perceived creation of riskier situations and safety decrement leading to general lack of safety in GA:

> *We're entering a very interesting time in the aviation industry, particularly in GA, where the best competent engineers are in the office, off the tools, and the least competent engineers are operating hands-on on the aircraft;* (P_2)

> And

> *Those that are trying to do it the right way can't afford to stay in business, so they fall out of business. Which means that all you end up with is are those that have cut the corners and you now have an unsafe engineer environment in General Aviation. End of story.* (P_4)

Safety-risk and concerns were also forecast based on a *loss of skills* within the industry, and the level of experience, *"the safety aspect [of concern] would be the experience level of the people"* (P_8), and the notion that there will continue to be no perceived consequences for trainees:

> *Back in my day, you were taught, you were given a competency, when you expressed that you knew what you were doing, you were examined and tested on that. You were then given that privilege. If you didn't know and you made a mistake, you got a thick ear. You were off your machine, you were out sweeping floors. There was a consequence for you not putting that effort in. These days, there's no consequence.* (P_10)

A series of problematic *workforce development* issues were also perceived to be a major practical and operational challenge going forward. An intergenerational *loss of knowledge and skills* from an ageing workforce was perceived to be a major industry affliction, where the acquisition of the knowledge required to identify problems was being lost:

> *If I could know half of what [my Dad] knows [about aviation maintenance engineering] I'd be doing well, you know? I think there's still a bit of a concern now that there's not that much experience coming out of the schools and apprentices and stuff* ; (P_3)

And

> *The sad thing is that the guys at my age are becoming less and less, and those young people aren't getting the experience from them.* (P_8)

A loss of applied skills and knowledge was also being attributed as a by-product of the increase in professionalism, *"[...] I think the downside to the professionalism is the loss of the knowledge how [sic] to fix things yourself"* (P_3). Directly related to these issues was the perception of a high *staff turnover* of people within general safety roles, *"basically, there's just enough staff employed to do what's required, and that's it"* (P_2), and a shortage of staff which meant that there not many apprentices in the industry and less opportunity to "pass the torch" onto younger people through *mentorship*:

> *There's not many apprentices coming through [...] I just loved working with older [engineers] because they could give me hints and guidance on where to go and what to do. I'm, me now as being one of those older people, don't have the opportunity to teach the young.* (P_8)

These concerns were further complicated by a belief that when the opportunities were arising, apprentices were not being supervised or mentored effectively, creating a haphazard learning experience, and opportunity-cost considerations were impacting this:

> *The apprentices are just having to learn by trial and error and on their own. They're coming out very insecure and very poorly trained [...] As bigger companies try and keep costs down, those budgets that training apprentices, they don't have those budgets for extra people on the floor to mentor and to guide and to hand over those skills to the apprentices.* (P_2)

Following on from the last point, the *financial burden* in the GA sector, and financial considerations in employment, was a final key practical and operational challenge. The organisational structure of GA was perceived to have changed in a way that disadvantaged small operators, and the new system was increasing cost with the *safety improvement increasing cost* and adding to the financial burden:

> *Small operators are real disadvantaged. The days of having a small operator, be it a charter company or be it a maintenance organization with just the chief engineer, the owner, and a couple of LAMEs ... Those days are gone because you need so many senior persons now just to be able to meet requirements;* (P_2)

> *The main thing for me is money. That's the biggest negative at the moment is that you're always fighting it [...] For some reason, the General Aviation industry, people think that everyone should ... We should be a bank for them. [...] It's like, "well, we're a small business, we can't afford to do that." In the same respect, it's ... Then [maintenance engineers] have problems paying our people and our suppliers and, by law, we're supposed to be able to afford to operate;* (P_9)

And

> *The cost of parts have gone up because of the new system. The cost of maintenance has gone up because of the new system. Owners can't afford planes these days and when they do, they can only just afford them. However, they still want the plane maintained perfectly and they still have to have their plane yesterday.* (P_5)

Lastly, the age and cost of the aircraft was thought to be contributing to the financial burden; the expensive nature of planes coupled with their age, which despite a link for more work perpetuated the view that work was simply unprofitable:

> *Back when it was [the aircraft] only 10 years old, yeah, $150 is fine because it didn't need all this major work. Now, you need to be putting money aside for not just your mandatory component overhauls, but major refurbishment: air frame and avionics upgrades, paint interiors, things that have never been replaced in the aircraft's life;* (P_2)

> *We haven't made profit in two years. It's not possible to make a profit, but the customers still complain about the bill;* (P_5)

> And

> *The hardship now, really, is that there's a lack of money in the industry, not only, from an engineering perspective, but from the customer's perspective. Everyone's trying to scrimp and save.* (P_9)

Figure 4 depicts a bleeding brakes scenario created by a participant in the form of their own concept map/mind-map/list which involves many of the central themes and issues across the results of the study, such as competency/experience and manuals, but also shows how concerns and worries around time, cost, business viability, and regulation may telegraph into the work and impact safety and performance.

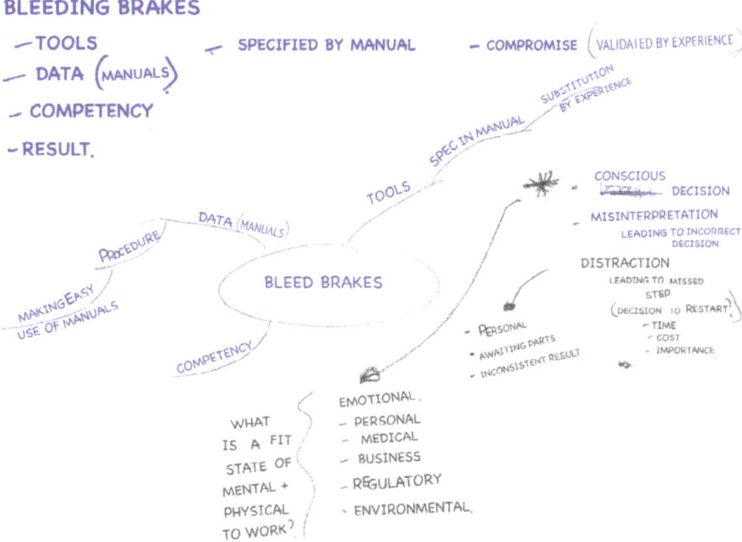

Figure 4. Example SITT scenario constructed as a mind-map by a participant depicting various factors and considerations perceived to be involved in bleeding brakes. Note: Handwriting has been replaced by a typescript to preserve anonymity of participant; size, colour and placement are the same.

3.6. General Discussion

The next sections provide a general discussion and reflection of findings based on the research questions, followed by a discussion of the strengths and limitations of the study, its implication, and potential future research directions.

3.6.1. Changes and Impacts in Regulation and to the Regulator Role to the GA sector

The results draw attention to an interconnectedness between the various components of the GA system, and illustrate how easily system behaviour can change overtime based on various feedforward and feedbacks loops associated with the regulatory/policy layers. Many of the findings reflect perceptions and statements on ways of working that carry safety implications, but are also a reflection of behaviours in an ostensibly complex and highly dynamic system. Proportional regulatory requirements, compliance, and safety oversight were among the main issues raised by participants (i.e., maintenance engineers/technicians). This set of issues are vital for operation of the GA industry, primarily due to the sustainment challenges brought by ageing fleets.

The results echo the industry's wish to balance safety and cost in a way that will allow the GA to remain profitable. Similar efforts have taken place in Europe and the US, thus, this finding indicates the international connectedness of the Australian GA industry, as well as the high degree of awareness of Australian maintenance professionals around these matters. On the other hand, it also helps us understand the driving forces behind the recent adoption of the US (FAA) regulatory model for GA. The choice of the FAA regulatory model offers greater autonomy to the Australian GA sector, but only at the cost of creating disconnection with the EASA-based CASA regulatory framework.

The existence of perceived power distance issues between "regulated entities" (i.e., maintenance technicians) and the regulator is insightful. Power distance has been well-researched in many different contexts [33–35], and in this study, responses suggest that GA technicians do not necessarily see themselves as the regulator's "long arm" in the effort to safeguard airworthiness. This was reflected both directly and indirectly (i.e., reference to the educational role of the regulator) through a wide range of narratives. This is an important research finding, since power distance can have an impact on the level of regulatory compliance and the effectiveness of reforms and changes attempted by the regulator. Interestingly, the results suggest that authority gradients—which relate to perceived power hierarchies in decision-making within teams or groups—also exist in the ranks of maintenance technicians, between those working on the "floor" and those in maintenance management and administration posts. As a natural extension of power distance, authority gradients between specific teams have also been explored at length in different safety-critical contexts [22,36]. In the aircraft maintenance context, it is a common expectation that the more experienced maintenance technicians have more opportunities (and do) take up office-based roles, especially in larger organizations. However, authority gradients also draw attention to cultural impacts in dynamic teams that can also impact and create fissures in safety culture.

3.6.2. Practical and Operational Challenges in the GA Industry

The availability of funds for maintenance has always been a strident issue for the GA sector, and this was confirmed in findings. Interestingly, participants expressed numerous concerns in relation to the role of owners in sustaining the airworthiness condition of their aircraft, broadening the systems perspective. The relationship of technicians and aircraft owners was highlighted over several narratives, offering indications of safety discrepancies. This relationship, in turn, has a negative effect on the job satisfaction of technicians since they may perform their work (i.e., maintenance tasks) within a stressful environment. This view was supported in a previous related research [12,13].

Another key finding was related to the attractiveness of the GA sector as a career pathway for aircraft technicians. This was recognized by participants as a challenge, given the high demand for more technicians (as the ageing GA aircraft required more maintenance) and the ageing workforce at large. Moreover, there was a perception that this demand cannot be met by the current supply of newly qualified technicians, as they generally lack the necessary training, skills, and motivation. These highly critical views expressed by participants may reflect a generation gap mindset, but also draw attention to perceived issues and hindrances with recruitment and retention of the workforce.

3.6.3. Strengths and Limitations of the Study

A key strength of this study was the use of the methodology that underpinned the data collection. Integration and application of the SITT elicited rich and concrete examples of system-level behaviours from the technician and engineer (i.e., end-user) perspective associated with the regulator and regulation and other challenges facing everyday operations. This enabled participants to harness and articulate the technical complexity of their work succinctly, but also allowed them to share their first-hand views in a more vicarious manner.

While relatively modest, the sample size did not serve as a limitation of the methodology as data saturation within data was reached, however, the small sample does limit the ability to make broad generalizations across the GA sector. Although the perceptions received were common across many participants, they are nevertheless subjective and nuanced and require care with interpretation.

3.6.4. Implications and Future Research Directions

In 2019, CASA decided to rearrange the GA aircraft maintenance regulatory framework in Australia. Changes within aviation can happen relatively quickly, but building a long and lasting positive culture is a slow process. For this reason, the impact of any change (either positive or negative) will be witnessed in the years to come. This analysis of data collected during late 2016—prior to the extensive industry consultation process which CASA employed for this purpose—offers important empirical insights which may be of value during transition, and also in years to come for benchmarking the sector's cultural maturity. It is likely that the regulatory changes happening in the GA sector will not be as fruitful as expected if the issues identified in this study are not tackled, and if the understanding between the regulator and those who are regulated is not improved. Issues in the key *human* factors cannot be addressed through new regulation, or through forced compliance. This paper reports some of these issues and it is believed that further independent research and publications (from external researchers, not associated with the regulator or the regulated bodies), can assist in the overall effort. Given the larger systems-oriented processes uncovered in the findings, further research could seek to unpack these dynamics relationships, for example through application of specific systems mapping processes that identify and attribute the feedback and feedforward loops for different elements [37–39].

Lastly, while this paper has focused on the Australian GA sector, comparison of the GA safety performance at an international level may be helpful in garnering further insights of the underlying regulation, practice, and safety culture issues [40]. A comparative analysis of these safety statistical data and practice could provide an indication of the effectiveness of the policies now being pursued by the Australian regulator. Therefore, a follow up research and study in this space is likely to constitute a valuable retrospective evaluation of the issues identified and discussed in the present paper.

4. Conclusions

In Australia, the diverse profile of General Aviation does, indeed, pose challenges for effectively managing safety and performance in aircraft maintenance operations. Based on the perceptions of this study, it is unsurprising that this has been a targeted area for continuous improvement.

The findings highlight that for the most part, perceived changes in the role of the regulator and regulation within the sector are not only perceived to have impacted the behaviours of the system underpinning the sector, but also diminished operations around safety and invoked a general view of a negative safety climate.

There are many perceived practical and operational challenges in the GA sector that must be dealt with to sustain safe maintenance operations going forward, many of which require change and growth across various cultural dimensions, and not only in how maintenance engineers and technicians work, but also other areas of the system, including the regulator, policy and practice, customers/owners, and tertiary education and vocational training.

In nearly every regard, if a central theme was to be assigned to this study it is that the GA sector is an ostensibly complex system, with high stakes in safety, and would therefore benefit from systems thinking when seeking to facilitate change management.

Author Contributions: Conceptualization, A.N. and K.I.K.; methodology, A.N.; formal analysis, A.N.; investigation, A.N. and K.I.K.; validation, K.I.K.; data curation, A.N.; writing—original draft preparation, A.N. and K.I.K.; writing—review and editing, A.N. and K.I.K. All authors have read and agreed to the published version of the manuscript.

Funding: This research received no external funding.

Acknowledgments: The authors are extremely grateful to Kate Kingshott for her significant role in the collection of the data used to underpin this research, and are very thankful for her support during very early discussions of this study.

Conflicts of Interest: The authors declare no conflict of interest.

References

1. International Civil Aviation Organisation (ICAO). *Annex 6 to the Convention on International Civil Aviation, Operation of Aircraft, Part I—International Commercial Air Transport–Aeroplanes*, 11th ed.; International Civil Aviation Organisation: Montreal, QC, Canada, 2001.
2. Dekker, S. *Foundations of Safety Science: A Century of Understanding Accidents and Disasters*; CRC Press: Boca Raton, FL, USA, 2019.
3. CASA. Sector Risk Profile for the Small Aeroplane Transport Sector. Available online: https://www.casa.gov.au/sites/default/files/srp_small_aero_booklet.pdf (accessed on 15 June 2020).
4. Boyd, D.D.; Stolzer, A. Accident-precipitating factors for crashes in turbine-powered general aviation aircraft. *Accid. Anal. Prev.* **2016**, *86*, 209–216. [CrossRef]
5. Boyd, D.D. Causes and risk factors for fatal accidents in non-commercial twin engine piston general aviation aircraft. *Accid. Anal. Prev.* **2015**, *77*, 113–119. [CrossRef]
6. CASA. Safety Behaviours, Human Factors: Resource Guide for Engineers. Available online: https://www.casa.gov.au/sites/default/files/_assets/main/lib100215/hf-engineers-res.pdf (accessed on 3 April 2020).
7. ATSB. Aviation Occurance Statistics 2005 to 2014. Available online: https://www.atsb.gov.au/publications/2015/ar-2015-082/ (accessed on 23 January 2020).
8. BITRE. General Aviation Study—[Statistical Report—978-1-925531-77-0]. Available online: https://www.bitre.gov.au/sites/default/files/2019-11/cr_001_0.pdf (accessed on 2 January 2020).
9. Haynes. *The Theory and Practice of Change Management*; Palgrave: Hampshire, UK, 2002.
10. CASA. SMS for Aviation—A Practical Guide: Safety Assurance. Available online: https://www.casa.gov.au/safety-management/safety-management-systems/safety-management-system-resource-kit (accessed on 3 April 2020).
11. CASA. Part 43—Maintenance of General Aviation and Aerial Work Aircraft (CD 1812SS). Available online: https://consultation.casa.gov.au/regulatory-program/cd1812ss/ (accessed on 31 January 2020).
12. Naweed, A.; Kingshott, K. Flying off the handle: Affective influences on decision making and action tendencies in real-world aircraft maintenance engineering scenarios. *J. Cognit. Eng. Decis. Making* **2019**, *13*, 81–101. [CrossRef]
13. Kingshott, K.; Naweed, A. "Taxiing down the runway with half a bolt hanging out the bottom": Affective influences on decision making in general aviation maintenance engineers. In Proceedings of the 2018 Ergonomics & Human Factors Conference, Birmingham, UK, 23–25 April 2018.
14. Vaughan, D. *The Challenger Launch Decision: Risky Technology, Culture and Deviance at NASA*; Uni of Chicago Press: Chicago, IL, USA, 1997.
15. Naweed, A.; Chapman, J.; Trigg, J. "Tell them what they want to hear and get back to work": Insights into the utility of current occupational health assessments from the perspectives of train drivers. *Transp. Res. Part. A* **2018**, *118*, 234–244. [CrossRef]
16. Naweed, A.; Dorrian, J.; Rose, J. *Evaluation of Rail Technology: A Practical Human Factors Guide*; CRC Press: London, UK, 2013.

17. Naweed, A.; Balakrishnan, G.; Bearman, C.; Dorrian, J.; Dawson, D. *Scaling Generative Scaffolds Towards Train Driving Expertise, in Contemporary Ergonomics and Human Factors 2012: Proceedings of the International Conference on Ergonomics & Human Factors 2012*; Anderson, M., Ed.; CRC Press: Blackpool, UK, 2012; p. 235.
18. Naweed, A. Psychological factors for driver distraction and inattention in the Australian and New Zealand rail industry. *Accid. Anal. Prev.* **2013**, *60*, 193–204. [CrossRef] [PubMed]
19. Klein, G.; Calderwood, R.; Macgregor, D.G. Critical decision method for eliciting knowledge. *IEEE Trans. Syst. Man. Cybern.* **1989**, *19*, 462–472. [CrossRef]
20. Monk, A.; Howard, S. Methods & tools: The rich picture: A tool for reasoning about work context. *Interactions* **1998**, *5*, 21–30.
21. Checkland, P. *Systems Thinking, Systems Practice*; Wiley: Chichester, UK, 1980.
22. Naweed, A. Getting mixed signals: Connotations of teamwork as performance shaping factors in network controller and rail driver relationship dynamics. *Appl. Ergon.* **2020**, *82*, 102976. [CrossRef] [PubMed]
23. Naweed, A.; Rainbird, S.; Chapman, J. Investigating the formal countermeasures and informal strategies used to mitigate SPAD risk in train driving. *Ergonomics* **2015**, *58*, 883–896. [CrossRef]
24. Pabel, A.; Naweed, A.; Ferguson, S.A.; Reynolds, A. Crack a smile: The causes and consequences of emotional labour dysregulation in Australian reef tourism. *Curr. Issues Tour.* **2019**, *23*, 1598–1612. [CrossRef]
25. Naweed, A.; Rose, J. "It's a Frightful Scenario": A Study of Tram Collisions on a Mixed-traffic Environment in an Australian Metropolitan Setting. *Procedia Manuf.* **2015**, *3*, 2706–2713. [CrossRef]
26. Naweed, A.; Ambrosetti, A. Mentoring in the rail context: The influence of training, style, and practice. *J. Workplace Learn.* **2015**, *27*, 3–18. [CrossRef]
27. Braun, V.; Clarke, V. Using thematic analysis in psychology. *Qual. Res. Psychol.* **2006**, *3*, 77–101. [CrossRef]
28. Hofstede, G.; Milosevic, D. Dimensionalizing cultures: The Hofstede model in context. *Online Read. Psychol. Cult.* **2011**, *2*, 8. [CrossRef]
29. CASA. Cessna Supplemental Inspection Documents. Available online: https://www.casa.gov.au/aircraft/airworthiness/cessna-supplemental-inspection-documents (accessed on 2020 25 March).
30. Dekker, S.W.A. The bureaucratization of safety. *Saf. Sci.* **2014**, *70*, 348–357. [CrossRef]
31. Hollnagel, E. The nitty-gritty of human factors. In *Human Factors and Ergonomics in Practice: Improving System Performance and Human Well-Being in the Real World*; Shorrock, S., Wiliams, C., Eds.; CRC Press: Boca Raton, FL, USA, 2016; pp. 45–64.
32. Naweed, A.; Young, M.S.; Aitken, J. Caught between a rail and a hard place: A two-country meta-analysis of factors that impact Track Worker safety in Lookout-related rail incidents. *Theor. Issues Ergon. Sci.* **2019**, *20*, 731–762. [CrossRef]
33. Tan, W.; Chong, E. Power distance in Singapore construction organizations: Implications for project managers. *Int. J. Project Manag.* **2003**, *21*, 529–536. [CrossRef]
34. Cole, M.S.; Carter, M.Z.; Zhang, Z. Leader–team congruence in power distance values and team effectiveness: The mediating role of procedural justice climate. *J. Appl. Psychol.* **2013**, *98*, 962. [CrossRef]
35. Rieck, A.M. Exploring the nature of power distance on general practitioner and community pharmacist relations in a chronic disease management context. *J. Interprofessional Care* **2014**, *28*, 440–446. [CrossRef]
36. Naweed, A.; Dennis, D.; Krynski, B.; Crea, T.; Knott, C. Delivering simulation activities safely: What if we hurt ourselves? *Simul. Healthc.* **2020**, in press. [CrossRef]
37. Leveson, N. A new accident model for engineering safer systems. *Saf. Sci.* **2004**, *42*, 237–270. [CrossRef]
38. Read, G.J.; Naweed, A.; Salmon, P. Complexity on the rails: A systems-based approach to understanding safety management in rail transport. *Reliab. Eng. Syst. Saf.* **2019**, *188*, 352–365. [CrossRef]
39. Salmon, P.M.; Read, G.J.; Walker, G.H.; Goode, N.; Grant, E.; Dallat, C.; Carden, T.; Naweed, A.; Stanton, N.A. STAMP goes EAST: Integrating systems ergonomics methods for the analysis of railway level crossing safety management. *Saf. Sci.* **2018**, *110*, 31–46. [CrossRef]
40. De Voogt, A.; Chaves, F.; Harden, E.; Silvestre, M.; Gamboa, P. Ultralight Accidents in the US, UK, and Portugal. *Safety* **2018**, *4*, 23. [CrossRef]

Review

Development of the Minimum Equipment List: Current Practice and the Need for Standardisation

Solomon O. Obadimu [1], Nektarios Karanikas [2] and Kyriakos I. Kourousis [1,*]

[1] School of Engineering, University of Limerick, V94 T9PX Limerick, Ireland; solomon.obadimu@ul.ie
[2] School of Public Health and Social Work, Queensland University of Technology, Brisbane QLD 4000, Australia; nektarios.karanikas@qut.edu.au
* Correspondence: kyriakos.kourousis@ul.ie; Tel.: +353-61-202-217

Received: 10 December 2019; Accepted: 15 January 2020; Published: 17 January 2020

Abstract: As part of the airworthiness requirements, an aircraft cannot be dispatched with an inoperative equipment or system unless this is allowed by the Minimum Equipment List (MEL) under any applicable conditions. Commonly, the MEL mirrors the Master MEL (MMEL), which is developed by the manufacturer and approved by the regulator. However, the increasing complexity of aircraft systems and the diversity of operational requirements, environmental conditions, fleet configuration, etc. necessitates a tailored approach to developing the MEL. While it is the responsibility of every aircraft operator to ensure the airworthiness of their aircraft, regulators are also required to publish guidelines to help operators develop their MELs. Currently, there is no approved standard to develop a MEL, and this poses a challenge to both aviation regulators and aircraft operators. This paper reviews current MEL literature, standards and processes as well as MEL related accidents/incidents to offer an overview of the present state of the MEL development and use and reinstate the need for a systematic approach. Furthermore, this paper exposes the paucity of MEL related literature and the ambiguity in MEL regulations. In addition, it was found that inadequate training and guidance on the development and use of MEL as well as lack of prior experience in airworthiness topics can lead to mismanagement and misapplication of the MEL. Considering the challenges outlined above, this study proposes the combination of system engineering and socio-technical system approaches for the development of a MEL.

Keywords: minimum equipment list; aviation; aircraft; safety; airworthiness

---

1. Introduction

Airworthiness relates to the ability of an aircraft to conduct safe operations, and aircraft maintenance activities comprise its backbone [1]. Aircraft must be maintained and certified under regulatory standards published by regional authorities to ensure airworthiness in every private or commercial aircraft operation and achieve acceptable flight and ground safety levels while ensuring dispatch reliability. Under this mandate, in collaboration with aircraft manufacturers and aviation regulatory bodies, the Master Minimum Equipment List (MMEL) and Minimum Equipment List (MEL) were introduced in the late 1960s [2]. The MMEL and MEL are documents with a list of aircraft components or systems that may be inoperative for aircraft dispatch [3,4]. The former is developed by the aircraft manufacturer, and the latter is based on the MMEL but further customised by each airline depending on its distinct operational characteristics and needs. MMEL and MEL are reviewed, rejected or approved by the corresponding National Aviation Authority (NAA).

However, although several parties and professionals are involved in the development and review of the MMEL, the MEL still lacks a similarly standardised process. Typically, the MEL per operator and aircraft type is approved by the respective competent authority after it has been compiled according to generic guidelines [3–5] which describe what must be achieved but provide only a little guidance on how to develop a MEL. Although the MMEL can serve as the basis to build MEL, under a systems approach, the existence and performance of each component, as well as the combined effects of various malfunctioning components or systems, can resonate and lead to adverse situations that had not been anticipated when examining the performance of the former separately from their environment [6]. Thus, mere reliance on MMEL, which typically refers to behaviours of individual components and subsystems under assumed conditions, might not suffice to publish a MEL proper for the operational environment of each air carrier.

Considering the above, the overall aim of our study was to examine the current situation around MEL and suggest whether a more standardised framework is necessary. In the next section of the paper, we present an overview of the current MMEL/MEL development process and respective standards, and we refer to the associated topics of reliability, safety/acceptable level of safety, environment and human factors. The paper continues with a review of MEL-related studies and an analysis of MEL-related accidents and incidents to detect the types of relevant problems/issues and identify possible gaps in the MEL development and application process. After a discussion of the overall picture, our paper concludes with recommendations about the application of system engineering and socio-technical system approaches to the development of MELs.

2. Master Minimum Equipment List (MMEL)/Minimum Equipment List (MEL)

Aircraft are designed with highly reliable equipment. Nevertheless, failures could occur at any time resulting in an accident/incident or simply flight cancellations and delays. The main objective of the Master Minimum Equipment List (MMEL) is to " ... reconcile an acceptable level of safety with aircraft profitability while operating an aircraft with inoperative equipment" [3]. MMELs/MELs are used to examine the release of an aircraft with inoperative equipment for flight. Their aim is to permit operation for a specific period under certain restrictions pending replacement or repair of the faulty item. However, the repair must be carried out at the earliest opportunity [2–4].

Before the introduction of MEL, the permission or not to operate with inoperative or underperforming systems/components was more a topic of negotiations between the operator and the regulator. Each regulator was forming its judgment and evaluation based on the competence of its staff, personal experiences, and information from previous cases depending on the type of aircraft under assessment [2]. This led to operators claiming favouritism when they had discovered that the list of permissively inoperative components and systems of aircraft of the same type belonging to another airline was less restrictive. The regulation and management of MELs were institutionalised in the 1960s [2].

According to the International Civil Aviation Organisation (ICAO) [7], the overall goal of a MEL is to describe when an aircraft with inoperative components or systems is still airworthy and authorised for dispatch. Airbus [8] describes MEL as a document based on the MMEL and developed by the air operator to optimise flight planning and dispatch as well as the operator's profitability while maintaining an acceptable level of safety. Nowadays, the general principles governing the compilation, approval, maintenance, and monitoring of MEL are the following [3–5]:

- The MEL is based on the MMEL, but the former must be more restrictive than the latter because MMEL is generic.
- The MEL must be produced by the operators and approved by the respective NAA.
- The MEL must be customised/tailored to account for various environmental conditions and operational contexts. This means that approving NAAs consider the environmental conditions (e.g., operating temperature and humidity) as well as the operators' scheduled destinations before a MEL is approved. For example, the MEL for an airline operator in the United Kingdom (UK) will be different from airlines in China who operate the same aircraft (e.g., prolonged exposure to cold of aircraft systems/components in the UK compared to exposure to dust and sand in China).
- The MEL allows operators to optimise aircraft dispatch reliability. The use of a MEL reduces aircraft downtime and increases airlines' profit without compromising safety.

- Each of the item/equipment in the MEL has conditions for dispatch. An aircraft cannot be dispatched until the category of the equipment/item in question is confirmed from the MEL.
- Equipment not listed in the MEL are automatically required to be operative for the dispatch of an aircraft.

Various professionals are involved in the process of MMEL [3,5]. Interactions within and across (sub)systems are extensively analysed, ensuring that multiple failures would not lead to an unsatisfactory level of safety by considering the impacts of critical failures and/or unserviceable items on flight safety, crew workload and operations. Table 1 illustrates the differences in the MMEL approval process between the Federal Aviation Authority (FAA) and the European Aviation Safety Agency (EASA).

Table 1. Differences between the Federal Aviation Authority (FAA) and the European Aviation Safety Agency (EASA) MMEL development and approval [5,9–11].

FAA	EASA
Final MMEL is developed by FAA	MMEL is developed by the manufacturer
Flight Operations Evaluation Board is the primary point of contact	Flight Standards Department is the primary point of contact
Proposed Master Minimum Equipment List is submitted for review	A full MMEL is submitted for review
Approved and published by FAA	Recommended by EASA's Operations Evaluations Board, approved by the NAA and published by the manufacturer

2.1. Methods of MMEL Justification

Before creating an MMEL, a thorough analysis must be conducted quantitatively and qualitatively to justify whether a component/system should be included in the list. The main tools used for these analyses are the Failure Mode Effect Analysis (FMEA) and Fault Tree Analysis (FTA) [12,13]. While these tools are meticulous and versatile for performing safety assessments, there are also limitations associated with their use. Tables 2 and 3 highlight the advantages and limitations of FMEA and FTA correspondingly [14–16].

The application of FMEA and FTA is based on results from qualitative and quantitative analyses and considers any optional and redundant equipment to inform decisions and develop a MMEL. Qualitative analysis is carried out before quantitative analysis [3,4,13]. It includes an evaluation of the effects of inoperative or underperforming components and systems on aircraft operation, flight crew workload and passenger safety to assess the achievement of an acceptable level of safety for dispatch [4]. Additionally, the qualitative analysis must ensure that the combined impact of multiple inoperative pieces of equipment will not lead to a catastrophic/hazardous failure [4]. Quantitative analysis supplements qualitative analysis [3,4] and is performed for items/equipment/components that are characterised as critical to the safe operation of the aircraft [4]. Furthermore, additional analysis may be required to analyse the rectification interval of each component or system [13]. This type of analysis adopts the System Safety Assessment (SSA) process which is based on the quantitative results from the FTA or FMEA techniques [12,13].

If an item is over the minimum required for safe operations in a particular flight route/condition or the aircraft could be operated under restrictions, inoperativeness of the specific item may be accepted and approved for inclusion in the MMEL. For example, the flight data recorder system in a Bell 412 Helicopter may be out of service for a limited time [17]. In addition, although the number of items with identical functionality installed on an aircraft depends on the manufacturer, operating a piece of equipment in the optional category is subject to the satisfaction of the NAAs that an acceptable level of safety would be maintained. If the functions of the system/item under assessment can be substituted by an alternative system/item with similar functions, then, it would be accepted for MMEL inclusion on a redundancy basis. The condition is that the alternative system would provide an acceptable level of safety as long it is confirmed operative. However, redundancy cannot be claimed for the inclusion of an item in the MMEL if all items/equipment are required to be operative based on the aircraft's type certificate. For instance, in Bell 412, two air data interference units are installed on the aircraft where one may be inoperative provided the second unit is fully operative for flight [17].

In cases that MMEL allows inoperative items, the aircraft can be dispatched to prevent aircraft grounding situations subject to maintenance or replacement of the affected component or system within the time frame specified in the MMEL. Nonetheless, to maintain the same level of reliability certain restrictions might apply (e.g., transferring functions to another fully operative system, flight limitations, night/day operations restrictions) to ensure that safety is not compromised.

Table 2. Advantages and limitations of Failure Mode Effect Analysis (FMEA).

Advantages	Limitations
FMEA provides a systematic approach in assessing the performance of relatively simple systems/components as it provides a systematic approach to system safety analysis.	FMEA does not guarantee the identification of all critical failure modes of a component/system under assessment when there is a lack of information/knowledge/experience.
FMEA offers flexibility in system safety analysis because of its ability to examine a system's failure modes and their safety impacts on a system/subsystem level or on a component level.	FMEA does not consider the human factors element; therefore, it cannot be used as a stand-alone safety system analysis tool.
FMEA complements other safety assessment tools (e.g., Fault Tree Analysis and Markov Analysis) as it provides source data for critical items/components of a system.	While the FMEA can be very thorough, it does not consider external system/component threats during analysis (e.g., multiple failures and common cause failures).
FMEA considers all possible failure modes and impacts on system operation, reliability, safety, and maintainability.	FMEA needs continuous management to keep it up to date. It takes time and is expensive to generate.
FMEA identifies both critical single failure events and latent failures.	FMEA does not consider multiple failure analysis within a system.

Table 3. Advantages and limitations of Fault Tree Analysis (FTA).

Advantages	Limitations
FTA identifies all basic events and describes their relationship within a system.	FTA requires a thorough understanding of the design, and this might be a challenge especially when the design is new.
FTA permits the evaluation of hypothetical events to determine their potential impacts on the top event.	FTA is tailored to a top event; thus, it includes only failures/events concerning the top event. Besides, the contributing events are not exhaustive as they are based mainly on the analyst's judgement.
FTA forecasts potential failures in a system's design, thus identifying areas of improvement within a system and enabling safety improvements.	FTA is not 100% accurate because it is based on an estimate/perception of reality. In addition, it does not consider maintenance and periodic testing of a system/subsystem.
FTA can be used during the design and operational phase of a system.	FTA depends mainly on the top event; thus, if incorrectly defined, the FTA will be incomplete/incorrect.

2.2. Factors Considered in MMEL/MEL Development and Justification

This section summarizes the principal factors which are collectively mentioned by Airbus [3], EASA [4], and UK CAA [5] and exert a major influence on the development of a MEL and affect decisions to include or exclude items from the list.

2.2.1. Reliability

Despite the technological advancements over time, the demand for highly reliable and performing systems has increased due to the complexity of modern systems [18,19]. Reliability of a component/system plays a key role during the MMEL development process as it helps to analyse and predict possible failure causes through the application of engineering knowledge [19]. Reliability plays a role in both MMEL and MEL. Reliability analysis is performed during the MMEL development and justification by using the tools and techniques mentioned in Section 2.1 above, and sufficient reliability must also be ensured for MEL items during operations.

According to Airbus [3], the MMEL contributes to the operational reliability of airline operators as it optimises dispatch reliability by reducing aircraft downtime. To maintain a MEL item's reliability, servicing/scheduled maintenance is required [20], and MEL items must be inspected at predetermined time intervals prescribed by the manufacturer or regulator.

2.2.2. Safety/Acceptable Level of Safety

Acceptable Level of Safety (ALoS) is primarily related to aircraft accident prevention, and its definition depends on the region, regulatory framework, safety records, etc. During the MMEL development process, crew performance and workload must be considered along with the effects of system/component failures on landing and take-off and their impact on the aircraft and its occupants [4]. In the context of MEL, ALoS can be maintained by adjusting aircraft operational limitations, transferring a system or component's function to another functioning system or a system that provides the information needed or performs similar functions as long as the crew workload changes are acceptable, and the training provided to the crew remains sufficient and relevant [4].

Another way of maintaining an ALoS during MEL application is by developing maintenance actions based on the MMEL (e.g., deactivating and securing the system/component concerned). For instance, for the Bombardier CL600 concerned, to maintain an ALoS on a lavatory fire extinguishing system when it is inoperative a placarding procedure is carried out: Associated lavatory door is locked CLOSED and placarded, "INOPERATIVE–DO NOT ENTER" [21]. Another example regarding the Airbus A320 is the Emergency Locator Transmitter (ELT) which broadcasts signals in the event of an accident; to maintain the ALoS, the inoperative ELT must be deactivated and repaired within 90 days [22].

2.2.3. Environmental Factors

MMEL development cannot be complete without the assessment of the varying environmental conditions in which aircraft would operate. Commercial aircraft operate between different regions; thus, their availability must not be impacted by the environment since designing an aircraft for a particular region would not be cost-effective [23]. Therefore, designers must obtain adequate knowledge of the environment so that to design aircraft that can withstand varying environmental conditions [24] while observing various local and international environmental standards and regulations.

Furthermore, concerning MMEL development and justification, there must be measures to reduce environmental impact on aircraft dispatch, e.g., impact of rain and icing conditions. With provisions such as effective anti-ice and wiper systems, aircraft dispatch is not limited/restricted in these conditions. In addition, it is important to consider the aircraft's flight envelope to specify the aircraft's maximum and minimum altitudes quantitatively when systems or components underperform or are out of service and define dispatch conditions (i.e., GO–IF items) to facilitate aircraft operations.

2.2.4. Human Factors

The goal of system safety assessments is to enhance a system's reliability, performance and safety by considering the persons (e.g., flight crew, ground crew and engineers) that will interact with the system. Reducing the effects of possible human errors must be a proactive and systematic activity considered during the design phase of a system [15]. Therefore, every system analysis must encompass human factors, particularly during the MMEL development and justification process. For instance, the application of sound ergonomic principles in the design of cockpit instrumentation, displays and controls can alert flight crews of any technical failures and provide timely and effective information (e.g., appropriately sized and readable figures) [15].

Consequently, as part of justifying the inclusion of an item in the MMEL, human factor analysis is conducted to anticipate the operation of cockpit systems/computers in varying operational conditions. However, it must be noted that the management of human factors is equally important throughout the lifetime of the aircraft during which MEL applies. Hence, it is imperative to understand the impact of an inoperative MEL item on crew performance when developing the lists [3]. For instance, Job Task Analysis [25] is a technique that can be conducted before carrying out any MEL-related task to identify the required resources (e.g., knowledge, certifications, and experience) for the successful execution and completion of the task.

Furthermore, worksheets and logbooks should be used to communicate any component failure during post-flight or pre-flight inspections. Logbook entries are the starting point for assessing defects of components included in a MEL [8] and help the ground crew understand the causes of failures. This practice facilitates communication between pilots and engineers and ensures that they are aware of MEL tasks that have been completed or are pending. Moreover, the consistent use of logbooks encourages teamwork and improves coordination, especially between work shifts.

3. Literature Review

3.1. Utility of MEL

Munro and Kanki [26] described the MEL as an operational document for air operators, which has direct implications on the safety and airworthiness of an aircraft. Pierobon [27] shared a similar opinion by naming MEL as a barrier against aircraft dispatch with unauthorised defects. Grüninger and Norgren [28] believe that MEL serves two different purposes. Its first purpose is to identify aircraft systems/components that could be inoperative while keeping the aircraft airworthy for dispatch. The second purpose of MEL is to serve as a reference document for engineers and pilots to conduct MEL-specified tasks/procedures before aircraft dispatch. Interestingly though, Hertzler [29,30] suggested that the name MEL is unsuitable for the specific document as it could be possibly interpreted as a list of equipment/items/systems that must be installed on an aircraft. Thus, he suggested a name change to "permissibly inoperative instruments and equipment or stuff that doesn't have to work".

Following a report published by the Accident Investigation Board of Norway [31], Herrera, Nordskaga, Myhreb and Halvorsenb [32] researched the impact of changes and developments in the Norwegian aviation industry. Amongst other findings, the authors found that, although MEL provides information about the overall status and serviceability of an aircraft, it was not considered as a parameter in the holistic assessment of safety and a leading indicator of safety performance. Consequently, Herrera et al. [32] recommended the development of a risk assessment model that would integrate safety indicators such as the MEL to facilitate the identification of maintenance-related airworthiness and safety issues by air operators proactively.

3.2. Differences in MEL Standards and Requirements

In his review, Feeler [33] confirmed differences in MEL standards worldwide. For instance, he compared MEL standards for corporate/business jets published by Transport Canada (TC) and the Federal Aviation Administration (FAA), and he observed that MEL provisions and applications differ in these regions. In Canada, both business and commercial operators can operate without an approved MEL, and the decision to fly is ultimately based on the pilot in command by considering, amongst other factors, applicable Airworthiness Directives and aircraft equipment/systems required for the intended flight route and conditions (e.g., day or night flight, operating under visual flight rules or instrument flight rules) [34,35]. Even where there is an approved MEL, the ultimate dispatch decision is still made by the pilot in command [33,34].

However, in the United States, non-commercial, business/corporate operators (a.k.a. Part 91 operators) are not permitted to operate an aircraft without a MEL [33]. Nonetheless, compared to their commercial counterparts, Part 91 operators enjoy some leniency. For example, commercial operators or operators with MELs approved under the Federal Aviation Regulations (FAR) 135, 129, 125 or 121 must comply with the repair intervals specified in MMEL/MEL, whereas Part 91 operators are not obligated similarly. Hertzler [29,30] compared the use of MEL under two distinct types of regulations, FAR Part 135 and 91, and, amongst others, found that MEL is not a technically approved document for Part 91 operators because they are authorised to use the MMEL as MEL, subject to FAA approval. Moreover, for the same category of operators concerned, compliance with deferral categories/repair intervals specified in MMEL was not mandatory. It is noted that, according to the advisory circular published by FAA [36], MELs approved under Part 135 apply even when the operator conducts operations under Part 91. According to FAA [36] "to provide relief to operators under the MEL concept, some operators may find it less burdensome or less complicated to operate under the provisions of FAR 5 91.213(d)".

On the other hand, in Australia, aircraft are not allowed to operate without an approved MEL or a manual for permissibly unserviceable components/systems under the provisions of CAR 37 [37]. In addition, in the United Kingdom (UK), aircraft are not permitted to commence a flight with inoperative equipment governed by Commission Regulation (EU) No 965/2012 [5,38]. Therefore, although the institutionalisation of MMEL/MEL has been promoted since the 1960s [2], some countries still exempt operators from the scheme and the aviation industry lacks a uniform approach.

3.3. Issues in MEL Development and Application

In the late 1980s, FAA launched a special inspection program to evaluate compliance of commuter air carriers with FARs with the participation of fifteen FAA inspection teams, consisting of six airworthiness inspectors each. A total of 35 air carriers were thoroughly inspected with a focus on 13 different areas, the MEL included [39]. The teams identified a total of 87 findings relevant to MEL, and they highlighted the following ones:

- Aircraft had been dispatched with inoperative systems/equipment not covered/permitted by the MEL.
- Cases of late rectifications of MEL items; air carriers operated MEL items/equipment for extended periods.
- Unrevised MELs or MELs less restrictive than MMELs.
- Inadequate placarding procedures as required by MELs.
- Inappropriate use of MELs (e.g., use of an Airbus A319's MEL on an A320 model or Bell 412's MEL on AgustaWestland AW139).

The overall conclusion of the FAA airworthiness inspectors was that there was a need for MEL compliance training as management personnel were not familiar with the MEL. Interestingly, over 30 years later, similar concerns were raised by Airbus [8] whose report restated the correct application/use of the MEL with a focus on compliance with its provisions as well as the principles and best practices when deferring and dispatching aircraft according to MEL. Furthermore, Pierobon [27] stressed the need for pilots and engineers to have prior knowledge and experience in airworthiness management for proper interpretation of the MEL document because the short familiarisation training received by pilots is insufficient. For instance, as part of the MEL application process in Canada and Australia, there is a requirement for operators to have MEL training programmes in place before approval and commencing operations with the MEL [37,40]. On the other hand, in the USA and UK, MEL training requirements are not defined in their respective guidance documents [5,36,41].

Similar findings were revealed by the study of Munro and Kanki [26], who reviewed 1140 maintenance reports issued between 1996 and 2002 and detected 143 reports relevant or related to the use of MELs. Their research identified improper deferral of MEL defects, failure to accomplish MEL specified tasks due to errors of omission, placarding issues and unrecorded MEL defects in technical logbooks. However, Munro and Kanki [26] also revealed contributory factors to MEL-related incidents including time pressure, unclear MELs, lack of familiarity, misinterpretation of the MEL, and communication flaws regarding the applicability of the MEL to aircraft status.

Moreover, Pierobon [27] stated that there is no standard MEL development methodology. After collecting opinions from industry professionals, especially those experienced in the MMEL/MEL process, he concluded in the necessity for NAAs to publish more specified guidelines to help air operators develop their MELs. This position and urge agree with the work of Feeler [33]—as presented in Section 3.2 above, the concerns of Hertzler [29] about the difficulty in interpreting the FAR Part 91.213 MEL regulations, and the observations of Yodice [42] regarding the ambiguity in FAA MEL regulations. To overcome these inherent challenges, the professionals interviewed by Pierobon [27] recommended the following improvement points:

- Delegating the MEL responsibility to persons experienced in MEL development and understanding the methodologies behind the development and justification of the MMEL.
- Appointing and training staff to specialise in the MEL authoring and review processes.
- Outsourcing MEL to knowledgeable and experienced professionals to ensure consistency in the MEL development and review process.
- Mandating the customisation of MELs to each operator per different aircraft model (i.e., reject mere duplications from the MMEL).
- Provision of adequate guidance from NAAs through a clear and detailed framework or methodology.

3.4. Published Studies on MEL-Related Accidents

Grüninger and Norgren [28] analysed the Spanair's McDonnell Douglas MD-82 fatal crash. The aircraft crashed shortly after take-off because the flaps/slats were not deployed due to a series of omissions/mistakes [43]. The investigation concluded that the take-off warning system did not activate, leaving the crew clueless about the incorrect configuration of the aircraft. The flight had been dispatched according to the MEL due to a faulty ram air temperature probe heater. Albeit the 'ground sensing relay' controls for both the take-off warning system when on the ground and the ram air temperature probe heater when airborne, the MEL did not require a detailed inspection to determine if the source of failure was, in fact, the inoperative temperature probe or a defective ground sensing relay. Had the MEL specified this, then perhaps the inoperative take-off warning system, which is a 'No Go' item if faulty, could have been discovered. Furthermore, the report also highlighted other instances of MEL misuse days before the accident; in one case, the crew had used the MEL during taxiing while it should be consulted only on ground before taxiing or take-off, and, in another case, the crew used the MEL without consulting with technicians/engineers.

Grüninger and Norgren [28] asserted that the interconnectedness and complexity of modern aircraft systems require a detailed understanding of the failure modes of each component/equipment within each system because a system's malfunction can be caused by several failures, but a single point of failure can also have several different effects. The authors above pointed out that the MMEL does not cover all conceivable scenarios during the operational phase of an aircraft, and they stressed the importance for engineers and crew members to maintain their 'mindfulness' since they comprise the last line of defence when dispatching an aircraft under the provisions of a MEL. Nonetheless, as Thomas Fakoussa, founder of Awareness Training Fakoussa (cited in Pierobon [27]) articulated, even with the MEL, pilots need advanced troubleshooting skills to tackle failures/defects under the provisions of the document. This is because flight crews are more aware of the type of operation and condition ahead of them (weather, terrain, region, etc.) as well as the required components/systems for aircraft dispatch.

Pierobon [27] analysed the Air Canada Boeing 767 event in 1983 where the aircraft was dispatched with inoperative fuel tank gauges and ran out of fuel while airborne [44]. Although the aircraft landed safely, the investigation report states that the captain had "consulted the MEL in a very cursory way" before the flight [44], suggesting that the MEL was not thoroughly reviewed by the pilot. Another concern raised by the investigators was the fact that the maintenance control centre on another occasion had granted an aircraft release against the restrictions of the approved MEL. However, the MMEL should not be consulted once a MEL has been customised for an aircraft; before this accident, an illegal relief had been granted based on the MMEL [44]. Pierobon [27] believes that the dispatch was based on (mis)perception rather than the use of the MEL, which must be the ultimate decision-making tool for aircraft dispatch for both pilots and engineers.

4. Review and Analysis of MEL-Related Events

Considering the standards, guidelines, reports and literature reviewed above, the authors aimed to complement the work of Munro and Kanki [26] and examine the types of MEL-related issues emerging from aviation safety events to detect necessary intervention areas. It is noted that the goal of the research was not to derive epidemiological data and rates across and within the regions of the investigation agencies. Instead, we aimed to map types of MEL issues that have contributed to incidents and accidents, generate an overview and compare the findings against relevant literature to detect possible development gaps in the particular area. The first step was to identify the sources of safety investigation reports and define proper keywords to conduct the search. The criteria for selecting sources were the availability of reports online and publicly in the English language from regional agencies that are responsible for a large volume of aviation operations. The databases identified were the ones maintained by National Transportation Safety Board (NTSB) of the United States, Transportation Safety Board of Canada (TSBC), Air Accidents Investigation Branch (AAIB) of the United Kingdom and Australian Transport Safety Bureau (ATSB). All of these repositories relate to the respective country's registered aircraft and include occurrences that have been investigated by the respective agencies.

Relevant incidents and accidents were identified by utilising the search option on the AAIB and NTSB websites and using 'Maintenance' and 'Minimum Equipment List' as keywords. This way, the research strings covered also MMEL-related records since the term MEL is a subset of MMEL. The keywords above were also used on the ATSB and TSBC websites, but the search did not result in a substantial number of reports to review; TSBC's website produced 0 and 4 reports, respectively, and ATSB's website produced 0 and 1 reports correspondingly. Consequently, considering research time limitations, we decided to review the 400 most recently published incident/accident investigation reports on both ATSB and TSBC websites.

The search described above resulted in 1323 investigation reports, the synopsis/summary and findings/conclusions of each were studied to identify and analyse MMEL/MEL related events. The analysis included reports where MEL-related issue was identified regardless of its attribution as a contributory factor or not to each event. In addition, cases where unreported or unrecorded defects contributed to the events were reviewed and analysed. The latter cases were consulted based on the view of Airbus [8] on the importance of recording defects to, amongst others, allow the retrofit of the MEL and assess their criticality in conjunction with other possible system and component malfunctionings or failures.

The filtering process described above resulted in the identification of 52 investigated events, 42 of which were directly related to MEL and ten regarded maintenance issues not covered by MEL but indirectly and likely affecting safety. In addition, although the search strategy followed is seen as comprehensive enough to generate a representative sample of relevant safety investigations, there might be reports that were unintentionally excluded from this study (e.g., input/typing errors of operators when populating the fields of the databases searched) as well as MEL-related events that have happened in other regions. Nonetheless, we believe that the final sample analysed represents the best-case picture adequately when considering the intensive safety initiatives of the specific agencies and the overall developments in the aviation industry of the respective regions.

Table 4 presents the results of each of the search steps, and Table 5 summarises the classification of the analysed events. The full list of the reports included in our study is available to the reader upon request to the corresponding author of the paper.

Regarding the accident/incident reports reviewed, AAIB and NTSB provided a detailed discussion of the MEL-related issues identified via a dedicated MEL section/paragraph in their reports, especially in the cases where MEL-related issues directly contributed to the particular events On the other hand, although ATSB and TSBC highlighted MEL-issues in different sections of their reports such as findings, analysis and conclusions, we did not find dedicated parts in the reports addressing MEL-related findings cumulatively. Nonetheless, all the reports reviewed provided enough information to enable the authors of this paper understand the situation and circumstances around the MEL-relevant events. On a side note, only 8 of the 29 reports including MEL-related contributory factors addressed the respective findings through specific recommendations. Although it was outside the scope of our research to evaluate the quality of the investigation reports and the degree of coverage of MEL-related issues through targeted measures, the picture above indicates that investigating agencies did not focus on the resolution of such problems even when they were detected in the course of investigations.

As Table 5 suggests, 50% of the MEL-related events regarded aircraft dispatch against MEL requirements, followed by cases where operations were conducted before MEL was approved (about 17%); this suggests non-compliance with MEL to the level of 67%. Cumulatively, the issues concerning MEL development (i.e., exhaustiveness, completeness and clarity) accounted for 21% of the cases, whereas cases relevant to human error/decision-making when applying MEL were jointly at the level of 12% of the events analysed (i.e., delayed rectifications or misused MEL item).

Table 4. Results of the search strategy and filtering of reports.

Investigation Agency	Search Strategy	Time Period Covered	Number of Reports Identified/Analysed	Number of Reports Relevant to This Research	MEL-Related Issues (Contributory)	MEL-Related Issues (Non-Contributory)	Indirect MEL-Related Issues
Accidents Investigation Branch, UK (AAIB)	Keyword: 'Minimum Equipment List'	2004–2018	106	5	5	0	0
	Keyword: 'Maintenance'	2002–2018	179	8	1	0	7
National Transportation Safety Board (NTSB)	Keyword: Minimum Equipment List	1982–2016	83	26	19	6	1
	Keyword: Maintenance Report status: Factual	1994–2019	155	0	0	0	0
Transportation Safety Board of Canada (TSBC)	400 most recently published incident/accident final reports	2005–2018	400	9	1	6	2
Australian Transport Safety Bureau (ATSB)	400 most recently published incident/accident final reports	2012–2018	400	4	3	1	0
		Total	**1323**	**52**	**29**	**13**	**10**

Table 5. MEL events categories.

Code	Category	Description	Result
UNJMEL	Unauthorised and unconfirmed MEL dispatch	Events where aircraft were dispatched against MEL specified requirements, e.g., flying below MEL specified altitude.	21
OPWMEL	Operating without a MEL	Events where operators dispatched aircraft without an approved MEL, e.g., operating an aircraft prior to developing a MEL.	7
UNAMEL	Unanticipated failures during MMEL development and justification	Events where failures were not foreseen during the MMEL justification phase, e.g., unanticipated Electronic Centralised Aircraft Monitor (ECAM) failure.	5
UNSCMEL	Ambiguous and incomplete MEL due to unanticipated scenarios during MEL development	Events where MEL scenarios were not foreseen during MEL development, e.g., insufficient MEL maintenance procedures.	4
LMEL	Late rectification of MEL items	Events where MEL related defects/deficiencies were not rectified on time in accordance with MEL specified intervals, e.g., Late replacement of inoperative thrust reversers.	4
INAPMEL	Inappropriate use of a MEL item	Events where MEL items were misused, e.g., Use of MEL lockout pins for the deactivation of the thrust reversers when it was not required.	1
DHIS	Defects recording/Handover issues	Events where defects were not reported or recorded in the aircraft's technical logbook as required.	10

5. Discussion

Overall, the literature reviewed suggests that aviation professionals are concerned about the current state and application of MELs. Notably, all positions highlighted the importance of MELs and point to the utility of the MEL as a balancing factor between safety and operations where aircraft can be dispatched with inoperative equipment as long as safety is not compromised. However, it has been postulated that a holistic approach is required to streamline the development of a framework/methodology to support the development, maintenance and monitoring of MELs [27,32]. Regulatory authorities and aircraft manufacturers are expected to offer to operators more detailed MEL-related guidance and specific tools along with requirements for respective training programmes.

Given the increasing complexity of aircraft systems, coupled with issues identified in the literature such as different MEL standards worldwide and cases of MEL mismanagement and misapplication [26,29,33], it is important to reinstate the need for standardisation and reinforcement. The issues identified in the literature and revealed through our review of the accidents and incidents above could pose a serious problem and are still prevalent despite MEL was introduced in the 1960s. The urge for MEL standardisation followed by targeted interventions to ensure its consistent and substantive application has become quite undeniable according to aviation researchers and professionals such as Pierobon [27], Hertzler [29] and Yodice [42]. Although literature and previous studies do not argue that the MEL framework should be entirely reformed, its standardisation is expected to ameliorate current issues and support proper and justified customisation of MEL to the operational profile and needs of airlines, minimisation of ambiguity in its implementation and enrichment of respective training. Furthermore, harmonisation of MEL development will allow valid comparisons of practices followed across regions and operators and offer to airlines and NAAs a common reference baseline for knowledge exchange as well as possibilities for continuous review, update and improvement of shared MEL-related processes.

Furthermore, it seems that there is an assumption that the factors/parameters considered during the MMEL development are directly applicable to the MEL, while this may be valid to some extent as MMELs are developed for operators to use as a guide for their MEL development, it is important to note that the MMEL alone might not be entirely suitable and adequate for every operator. MMEL professionals attempt to anticipate the worst possible effect of systems failure, but they may not anticipate all probable scenarios and failure modes that can emerge during operations and stem from the complexity of aircraft systems and their interactions with humans and the environment [28].

The above was also confirmed during our analysis of MMEL/MEL related events under the category UNAMEL where professionals sometimes did not consider the history of failure of an equipment/system during the justification phase. Although it can be argued that the events under this category were random, their occurrence highlights the need for operators to customise their MELs to their environment and type of operations that can affect system/component performance rather than just duplicating the master MMEL document which is based on different datasets of failures and performance. In addition, despite most of the events analysed in our study were not fatal or catastrophic, the outcome severity of any future event cannot be guaranteed, especially when flight crews are unaware or unfamiliar with the problem and cannot exert full control over the unfolding situation [45]. The Spanair's crash studied by Grüninger and Norgren [28] was linked to an unanticipated MEL failure.

Moreover, it is interesting that, even under the current regulatory mandates around MEL, the importance of the latter might not have been understood completely across the aviation industry as indicated by the high frequency of non-compliant cases. For operators with approved MELs, it was observed that, in several instances, aircraft were dispatched with known inoperative equipment or defects even though the operators had MELs in place (category UNJMEL). In addition, all the events where operators dispatched aircraft without an approved MEL (category OPWMEL) regarded the US region. Most of the operators falling under the latter category were FAR 91 operators or regarded operations conducted under FAR 91. This confirms Hertzler's [29,30] call to operators to apply for

MELs under FAR 135 because FAR 91 operators are the most neglected in terms of MEL oversight. The latter enjoy some leniency and do not utilise the MEL concept compared to FAR 135 operators where compliance with MEL and applicable MEL intervals are mandatory as mentioned in Section 3.2 above.

Another issue identified during the review of MEL events was the late rectification of MEL items (category LMEL). Indeed, EASA [4] and Airbus [3] stress the importance of repairing or replacing an inoperative item at the earliest opportunity and not at the most convenient time for an airline. However, although someone could argue a possible relationship of these cases with human error (e.g., lapses or slips) and non-compliance, these events can also be attributed to a lack of understanding of operators about the intended objectives and philosophy of MEL. The latter, instead of being approached as a constraint to operations, should be viewed as a risk management tool that can help in evaluating operational risks and specifying procedures in maintaining safety margins. Nonetheless, we did not identify literature suggesting any direct links between the MEL and the risk management framework of companies.

Furthermore, the cases associated with misinterpretations of MELs (category UNSCMEL) accord with the findings of FAA cited in Pope [39] and Munro and Kanki [26]. As stressed in the literature reviewed, the clarity of MELs and their related regulations along with MEL designated roles within airlines would facilitate the MEL review and development process and improve the reliability of MEL application [27,29,42]. Additionally, air operators need to train their pilots, engineers and aircraft dispatchers on MEL-related operational and maintenance requirements. Based on the nature of events under the specific category, it can be argued that adequate training could have led to anticipation of scenarios within the operator's operational environment and could have played a positive role. Furthermore, those currently involved in the MEL process might have little or no experience in airworthiness management or competencies and skills in MMEL/MEL. Being type-rated on an aircraft does not necessarily mean that an engineer or pilot is able to fully understand the parameters/factors surrounding the development and application of the MEL and interpret it correctly. Such a situation might lead to adverse events like the ones studied by Grüninger and Norgren [28] and Pierobon [27].

Regarding the ten cases indirectly related to MEL (category DHIS), Airbus [8] highlighted that a logbook entry is the starting point for assessing MEL-related defects/deficiencies. Perhaps, in conjunction with the remarks stated above about proper training, engineers and pilots might not have understood the criticality of registering technical works and problems in logbooks. Undocumented maintenance, unrecorded/unreported defects and improper handover will reduce the information available to pilots, maintenance staff and engineers in making informed decisions about the status and serviceability of aircraft.

Finally, the traditional approaches highlighted in Section 2.1 above have been criticised because they do not consider visibly and methodologically the human interactions with systems [6,46] which are inextricable parts of aircraft operations and are closely related to the development and application of MEL. Due to the interconnectedness of elements and processes that increase the complexity of modern systems, there is a need for more holistic and nonlinear frameworks to system safety analysis. Recent socio-technical systems engineering approaches, which are built upon systems theory, consider the interactions and interdependencies between human and technology [6,47] and have introduced tools and techniques to tackle the limitations of traditional approaches. For example, Leveson [6] has proposed the Systems Theoretic Process Analysis (STPA) technique, Hughes et. al. [46] recommends the Systems Scenarios Tool (SST), while Mumford [48] introduced the Effective Technical and Human Implementation of Computer based System (ETHICS) tool. Although each approach is accompanied by limitations in its endeavour to understand and deal with complexity, these tools suggest a more structured path to socio-technical systems modelling and offer a dynamic approach to systems safety engineering. While such techniques are relatively new compared to FMEA and FTA and perhaps more resource-demanding in their application, they are promising in overcoming the limitations highlighted in Tables 2 and 3 above and, apart from the proximal technical and human components of aircraft operation systems, could also account for various complex roles aviation stakeholders hold in the MEL development process and consider contextual parameters (e.g., specific NAAs policies and strategies, cultural and societal factors).

6. Conclusions

In this study, we reviewed the current situation around the development and application of MEL as well as literature referring to respective challenges, and we performed an analysis of MEL-related events. Overall, the results of our analysis as presented in Table 5 agree with previous work and suggest the need to revisit the processes related to MEL as demonstrated by the several issues mentioned in the safety investigation reports reviewed and related to unauthorised MEL dispatch, ambiguous MELs and airlines operating without a MEL. The lack of a systematic and uniform approach to MEL, apart from depriving the aviation industry of a standardised and reliable application of MEL, might have led to an underestimation of its importance and criticality for safe operations. The ambiguity detected in MEL-related standards along with the diversity of approaches to MEL in various regions might have contributed to building perceptions which suggest, on the hand, that the MEL is a quick-to-achieve compliance requirement by solely replicating or slightly changing MMELs, and, on the other hand, that individual defects emerging in day-to-day operations can be dealt only subjectively.

However, MEL is not just about the aircraft; it is about the aircraft in service operated by humans within a specific environment. While manufacturers try to anticipate varying environment conditions when compiling MMELs, the latter alone are usually inadequate for the development of MELs. As stressed by Leveson [6], system failures do not occur only as a result of random and individual component malfunctions; instead, we must consider interactions between socio-technical system agents (i.e., technology, end-users, organisations, regulators, society and environment) within and across system levels, which are often neglected during current MEL practice. Therefore, in line with the views of Leveson [6], the MEL development process should be viewed as a dynamic process involving the NAAs, air operators, pilots, engineers and flight dispatchers. The work of Karanikas and Chatzimichailidou [49] who suggested a combined qualitative and quantitative approach to compare system configurations encapsulates the newly introduced Systems-Theoretic Process Analysis (STPA) technique [6] and is an example of how system analysis could consider (1) non-binary behaviours of system agents, (2) the degree of influence of each agent on system performance, and (3) the criticality of each agent. The particular approach, as well as any other approach that encompasses systems engineering and socio-technical principles, could comprise the basis for a holistic and systematic methodology for MEL development and application and address the weaknesses of currently used techniques as presented in Section 2.1 of this paper.

Funding: This research received no external funding.

Conflicts of Interest: The authors declare no conflict of interest.

References

1. King, F.H. *Aviation Maintenance Management*; South Illinois University Press: Carbondale, IL, USA, 1986.
2. Hessburg, J. *Air Carrier MRO Handbook*; McGraw-Hill Professional: New York, NY, USA, 2001.
3. Airbus. Getting to Grips with MMEL and MEL. Airbus. 2005. Available online: http://www.smartcockpit.com/docs/getting-to-grips-with-mmel-and-mel.pdf (accessed on 13 December 2018).
4. EASA. *Easy Access Rules for Master Minimum Equipment List (CS-MMEL)*; European Aviation Safety Agency (EASA): Cologne, Germany, 2018.
5. UK CAA. *CAP 549 Master Minimum Equipment Lists (MMEL) and Minimum Equipment Lists (MEL)*; UK Civil Aviation Authority: London, UK, 2010.
6. Leveson, N.G. *Engineering a Safer World: Systems Thinking Applied to Safety*; The MIT Press: Cambridge, MA, USA, 2011.
7. ICAO. *International Standards and Recommended Practices: Operation of Aircraft Annex 6 to the Convention on International Civil Aviation*; International Civil Aviaton Organisation: Montreal, QC, Canada, 2010.

8. Airbus. A Recall on the Correct Use the MEL. Airbus. 2018 Airbus. A Recall on the Correct Use the MEL. Available online: https://safetyfirst.airbus.com/a-recall-on-the-correct-use-of-the-mel/ (accessed on 13 December 2018).
9. FAA. Chapter 4 Configuration Deviation Lists (CDL) and Minimum Equipment Lists (MEL). In *Volume 4 Aircraft Equipment and Operational Authorizations*; Federal Aviation Administration: Washington, DC, USA, 2009.
10. EASA. *Oeb Administrative and Guidance Procedures*; European Aviation Safety Agency: Cologne, Germany, 2010.
11. FAA. Chapter 2 Technical Groups, Boards, and National Resources. In *Volume 8 General Technical Functions*; Federal Aviation Administration: Washington, DC, USA, 2017.
12. Zhenyu, W. Dispatch reliability oriented MMEL formulation technology. In Proceedings of the 30th ICAS Congress, Daejeon, Korea, 25–30 September 2016.
13. Wang, P. *Civil Aircraft Electrical Power System Safety Assessment: Issues and Practices*; Elsevier: London, UK, 2017.
14. Vincoli, J. *Basic Guide to System Safety*; Van Nostrand Reinhold: New York, NY, USA, 1993.
15. Kritzinger, D. *Aircraft System Safety Assessments for Initial Airworthiness Certification*; Elsevier: London, UK, 2017.
16. Ren, H.; Chen, X.; Chen, Y. *Reliability Based Aircraft Maintenance Optimization and Applications*; Elsevier: London, UK, 2017.
17. Bell 412 MMEL. U.S. Department of Transportation. Available online: http://fsims.faa.gov/wdocs/mmel/bh-212-412_rev%209.pdf (accessed on 3 January 2019).
18. Coit, D.W.; Zio, E. The evolution of system reliability optimization. *Reliab. Eng. Syst. Saf.* **2018**, *192*, 106259. [CrossRef]
19. Misra, K.B. Reliability Engineering: A Perspective. In *Handbook of Performability Engineering*; Springer: London, UK, 2008; Volume 36. [CrossRef]
20. Kinnison, H.A.; Siddiqui, T. *Aviation Maintenance Management*, 2nd ed.; The McGraw-Hill Companies, Inc.: New York, NY, USA, 2013.
21. Bombardier CL600 MMEL. Master Minimum Equipment List (MMEL). Available online: http://fsims.faa.gov/wdocs/mmel/cl-600-2b19_rev%2019.pdf (accessed on 5 January 2019).
22. Airbus A320 MMEL. Master Minimum Equipment List. Available online: http://fsims.faa.gov/wdocs/mmel/a-320%20r21.pdf (accessed on 7 January 2019).
23. Moir, I.; Seabridge, A. *Aircraft Systems: Mechanical, Electrical and Avionics Subsystems Integration*; John Wiley and Sons: Chichester, UK, 2008.
24. Kundu, A.K.; Price, M.A.; Riordan, D. *Theory and Practice of Aircraft Performance*; Wiley: Hoboken, NJ, USA, 2016.
25. Johnson, W.B.; Maddox, M.E. A model to explain human factors in aviation maintenance. *Avionics News*, March 2007; pp. 38–41.
26. Munro, P.; Kanki, B. An analysis of ASRS maintenance reports on the use of minimum equipment lists. In Proceedings of the 12th International Symposium on Aviation Psychology, Dayton, OH, USA, 14–17 April 2003.
27. Pierobon, M. Situational Awareness for MELs—Minimum equipment lists need maximum vigilance. *Aerosaf. World* **2013**, *8*, 32–37.
28. Grüninger, M.R.; Norgren, C. MEL—A risk management tool. *Bart Int.* **2013**, *146*, 68–69.
29. Hertzler, J. Operating an AIRCRAFT: With Inoperative Instruments or Equipment. Available online: https://www.aviationpros.com/home/article/10387201/operating-an-aircraft-with-inoperative-instruments-or-equipment (accessed on 15 July 2019).
30. Hertzler, J. Do I Need to Have an MEL? Available online: https://www.aviationpros.com/home/article/10383380/do-i-need-to-have-an-mel (accessed on 15 July 2019).
31. AIBN. *Safety in Norwegian Aviation during the Process of Change*; Accident Investigation Board Norway: Lillestrøm, Norway, 2005.
32. Herreraa, I.A.; Nordskaga, A.; Myhreb, G.; Halvorsenb, K. Aviation safety and maintenance under major organizational changes, investigating non existing accidents. *Accid. Anal. Prev.* **2009**, *41*, 1155–1163. [CrossRef] [PubMed]
33. Feeler, R. MELs for corporate and business aircraft guide deferred-mainteance decisions. *Flight Saf. Found. Aviat. Mech. Bull.* **2002**, *50*, 1–9.

34. TC. Commercial and Business Aviation Advisory Circular (CBAAC) No. 0107R. Available online: https://www.tc.gc.ca/en/services/aviation/reference-centre/pre-2007/commercial-business-aviation-advisory-circulars/cbaac-0107R.html (accessed on 15 July 2019).
35. TC. Canadian Aviation Regulations: SOR/96-433. Available online: https://lois-laws.justice.gc.ca/PDF/SOR-96-433.pdf (accessed on 15 July 2019).
36. FAA. *Advisory Circular: Minimum Equipment Requirements for General Aviation Operations under FAR PART 91. U.S.*; Department of Transportation, Federal Aviation Administration: Washington, DC, USA, 1991.
37. CASA. *Minimum Equipment List: CAAP 37-1(5)*; Australian Government Civil Aviation Safety Authority: Canberra, Australia, 2016.
38. EASA. Commission Regulation (EU) No 965/2012 of 5 October. 2012. Available online: https://eur-lex.europa.eu/LexUriServ/LexUriServ.do?uri=OJ:L:2012:296:0001:0148:En:PDF (accessed on 15 July 2019).
39. Pope, J.A. Special inspection of commuters: Evaluations are only as constructive as the appropriateness of the industry's response to them. *Flight Saf. Found. Accid. Prev.* **1989**, *46*, 1–6.
40. TC. *TP 9155E—Master Minimum Equipment List/Minimum Equipment List Policy and Procedures Manual*; Transport Canada Civil Aviation Communications Centre (AARC): Ottawa, ON, Canada, 2006.
41. FAA. Code of Federal Regulations Sec. 91.213: Part 91 General Operating and Flight Rules. Available online: http://rgl.faa.gov/Regulatory_and_Guidance_Library/rgFar.nsf/FARSBySectLookup/91.213 (accessed on 17 July 2019).
42. Yodice, J. A Pilot's Dilemma: Inoperative Instruments or Equipment. Available online: https://airfactsjournal.com/2018/09/a-pilots-dilemma-inoperative-instruments-or-equipment/ (accessed on 15 July 2019).
43. CIAIAC. *Report A—032/2008 Accident Involving a McDonnell Douglas DC-9-82 (MD-82) Aircraft, Registration EC-HFP, Operated by Spanair, at Madrid-Barajas Airport, on 20 August 2008*; Civil Aviation Accident and Incident Investigation Commission: Madrid, Spain, 2011.
44. TC. *Investigating the Circumstances of an Accident Involving the Air Canada Boeing 767 Aircraft C-GAUN That Effected an Emergency Landing at Gimli, Manitoba on the 23rd Day of July 1983*; Minister of Supply and Services: Ottawa, ON, Canada, 1985.
45. Karanikas, N.; Nederend, J. The controllability classification of events and its application to aviation investigation reports. *J. Saf. Sci.* **2018**, *108*, 89–103. [CrossRef]
46. Hughes, H.P.N.; Clegg, C.W.; Bolt, L.E.; Machon, L.C. Systems scenarios: A tool for facilitating the socio-technical design of work systems. *Ergonomics* **2017**, *60*, 1319–1335. [CrossRef] [PubMed]
47. Baxter, G.; Sommerville, I. Socio-technical systems: From design methods to systems engineering. *Interact. Comput.* **2010**, *23*, 4–17. [CrossRef]
48. Mumford, E. *Designing Human Systems for New Technology—The ETHICS Method*; Manchester Business School: Manchester, UK, 1983.
49. Karanikas, N.; Chatzimichailidou, M.M. *The COSYCO Concept: An Indicator for Comparing System Configurations*; Amsterdam University Press: Amsterdam, The Netherlands, 2018; Volume 1, pp. 154–174. [CrossRef]

Editorial

Airlift Maintenance and Sustainment: The Indirect Costs

Kyriakos I. Kourousis

School of Engineering, University of Limerick, V94 T9PX Limerick, Ireland; kyriakos.kourousis@ul.ie

Received: 30 August 2020; Accepted: 2 September 2020; Published: 2 September 2020

Abstract: This article aims to present and discuss a set of technical matters affecting the maintenance and sustainment cost of military transport aircraft (airlifters). An overview of the military aviation technical support system is provided, in conjunction with a high level discussion on the life cycle cost. Four technical support pillars are defined as part of this analysis: supply, restoration and upgrade, engineering and regulatory compliance. A focused discussion on airlift sustainment factors, based on past experience, is used to identify technical considerations that can be used for the evaluation of new aircraft. A number of technical considerations which are key for cost purposes are identified and mapped against the defined technical support pillars, related to engineering and technical support and airworthiness management aspects. Important practical technical considerations are identified, discussed and critiqued under an independent lens. This article can stimulate discussion of the maintenance and sustainment costs of airlifters, both within military aviation operators and the defence industry community but also within the civil aircraft maintenance industry.

Keywords: aircraft maintenance; airworthiness; military aviation; airlift; cost

---

1. Technical Support in Military Aviation

The technical support system in the United States and European military aviation is typically structured as organisational (squadron level), intermediate (base level) and depot level maintenance. Looking at the different functions within these segments, one can identify technical activities broadly covering:

- Supply of material (aircraft components, parts, consumables, etc.) in support of maintenance tasks, at organisational (O) and intermediate (I) level, where the focus is on the military operational unit.
- Restoration and upgrade at depot (D) level, where overhaul takes place and, in general, any technical (maintenance) activity beyond the capability of the base and the squadron. This includes shop-level maintenance for components and engine overhaul.
- Engineering, in terms of services provided for resolving problems, designing, approving, assessing and implementing engineering changes.
- Regulatory compliance spans across all activities and accompanies oversight at the various levels of command. However, often, in military aviation, the role of the regulator is not completely independent of the chain of operational command [1], unlike civil aviation.

These technical support pillars (supply, restoration and upgrade, engineering and regulatory compliance) are intended to sustain the airworthiness and the operational capability of a military aircraft fleet. Operational readiness and utilisation (including training) interact with all facets of maintenance at all levels (I, O and D), since the objective of technical support is to supply mission ready aircraft. In turn, operational readiness has an impact and it is affected by grounding time. This simplified view of the military aircraft maintenance and sustainment system is illustrated in Figure 1.

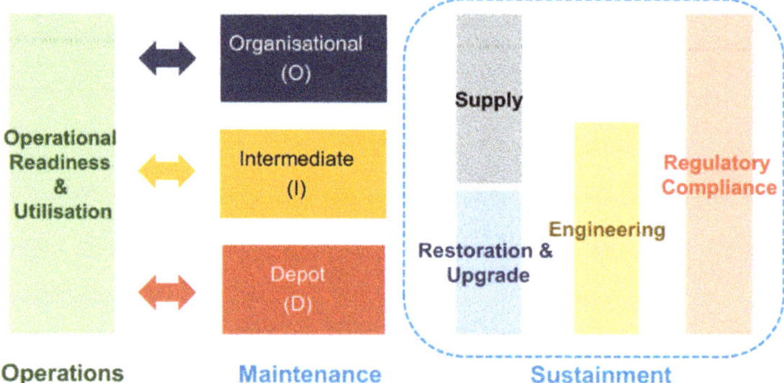

Figure 1. The different elements of a maintenance and sustainment (technical support) system for military aircraft and their interactions with operational readiness and utilisation.

2. Life Cycle Cost

As with every technical system, a military aircraft is designed, and expected to be operated, over a specified life cycle. An operator needs to ensure that the aircraft fleet remains healthy (airworthy) and operationally capable throughout its life. Complex maintenance and sustainment activities are performed to ensure that the aircraft systems operate reliably, structural integrity is assured and upgrades allow it to offer value to the operational capabilities of the defence force. The cost associated with maintenance and sustainment is continuously evolving, and it is generally expected to increase over time. The focus of the discussion is not how this cost evolves but the indirect costs associated with the ownership of military aircraft and particularly military transport aircraft (airlifters).

Military operators may set a lower (start) point and an upper (cut-off) point for the operating cost, both for monitoring/management and decision-making purposes. These two points define the acceptable operating margins for an aircraft fleet. The life cycle can be segmented, in broad terms, into four phases: the initial investment, the learning and maturing phase, the maturity phase and the end of life (disposal) of the aircraft (Figure 2).

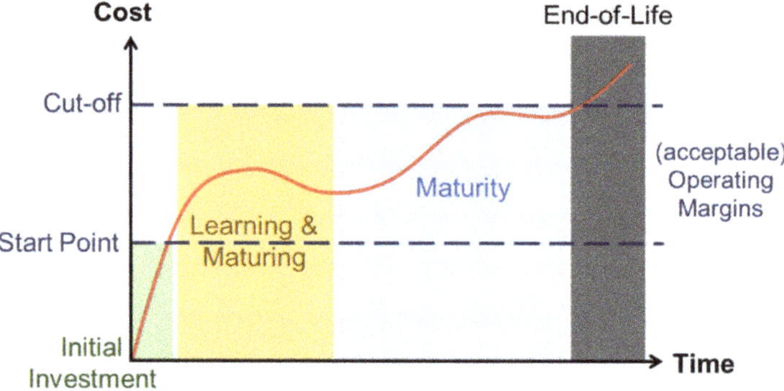

Figure 2. Life cycle, in phases, of military aircraft.

Entry to service requires an initial investment, which accompanies/stems from the procurement contract for new systems. The level of initial investment defines the start point in the cost curve, with costs in this phase associated with:

- Technical training for the engineering and technical staff;

- Development of maintenance capability, at different levels and depending on what or how an operator wishes to utilise the aircraft (i.e., configuration, mission profiles, etc.);
- Acquisition or preparation of the necessary infrastructure for maintenance, logistics support and operation of the fleet.

The next phase in the life of aircraft, which may be the most challenging in terms of managing some uncertainty around the performance of a new system, is that of learning and maturing. Inexperienced/smaller operators with fewer resources than larger defence forces may face additional challenges. For example, small operators typically do not have a sufficient level of depth in engineering expertise or in-house technical resources which would help them in resolving early-life challenges. In this phase, the operator needs to:

- Build up its engineering capability and expertise at various levels. This involves a direct or indirect investment in technical knowledge, acquired through training/retraining or obtaining experience from the manufacturer via technical support services. Strategies to retain this expertise in the long term are important, as engineering and technical staff engaged in this phase obtain valuable first-hand experience from their interaction with the manufacturer.
- Develop or adjust the logistics support. A technical supply system needs to be established for the new aircraft, in most cases, working in sync with follow on support (FOS) contracts and existing structures for other aircraft types of the defence force.
- Establish contract management structures and train staff to perform combined engineering and supply functions, in an integrated logistics support (ILS) or programme office environment. This set of activities is not only time consuming but also creates value and/or waste within the overall technical system.

The maturity phase may need more attention than the learning and maturing phase. As the technical system (aircraft) ages, it can become more resilient or predictable, but the ageing effects will start to have an impact on integrity or reliability at the same time. This can be a very interesting period for engineers, since older aircraft offer new challenges. Typically, this phase includes:

- Major maintenance tasks arising from the evolving maintenance programmes (often dictated by the manufacturers and/or operators).
- Major structural repairs mandated to rectify findings from inspections performed to monitor the effects of fatigue and corrosion.
- Enhanced focus on risk management since there is always a need for unobstructed operations. The aircraft fleet is required to operate under the same profiles it was procured for, rather than imposing operations restrictions or limiting its specifications to meet engineering/technical issues.
- Management of obsolescence, which presents additional challenges for the supply and operational capability of some of the systems of the aircraft.
- Aircraft upgrades, by default, large-scale engineering change activities, involving and complemented by heavy maintenance tasks. Upgrades introduce new features to the aircraft or improve the existing systems' functionality, integrity and reliability, i.e., structural fatigue mitigation, replacement of electrical wiring and other degraded components, etc.

When the upper (cut-off) point is reached (or exceeded in some cases), the decision-makers may place the aircraft at the of end-of-life phase, leading to disposal, storage or decommissioning. Often, these decisions are made on the basis of operational and/or political reasons.

3. Airlift Sustainment

A number of factors have a positive (increasing) contribution to the sustainment cost for military transport aircraft. Some of these factors are also applicable to other military aircraft types. These are discussed here.

3.1. Diverse and Non-Typical Operations

The nature of the military aircraft operations is dictating or imposing diverse and often non-typical (when comparing to civil transport aircraft) mission profiles and operating environment. This, in turn, has an effect on the cost of sustainment, since the maintenance programme needs to capture proactively any problems that may reduce the reliability of the aircraft systems.

3.2. Reliance on Offshore Maintenance Services

Heavy (depot-level) maintenance and, in some cases, intermediate-level maintenance, may have to be performed at overseas maintenance centres. This is the case when such capability has not been developed in-house (within the defence force organisation or at in-country aircraft maintenance organisations). Military operators have to rely on this set of services, which, in the long term, can have a substantial impact on sustainment cost. This includes out of country capabilities, offered by commercial entities. An example of specialised maintenance providers for airlifters is the Marshall Aerospace and Defence Group in the United Kingdom, which has extensive maintenance capabilities on the Lockheed Martin C-130 aircraft platform [2].

3.3. Disconnect between Military and Civil Airworthiness Requirements

Civil and military aviation are, by default, not aligned or consistent with each other, in terms of both regulations and practice. There are good reasons for this different treatment, primarily attributed to the nature of the operation of the military transport aircraft. This disconnect creates complexities and imposes defence-specific requirements for the preservation of the continuing airworthiness of the aircraft, even when the same aircraft type has dual (civil–military) certification. For example, military transport aircraft may have to be retrofitted with defence-specific equipment, such as chaff and flare dispensers (for self-protection purposes). This equipment cannot be certified under a civil regulatory framework and this activity adds cost for the military aircraft owner/operator.

3.4. Lack of Civil Type Certificate

Type certification of some airlifters does adhere to civil (i.e., the European Aviation Safety Agency, EASA, or the Federal Aviation Authority, FAA) regulations. The Airbus A400 M is a relatively recent example of a civil (EASA) certified airlifter [3]. However, for most types, this is not the case. This may have a smaller impact on cost, though modern safety regulations (and certification standards) for civil aircraft have progressively evolved to become more user-friendly, meaning higher efficiencies and improved reliability, resulting in lower long-term sustainment costs.

3.5. Legacy Structural Design

Design, especially structural, for many military transport aircraft have evolved (to a small or larger extent) from previous legacy versions. There are many examples, such as the Leonardo C-27J, which is an evolution of the FIAT G.222. Legacy design standards are, in general, less cost-effective than modern standards, as less efficient maintainability and reliability standards and practices are employed. This, in turn, has a toll on the structural maintenance and sustainment cost, since a higher volume and/or more frequent inspections and repairs may be required.

3.6. Ageing and Obsolescence

Due to the tendency to utilise military transport aircraft for longer than originally projected, ageing and obsolescence are becoming more profound challenges. Both have to be taken into consideration when evaluating for the acquisition of older (pre-owned), yet more affordable, airlifters. This, however, has to be examined in conjunction with the sought operational requirements, the operational tempo and the anticipated life cycle. An operator may utilise efficiently, within their budget, older aircraft with a good service history and a good structural condition, for five or even ten years. The recent example of the United States (US) Navy procurement of an ex-Royal Air Force (RAF) C-130J for $30 million (in comparison to the cost of $80 million for a new aircraft) is characteristic [4].

3.7. Ongoing Major Maintenance

Airlifters are large, complex platforms, with many components and parts highly susceptible to ageing effects: fatigue, corrosion, degradation and wear. Operators tend to use these aircraft under harsh conditions and environments. The manufacturers are aware of these issues, developing solutions which are reflected in the maintenance programmes. Maintenance programmes progressively become more thorough and demanding, dictating (higher cost) major inspection and maintenance tasks to sustain the airworthiness of aircraft being operated from their primary base and of those deployed elsewhere.

3.8. Upgrade

Upgrades are common for military aircraft, including airlifts. Avionics' and cockpit modernisation programmes, both for operational and technical reasons (some of them related to tackling obsolescence and reliability issues), are popular among operators. These kinds of programmes add value to older fleets by enhancing operational capabilities and/or extending service life. The Lockheed Martin C-130 serves as a good example, with the Hellenic Air Force C-130H and -B fleet avionics upgrade programme (AUP) in 2002 offering an indication in the costs involved—a 15 aircraft fleet upgrade at a cost of $6 million per aircraft [5].

These cost-raising factors are commonly observed in combination, especially in older fleets. The case of the Lockheed Martin C-130's structural integrity offers a good example. This aircraft type has served many defence forces around the world over the past fifty years, as well as civil operators offering contracted services to state organisations (a popular choice when cost-saving or exposure to risk is sought by the states). The entering of the Lockheed Martin C-130 into civil registers has placed the type under the scrutiny of civil aviation regulators, such as the Federal Aviation Authority (FAA) in the United States (US) and the European Aviation Safety Agency (EASA) in Europe. For example, a search in the EASA safety publications tool [6] reveals a number of primary structure-related Airworthiness Directives (ADs) for the Lockheed Martin C-130 (Model 382). These EASA ADs are linked with previously issued FAA ADs, which offer further details on the type and the estimated cost for major structural repairs and inspections required for the aircraft wings (i.e., center wing box, CWB, and outer wing). One can observe the significant labour and part replacement costs involved in complying with these ADs (i.e., the CWB replacement cost is estimated at $5 million and that of the outer wing at $8 million, both per aircraft) [7]. It is interesting to note that the newer C-130J model is required to undertake similar maintenance tasks to ensure the integrity of the wing structure. For example, in 2017, the Royal Air Force (RAF) decided to retain as operational 14 of their C-130Js until 2035. For this, it was required to replace the aircraft CWBs, at a total cost of $143 million [8]. This offers a flavour of the costs associated with legacy structural designs, heavy operational utilisation and the effect of heavy maintenance on the sustainment and affordability of older airlift types and fleets.

4. New Airlifter Technical Considerations

Assessing the long-term maintenance and sustainment costs of airlifters, i.e., 20, 25 or even 30 years from today, is important. There are various sources that can be used to inform the technical decision-makers, including the manufacturers, which can offer insight on the utilisation costs and publicly available research/industry reports, in conjunction with cost analysis and prediction models, i.e., [9]. For example, a report published by the RAND Corporation in 2013 provides, among others, a comprehensive analysis of the sustainment costs of US fleets of Lockheed Martin C-130 [10].

A "mix and match" strategy is especially important for small and diverse fleets, since these are more challenging and less cost-inefficient to run in comparison to more substantial size fleets. Any new aircraft type, including airlifters (given the much higher investment involved), should be evaluated against the existing technical support infrastructure, technical capability and fleet mix of the operator. An exercise evaluating different candidate airlift types can be very useful in this regard.

Developing and maintaining in-house technical support (and engineering) capability is essential not only for self-reliance purposes but also can be beneficial for sustainment cost-saving purposes. Efficient technical solutions, contributing to lower sustainment cost, can be sourced from defence engineers and technicians, and experience has shown that this can yield positive results for the operators. For example, aircraft fatigue monitoring programmes and inspection repair solutions are typically high-cost and value engineering capabilities. Moreover, the operator can interact in a more productive way with the aircraft manufacturer and external technical services' providers.

Learning from other users can be useful when evaluating new aircraft types. Experience of airlift users, especially for widely used types and models, offers valuable information on technical support matters. The sharing and exchange of technical information, data and findings, as well as practices that can have an effect on maintenance and sustainment, constitutes good practice for technical and engineering support purposes. This collective approach can also work for the benefit of operators when negotiating technical solutions with the aircraft manufacturer.

Dual certification and the airworthiness management framework can contribute (positively or negatively) indirectly to the maintenance and sustainment cost. This is related to the compliance requirements, since tailoring the certification standards (i.e., when an operator wishes to comply with non-typical requirements contained in widely used airworthiness codes) can increase the end-product cost. The European Military Airworthiness Requirements (EMARs) [11] were developed to bridge the gap between military-customised and civil certification requirements by adopting a common regulatory framework across different defence forces. Combining civil aviation regulatory structures and practice (where cost is an important element) with defence specific airworthiness requirements may offer to military operators the best of both worlds. In the case of airlifters, defence forces selecting dual-certified aircraft types have the opportunity to utilise these efficiently should such a hybrid airworthiness system be implemented. However, an operator can still operate dual-certified aircraft types, but it may be generally more onerous to manage technical support and ensure regulatory compliance when having to rely on bespoke contracts.

One option which military operators can also consider is civil-derivative or civil-certified airlifters. These aircraft types can be maintained under a civil or civil-based airworthiness framework. These aircraft can undergo maintenance and can be certified in EASA (or FAA) Part 145 aircraft maintenance organisations. This would offer cost benefits for the military operator and the defence and the civil aviation industry. In-country civil aircraft maintenance organisations would be able to offer their services to the defence force, expanding their business in the military aircraft maintenance sector. The same approach would also apply for the supply sector (spare parts and consumables), which can be sourced from a wider (non-defence specific) network/range of sources. Overall, this can have a positive effect on the sustainment cost of such (civil-derivative/civil-certified) airlifters.

The four pillars of technical support described in Figure 1 (supply, restoration and upgrade, engineering and regulatory compliance) can serve as a guide for identifying the technical considerations applicable to airlifters when evaluating the acquisition of new or used aircraft. These technical considerations have been mapped against the four pillars of technical support, presented in Figure 3 in a summarised way.

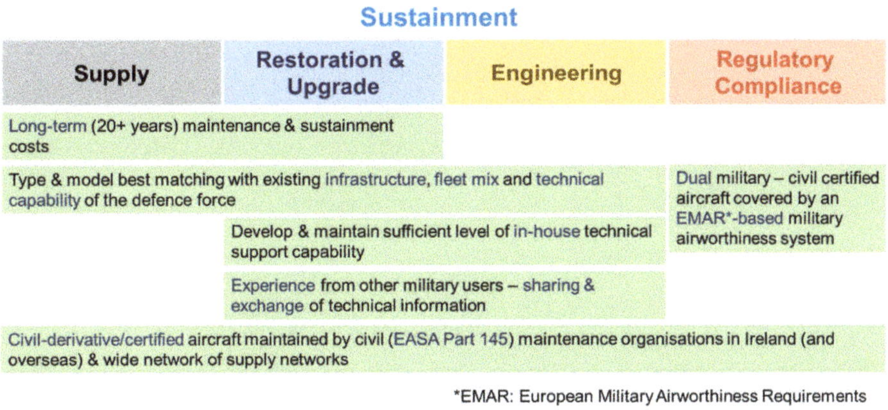

Sustainment

Supply	Restoration & Upgrade	Engineering	Regulatory Compliance
Long-term (20+ years) maintenance & sustainment costs			
Type & model best matching with existing infrastructure, fleet mix and technical capability of the defence force			Dual military – civil certified aircraft covered by an EMAR*-based military airworthiness system
	Develop & maintain sufficient level of in-house technical support capability		
	Experience from other military users – sharing & exchange of technical information		
Civil-derivative/certified aircraft maintained by civil (EASA Part 145) maintenance organisations in Ireland (and overseas) & wide network of supply networks			

*EMAR: European Military Airworthiness Requirements

Figure 3. Technical considerations for the evaluation of airlifters and mapping against the four pillars of technical support described in Figure 1.

5. Conclusions

This article has provided a general/broad view and discussion of technical considerations for the maintenance and sustainment of military transport aircraft (airlifters). In summary:

- Technical and engineering support can influence and influenced by the aircraft type and model choice;
- In-house capability and technical knowhow can be force-multipliers when it comes to achieving cost efficiency;
- Gaining experience from other aircraft users and creating synergies in exchanging technical information can be beneficial for maintenance and sustainment purposes;
- Airworthiness certification requirements for military aircraft are important considerations for older and newer aircraft types.

Funding: I have not received external funding.

Conflicts of Interest: I declare no conflict of interest.

References

1. Purton, L.; I Kourousis, K. Military Airworthiness Management Frameworks: A Critical Review. *Procedia Eng.* **2014**, *80*, 545–564. [CrossRef]
2. Marshall, Support Reimagined. Available online: https://marshalladg.com/capabilities/managed-services (accessed on 20 July 2020).
3. EASA Certifies the Airbus A400M. Available online: https://www.easa.europa.eu/newsroom-and-events/press-releases/easa-certifies-airbus-a400m (accessed on 12 June 2020).
4. Navy Flight Demostration Squadron to Receive "Fat Albert" Replacement. Available online: https://www.navair.navy.mil/news/Navy-Flight-Demonstration-Squadron-receive-Fat-Albert-replacement/Mon-06242019-1514 (accessed on 25 June 2020).
5. L-3 Communications' Spar Aerospace Awarded Contract by Hellenic Air Force to Upgrade C-130 Aircraft. Available online: http://www.defense-aerospace.com/article-view/release/9336/spar-wins-greek-c_130-upgrade-order-(mar.-26).html (accessed on 30 August 2020).
6. EASA Safety Publications Tool. List of Mandatory Continuing Airworthiness Information. Available online: https://ad.easa.europa.eu/ (accessed on 12 June 2020).
7. EASA Safety Publications Tool. US-2015-18-02: Wings—Center Wing Box (CWB) and Outer Wings–Replacement. Available online: https://ad.easa.europa.eu/ad/US-2015-18-02 (accessed on 30 August 2020).

8. Centre Wing Box Replacement Deal Supports RAF's Hercules. Available online: https://www.flightglobal.com/centre-wing-box-replacement-deal-supports-rafs-hercules/124819.article (accessed on 25 August 2020).
9. Lappas, I.; Bozoudis, M. The Development of an Ordinary Least Squares Parametric Model to Estimate the Cost Per Flying Hour of 'Unknown' Aircraft Types and a Comparative Application. *Aerospace* **2018**, *5*, 104. [CrossRef]
10. McGarvey, R.; Light, T.; Thomas, B.; Sanchez, R. *Commercial Intratheater Airlift, Cost-Effectiveness Analysis of Use in U.S. Central Command*; RAND Corporation: Santa Monica, CA, USA, 2013.
11. European Defence Agency. Approved MAWA Documents. Available online: https://www.eda.europa.eu/experts/airworthiness/mawa-documents (accessed on 30 August 2020).

Article

A Systematic Methodology for Developing Bowtie in Risk Assessment: Application to Borescope Inspection

Jonas Aust * and Dirk Pons *

Department of Mechanical Engineering, University of Canterbury, Christchurch 8041, New Zealand
* Correspondence: jonas.aust@pg.canterbury.ac.nz (J.A.); dirk.pons@canterbury.ac.nz (D.P.); Tel.: +64-33-695-826 (D.P.)

Received: 5 June 2020; Accepted: 25 June 2020; Published: 29 June 2020

Abstract: Background—Bowtie analysis is a broadly used tool in risk management to identify root causes and consequences of hazards and show barriers that can prevent or mitigate the events to happen. Limitations of the method are reliance on judgement and an ad hoc development process. Purpose—Systematic approaches are needed to identify threats and consequences, and to ascertain mitigation and prevention barriers. Results—A new conceptual framework is introduced by combining the Bowtie method with the 6M structure of Ishikawa to categorise the threats, consequences and barriers. The method is developed for visual inspection of gas turbine components, for which an example is provided. Originality—Provision of a more systematic methodology has the potential to result in more comprehensive Bowtie risk assessments, with less chance of serious omissions. The method is expected to find application in the broader industry, and to support operators who are non-risk experts but have application-specific knowledge, when performing Bowtie risk assessment.

Keywords: Bowtie analysis; risk assessment; safety; MRO; visual inspection; cause–consequence analysis; barrier model; 6M; Ishikawa; aircraft maintenance

---

1. Introduction

The barrier method of risk assessment, more commonly called Bowtie analysis, has been widely adopted in multiple industries. The key concept encapsulated in the method is that of preventative barriers that prevent a hazardous outcome (the 'top event') from occurring, and recovery processes that limit the escalation of that event into a larger catastrophe. It is a composition of a fault tree, event tree and barrier concept. The method is especially good at visually representing the event chains from the root cause to the consequence and identifying barriers that are in place, missing or ineffective. Industries in which the Bowtie method is particularly popular include oil and gas [1,2], aviation [3–7], transportation [8–10], chemical and process [11,12], mining [2,13], IT [14–16], and medical [17–19].

In the aviation industry, the safety of the passengers and crew is of utmost priority. To ensure this, maintenance, repair and overhaul (MRO) plays a crucial part. It includes the frequent inspection of the aircraft and its engines after a certain amount of flight hours or cycles, or after an unexpected event occurred, such as a bird strike. In both cases, different means of visual inspection are applied. The aircraft engine is mainly inspected via borescope inspection (BI) and if required via subsequent piece part inspection (PPI). Since the results of such inspections are crucial for the aircraft airworthiness and passenger safety, it is important to understand the inherent risks of the process. A high-level MRO process with the different borescope inspection procedures and related risks is presented in Figure 1.

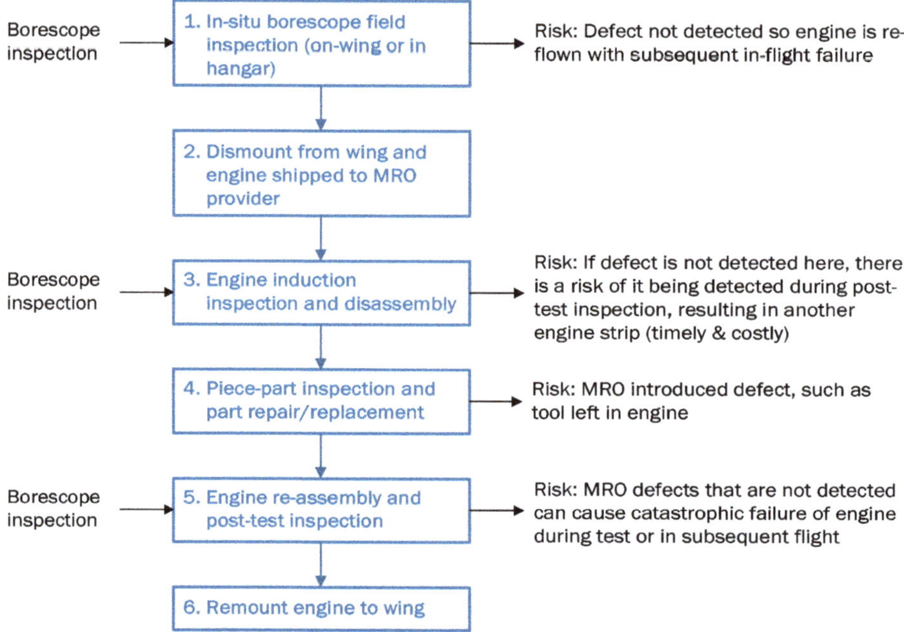

Figure 1. Maintenance, repair and overhaul (MRO) process with borescope inspection procedures and risks.

Previous work showed that Bowtie is a useful tool for such risk assessments, but has some limitations, such as the process of constructing a Bowtie being ad hoc and arbitrary [7]. It highly depends on expert knowledge and the personal preferences and outlook of the analyst [4,20–22]. This is problematic because a risk analyst will have different technical and operational insights to a technician. There is a risk of missing important threats, consequences, and barriers. This paper offers a solution by introducing a conceptual framework for a more systematic Bowtie risk assessment for manufacturing and maintenance operations. It achieves this by an integration with the 6M cause-and-effect methodology from Ishikawa [23].

2. Review of Bowtie Development and Structures

2.1. Existing Approaches Constructing a Bowtie Diagram

There are no standards for developing Bowtie diagrams, which results in a variety of different representations and interpretations [22]. However, a generally accepted and widely used approach for constructing a Bowtie diagram is presented in the following. This process aligns with the minimum requirements for a safety management system (SMS) and safety risk assessment introduced by the International Civil Aviation Organization (ICAO) [21]. Since the Bowtie methodology originates from the fault tree and event tree analysis, the diagram could be directly derived from these. In practice, however, the diagram is commonly developed based on brainstorming sessions [24].

A Bowtie diagram may be constructed using a bottom–up or top–down approach [4,25,26]. The latter starts with identification of the hazard, which sets the scope and context of the risk assessment [19,24]. As per the ICAO Safety Management Manual, a hazard is defined as "condition, object, process or activity with the potential of causing harm or damage, including injuries to personnel, damage to equipment, properties or environment, loss of material or reduction in ability to perform a prescribed function" [27].

The next step is defining the top event, which describes the release or loss of control over the hazard. "It has not caused any damage or negative impact yet, but can lead to undesired outcomes if all prevention barriers fail" [7]. The terminology of the 'top event' originates form the fault tree analysis. The top event forms the centre of the Bowtie diagram and links the fault tree and event tree. It can be caused by one or multiple threats.

Threats describe causes that can lead to the release of the top event, if all preventative barriers on the threat branch fail. They derive from fault tree analysis (FTA). Once identified, the threats are drawn as branches to the left of the top event.

The release of the hazard can lead to one or multiple consequences. These consequence branches are drawn to the right of the top event. Consequences are potential events or chain of events having a negative impact such as loss of control, damage, or harm. They originate from the event tree analysis (ETA).

Barriers, also referred to as 'controls' or 'layers of protection', are a means of prevention or mitigation for any negative outcome and can reduce the occurrence likelihood of the latter. Sklet [28] defined safety barriers as "physical and/or non-physical means planned to prevent, control, or mitigate undesired events or accidents". Depending on their purpose, barriers can be either on the left or on the right side of the Bowtie diagram. Prevention barriers are placed on the threat branches between the causes and the top event. Their function is to prevent the top event and ultimately the release of the hazard [12,18,21]. In contrast, mitigation barriers, also called recovery or protective barriers, aim to reduce the likelihood or minimise the severity of the consequences [29,30]. Thus, these barriers are positioned on the consequence branches between the top even and negative outcomes.

Barriers are not entirely effective or may not be permanently effective. Conditions that have the potential to adversely affect the effectiveness of a barrier are called escalation factors [31]. These factors are depicted as sub-branches from the main barrier path in the Bowtie diagram. To prevent the escalation factors from leading to barrier failure, additional controls, also called escalation factor barriers, are put in place [32]. These are drawn on the sub-branch of the escalation factor they are trying to prevent or mitigate. A generic Bowtie with all its elements is shown in Figure 2 below.

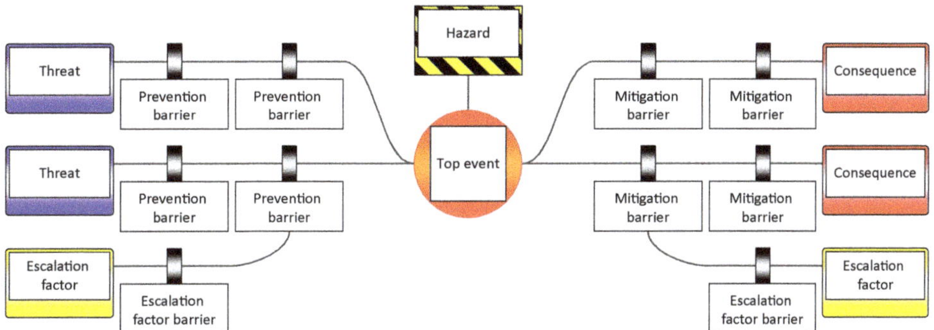

Figure 2. Generic structure of a Bowtie diagram.

As previously shown, the Bowtie method has elements of fault and event trees, albeit without the quantification or Boolean logic. There have been occasional efforts to re-introduce those features into Bowtie, e.g., in cyber security [14] and the process industry [33]. However, quantification and formalisation of the logic still suffers from the limitation of requiring estimates of probabilities—the provenance of which is as difficult as it originally was for FTA. This is particularly difficult if no historic data is available and must be estimated.

2.2. Categorisations Applied in Bowtie Analysis

Culwick et al. proposed two Bowtie structures [17]. One is a generic structure for general risk assessment, and the other one is an application-specific structure for malignant hyperthermia (MH) susceptibility. The focus for the following discussion is on the generic structure, as this might be transferable to other applications and industries. The generic Bowtie structure introduced by those authors has prompts and examples. For the preventive barriers, these include assessment, optimisation, preparation, planning, checklists, and forcing strategies. Examples given for the barrier controls are monitoring, vigilance, detection, and correction. As a means of recovery, the authors mentioned crisis management, resource and expertise, diagnosis and treatment as possible barrier categories. However, there was no structure or prompts provided for threats and consequences. It was proposed to organise the consequences from the top of the diagram to the bottom based on their severity reaching from 'no harm' to 'severe harm' respectively. The authors recommended that the Bowtie shall be "constructed by a group of individuals who have an interest in managing the particular hazard" [17]. This raises the question of whether or not an interest in the hazard is sufficient for creating a valid and comprehensive Bowtie risk assessment or if it would be better to have an expert or a group of experts, ideally with an experienced Bowtie facilitator, performing the risk assessment as suggested by CAA and ASEMS [4,22]. In addition to the generic Bowtie elements, the authors recommended consideration of factors influencing the efficacy of controls, namely patient factors, procedural factors, system factors, human factors, and chance factors. These factors are only suitable for the medical industry and were only conditionally transferable to other industries.

Hamzah developed a Bowtie based on a risk assessment matrix [34]. The matrix categorised the consequences into four categories, namely people, assets, environment, and reputation. These categories were used to evaluate the severity of consequences and, subsequently, to determine the risk of the hazard. However, this structure was not used for the development of the Bowtie diagram, but for the severity assessment and quantification.

Another approach was taken by Maragakis et al. [21], who did not provide a categorisation, but a hazard checklist deriving from past events and experience to help quickly identifying hazards. This list grouped hazards into the following categories: natural, technical, economical, ergonomic, and organisational hazards.

CAA UK used the 'Significant Seven' safety scenarios to categorise different top events and created a Bowtie for each. However, there was no categorisation suggested for threats, consequences and barriers [4,25].

As part of the "Basic Aviation Risk Standard (BARS) Program", the Flight Safety Foundation (FSF) performed a Bowtie risk assessment on offshore helicopter operation [35]. Similar to the Significant Seven by CAA UK, the threats were divided into eleven groups, namely: heliport and helideck obstacles, fuel exhaustion, fuel contamination, collision on ground, unsafe ground handling, controlled flight into terrain/water, aircraft technical failure, weather, loss of control, mid-air collision, and wrong deck landing. Furthermore, barriers were divided in organisational, flight operations, and airworthiness controls. This categorisation is particularly detailed and application specific, which makes the transferability to other industries somewhat difficult.

Barriers have been categorised based on their function, characteristics, and origins. Kang et al. [36] distinguished between physical and non-physical barriers and subsequently divided each group into three main categories, based on the work of Neogy [37] and Chevreau [38]. These include technological barriers, organisational barriers, and personnel barriers. Technological barriers can be further divided [18,29,32]. The division of risks into only two categories—'physical' and 'non-physical', or 'human behaviour' and 'technology related'—is insufficient for a broader categorisation as attempted here. Furthermore, an active and passive classification works well for barriers, but not for threats and consequences.

2.3. *General Categorisations and Classifications in Risk Management*

No systematic methodology is evident in the Bowtie literature, but there are some in the wider field of risk management. Some originate from the risk breakdown structure (RBS) focusing on project risks, while others derive from root cause analysis (RCA), a tool commonly used after a major, single-event problem occurred [39]. Still others focus on human factors, i.e., risks and conditions that cause humans to err.

2.3.1. Risk Categorisations

The most high-level and generally applicable categorisation of risks in any type of project, business and industry is based on the risk breakdown structure, a risk management tool, which has been broadly applied [40–42]. Hall and Hullet [40] proposed three 'universal risk areas', namely management risk, external risk, and technological risk, which differ from the four risk types proposed by Tanim et al. [43] comprising strategic, financial, operational, and compliance risks. These may be suitable for threats and particularly for consequences, but not so much for barriers.

The project management perspective of Lester [44] provides four main risk categories, namely organisational, environmental, technical, and financial, with further sub categories of technical, economic, environmental, operational, legal political, cultural, financial, commercial, resource, and security risks. It is one of the more detailed frameworks available.

The risk categorisation scheme of Chung and Zhu [45] includes operational, economic and environmental, strategic, technological and legal risks. This scheme was used to categorise company risks from news articles using a machine-learning algorithm. However, the categorisation emphasises management risks, less so the human involvement.

Industry specific approaches exist, such as a categorisation of airport construction risks into technical, logistical, economical, financial, legal, construction, commercial, social, natural, and legal [46]. In the pharmaceutical and health care industry, risks have been divided into facility, personnel, process, system, and product risks [47]. While these categories may well fit within the relevant industry, they may be less suitable for categorising risks in other areas—similar to the categorisation used in health care (see Section 2.2).

The PEST, also called the STEP framework, is a tool used in market research [48]. The acronym stands for Sociological, Technological, Economic and Political. This categorisation was later enhanced to PESTLE by adding a legal and environmental component. This framework is more common in strategic management or marketing rather than for risk assessment (for an exception, see [49]). This is because the PEST(LE) analysis helps to identify factors in the market that affect the development and viability of an organisation. However, these factors do not necessarily have to be risks or threats, but can also be opportunities. It is evident that many of the above categorisations have strong similarities with the PESTLE framework.

2.3.2. Threat and Root Cause Categorisations

In computer systems, threat can be classified into physical damage, technical failure, natural event, compromise of information, compromise of functions, and loss of essential service [50]. Jouini et al. [51] introduced another classification for threats in information systems. The two main categories are external and internal threats, with both having sub-categories of human, environmental and technological threats.

In Bowtie, the term threat is used to describe root causes. Taproot is a root cause tree dictionary that classifies causes into eight categories, namely equipment difficulty, management system, quality control, procedures, human engineering, communications, training, and work direction [52]. The International Air Transport Association (IATA) uses five categories for the accident root cause classification system including human, organisational, environmental, technical, and insufficient data [53]. This approach covers human factors only in the form of human engineering. Furthermore, equipment difficulty and insufficient data work well for categorising threats or accidents, for which the categorisation from

Taproot and IATA was developed. However, it may not work for barriers, since they should prevent or mitigate any negative outcome.

The most common categorisation for root causes originates from the cause and effect diagram, also referred to as 'Ishikawa' or fishbone diagram [23,47]. To help identifying the root causes in a manufacturing environment and to break them down, Ishikawa identified six categories starting with the letter M, hence the 6M method. The Ms stand for man (or mind power), machinery, materials, methods, and Mother Nature (or milieu) [47]. Variations exist on the basic idea, using different terminologies [54]. For instance, some used the term 'environment' instead of 'Mother Nature' or 'equipment' instead of 'machinery', hence leading to the 5M and 1E categorization [55,56], while others replaced the term 'manpower' with 'people' resulting in the 5M and 1P approach [57]. Some extended the categorisation to eight Ms by adding 'management', 'money', and 'maintenance' [58]. This categorisation originates from manufacturing and is therefore somewhat limited to its industry. However, the categories can be adjusted as needed.

Different categorisations have been developed to make the cause and effect diagram applicable to other industries. The 8P method (procedures, product, price, people, place, processes, policies, and promotion), for instance, is a common cause categorisation used in the marketing sector, while the 4S (surroundings, suppliers, systems, and skills) is well established in the service sector [55,56,59]. Both imply that a contextualisation may be required to apply a categorisation broadly.

2.3.3. Human Factors Categorisations

In aviation maintenance and inspection environments, errors can be categorised by their root causes including task, environmental, individual, organisational, and social factors [60]. However, most emphasis is on human factors, since it is generally accepted that humans have caused or contributed significantly to aviation incidents and accidents [61,62]. The two most common categorisations for human factors are the SHEL and PEAR models.

The SHEL model is a conceptual framework proposed by Edwards [63] to classify accident causes in aviation. The four categories are Software, Hardware, Environment, and Liveware. This concept was modified later by Hawkins, who added another 'liveware' component and presented the SHELL model [64,65]. Most recently, organisational factors were introduced by Chang and Wang making it the SHELLO model [66].

For the field of aviation maintenance, Johnson and Maddox developed the PEAR model, an acronym for People, Environment, Actions, and Resources [67]. It helps in categorising human factors and was first applied by Lufthansa.

Both PEAR and SHELL were developed for human risk factors in aviation. The SHELL model only focuses on interfaces within Human Factors (software–human, hardware–human, environment–human, and human–human) [27]. However, it does not include interfaces between the other factors (hardware-software, hardware-environment, and software-environment). Since we were looking for a categorisation covering risk factors and not only the interfaces between them, the SHELL model was not suitable. Similar, the PEAR model focused only on Human Factors and would not provide the universality needed for our purpose.

2.4. Limitation in the Methods for Bowtie Construction

While several approaches exist for Bowtie construction, there is no standard [22,33]. The two main issues are (i) the lack of a standard methodology for systematically identifying Bowtie elements, and (ii) the subjectivity of the process. The latter relates to the subjectivity of the brainstorming method. These issues are interrelated.

Existing methods focus on the diagram construction and bringing the elements into the characteristic bowtie shape, but not on the identification of these elements, i.e., threats, consequences and barriers. This identification is ad hoc. There is a need for a structured methodology to identify threats and consequences and to ascertain barriers without missing important ones.

There is also inconsistency regarding the hierarchy of the hazard and top event. Although it is commonly accepted that the Bowtie construction starts with identifying the hazard, followed by the top event, the hierarchy of these is inconsistent. Some risk assessments followed the top–down approach, whereby multiple Bowties with different top events for the same 'umbrella' hazard were developed, such as the Significant Seven Bowties by CAA UK [4,25]. Others followed the reverse approach and constructed several Bowties with different hazards for the same top event [26]. Still others analysed only one hazard with its top event, whereby the relationship was not problematic [26]. It was found that often the top event is a subjective and pragmatic choice of the risk analyst and that it is rephrased once the Bowtie diagram is completed [19]. This contributes to the inconsistency in the hazard and top event hierarchy.

Most Bowtie diagrams are developed using brainstorming sessions and hence are dependent on the expertise and personal view of the participants and the skills of the facilitator. Although brainstorming encourages creative unbounded thinking, it can be time consuming and sometimes chaotic, which results in an unstructured and incomprehensive approach, and is always subjective. Moreover, the sessions are prone to group dynamics, which can influence the assessment results and may lead to missing important risks [21].

3. Method

3.1. *Purpose*

The purpose of this research was to develop a systematic methodology for conducting Bowtie risk assessments. The desired attribute of such a methodology is to provide a structure to guide the analyst when identifying the barriers, so that no important threats, consequences or barriers are missed. The area under examination is maintenance engineering, and within that the aviation MRO and inspection operations, specifically visual borescope inspection of aero engine parts. We were especially interested in the risk of missing a defect on a turbine blade.

3.2. *Approach*

This work is of a theory-building nature. First, we reviewed the literature for candidate approaches to develop and construct a Bowtie diagram. Further, we reviewed different existing systems in the safety and broader risk management field that help identifying and categorising risks, hazards, threats, and root causes. These include general risk approaches, human factors models, and cause and effect classifications. From these we selected one approach, namely 6M, as the basis for the new Bowtie framework.

Next, we resolved the ambiguity in the hazard–top event coupling, and applied the 6M to the Bowtie process. In doing so, we contextualised it to the area under examination. We also addressed the structure of the escalation factors. We identified that strings of escalation factors were often common to different places in the Bowtie, and we proposed making these into modules for better representation.

The method was then applied to the specific case of visual borescope inspection of gas turbine components in an MRO environment. A total of 15 aircraft maintenance inspectors from the industry partner participated. The experience profiles were: five inspectors with more than twenty years of experience in visual inspection, seven inspectors who had worked for over ten years in the field, and three operators with up to ten years of experience. Their certifications included borescope operation and non-destructive testing (NDT). Each was observed independently for approximately 30 min during a real inspection process, and they were asked to articulate the risks and threats of the process, and what barriers were or could be in place to prevent negative outcomes. The specific instructions were:

- 'Please describe the inspection process you are performing and the challenges of each step.'
- 'What factors influence the inspection process and why are they safety critical?'
- 'What are the risks inherent in each process step?'
- 'What means of prevention or mitigation are or could be in place to prevent missing a defect during inspection?'

All the operators were familiar with Failure Mode and Effects Analysis (FMEA), and some were also familiar with the Bowtie methodology. Their verbal comments and insights were noted, and subsequently used by the primary author to construct the Bowtie diagrams shown in this paper. Ethics approval was obtained from the University of Canterbury (HEC 2020/08/LR-PS) and permission from the industry partner. The results and limitations of the new approach were then validated by discussion with the two highest certified borescope inspectors (who both held a level 2 certification in borescope inspection, which is the highest certification achievable in the industry), and a human factors and risk analyst for the organisation. They were presented with the Bowtie results and asked to comment. They confirmed the accuracy of the results, commented on the method (they were generally in favour), and made suggestions for improvements (primarily in the precise wording of threats and barriers). For the visualisation of the Bowtie diagrams, the software 'BowTieXP' revision 9.2.13 was used [68].

4. Results

4.1. Consistent Interpretation of Relationship between Hazard and Top Event

Regarding the inconsistency of usage of the hazard and top event, we propose the following categorisation.

Format A: In cases where the scope of analysis is limited to one situation, the correspondence between hazard and the top event is not problematic. If the Bowtie diagram is too large, it may be represented as multiple smaller diagrams, providing the hazard and top event are used in the same way. The case study presented in this paper follows this approach.

Format B: The top event may have multiple hazard dimensions. In which case the top event should be consistent across all the diagrams, while the hazard changes. The hazard thus corresponds to a different contextualisation for the same top event; see Figure 3.

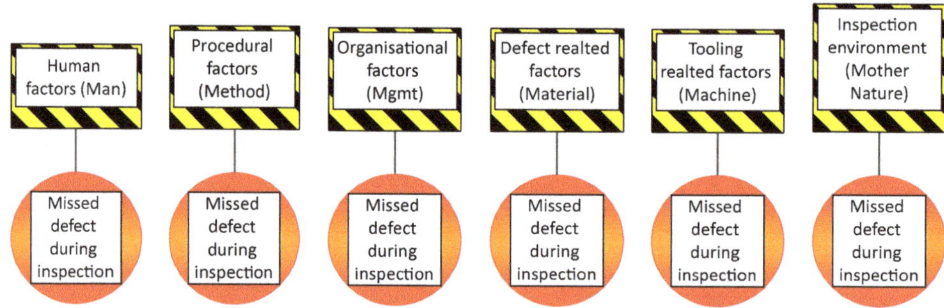

Figure 3. Consistent top event with multiple hazard dimensions.

Format C: A hazard may have different top event dimensions. Each dimension may represent a different step in a process. Figure 4 provides an example for human factors in different areas of the maintenance process.

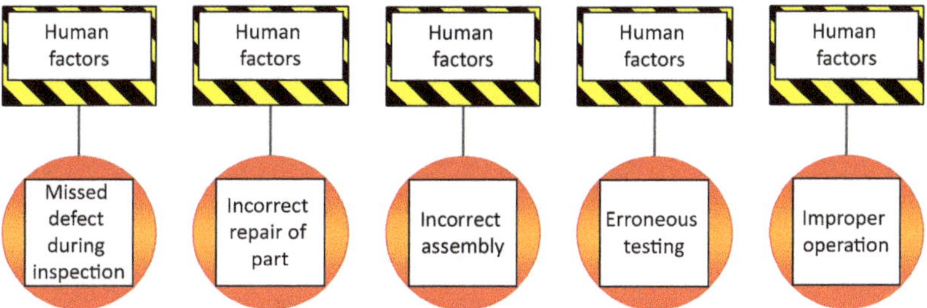

Figure 4. Hazard investigation with different top events.

We propose that the Bowtie analyst should decide beforehand how the problem is to be structured, and select one of the above formats, and then use it consistently throughout the analysis. This has the potential to avoid some of the inconsistencies seen in practice.

4.2. Proposal to Use 6M Structure

The challenge with selecting the most suitable categorisation was that Bowtie contains different risk elements including hazards, root causes (threats), consequences, barriers and controls. For each of these elements, different approaches and categorisations exist. Some are suitable to categorise risks (causes and effects), but less suitable for categorising barriers. Hence, we decided to choose a risk categorisation that works for both risks (threats and consequences) and means of risk prevention and mitigation (barriers).

The 6M approach was seen as most suitable for the planned research and case study for the following reasons. First of all, the 6M approach was chosen due to the familiarity of the Bowtie and the Ishikawa diagram, with both being a cause and effect diagram. The 6M has already been successfully applied to structure such a diagram [23]. Secondly, it has been successfully applied outside the manufacturing environment such as health care [69,70], management [71], and education [72]. Moreover, the 6M categorisation was well known at our industry partner, as the cause and effect diagram from Ishikawa is part of the measure phase of the DMAIC, a continuous improvement process often used in Six Sigma and know by industry practitioners [73]. The 6M structure can be used for both vertical categorisation of the threats and consequences, as well as horizontal categorisation of barriers.

The 6M aligns with other categorisations presented in the broader literature in Section 2.3 including the five accident categories used by IATA [53], and the four main risk categories by Lester [44]. Both are quite similar and a terminology adjustment, following the 'category starts with the letter M' concept, would make these categories match the selected 6M approach.

4.3. Integration of 6M with Bowtie (Contextualisation)

The MRO environment differs to the manufacturing environment, and hence the original 6M structure by Ishikawa required modification. The contextualisation was carried out in the light of organisational quality factors that influence the maintenance and inspection result. In a production environment, the measurement category commonly includes the inspection and measurement of the parts to check if they meet the quality requirements. However, in an inspection environment the 'measurement' category can overlap with 'machine' and 'method', since the inspection tool and followed processes would fit into all three categories. The measurement category was therefore less useful to apply in an inspection environment. Instead of the 'measurement' category, we included 'management', which is one of the additional categories from the 8M approach by Burch et al. [58] and aligns to the work by Gwiazda [74] and Vaanila [73]. This was carried out as it matches the area under investigation being the highly regulated aviation and maintenance industry (see below) [75,76]. Management refers to organisational and regulatory factors. The other category from the 8M we could have chosen was maintenance, but since we wanted to investigate risks in a maintenance environment, maintenance as a threat and a hazard at the same time would have caused problems with the consistency of the Bowtie structure.

In Table 1 below, we demonstrate how the categories and their interpretation may change according to the industry. This is demonstrated based on three examples, namely manufacturing, maintenance and health care. The latter was chosen to demonstrate the integration to a quite different industry.

Table 1. Contextualisation of the 6M categories to different industries.

Category	Manufacturing	Maintenance	Health Care
1st M	Method (workflow and production processes and procedures)	Method (maintenance processes and procedures)	Method (surgery or medical treatment procedures)
2nd M	Man (operator human factors)	Man (inspector human factors)	Man (personnel human factors)
3rd M	Mother Nature (Production environment and facilities)	Mother Nature (maintenance environment and facilities)	Mother Nature (hospital and GP facilities)
4th M	Machine (manufacturing machinery)	Machine (repair machinery and inspection tools)	Machine (surgery equipment and tools)
5th M	Material (manufactured product)	Material (maintained product)	Material (used product, e.g., medicine and aids)
6th M	Measurement (quality control and maintenance)	Management (MRO organisation and regulators)	Man (patient)

4.4. Threats and Consequence Structure Using 6M

4.4.1. Threat Structure in MRO

We started with integration of the 6M structure for the threats. A description of each category together with an example is given in the Table 2 below.

Table 2. The 6M categories for threats with description and example.

6M Category	Threat Description and Example
1. Machine-related threats	Machine or tools not working properly, e.g., faulty borescope
2. Mother Nature-related threats	Poor inspection environment, e.g., poor lighting
3. Man-related threats	Human error or failure, e.g., misinterpretation of the defect
4. Method-related threats	Lack of standard processes and procedures, e.g., incorrect, outdated or no standard working procedures
5. Material-related threats	Poor condition of the part, e.g., deposit on blade hides defect
6. Management threats	Poor operational management, e.g., time pressure leads to rushed inspection

4.4.2. Consequences for Different Stakeholders

There are two main stakeholders in this situation: (i) the MRO service provider, and (ii) the airline company and its passengers. The stakeholders are linked in a cascade of consequences [7]. A missed defect during borescope inspection can propagate from minor consequences for the MRO service provider and engine owner, towards catastrophic consequences for the airline, passengers and cabin crew; see Figure 5. For each link in the consequence chain, a new Bowtie risk assessment can be performed and a diagram drawn, tailored to the focus of the affected stakeholder.

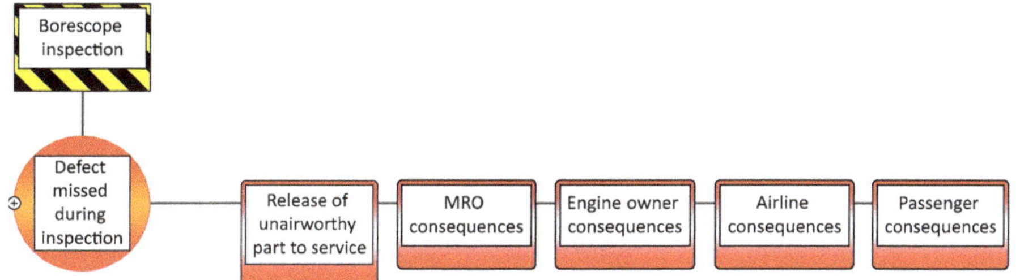

Figure 5. Cascading consequences with different stakeholders.

Immediate Consequences for the MRO Service Provider

The consequences for the MRO provider include additional costly and time-consuming repairs or improvement processes, and reputational damage. An overview of possible consequences is given in Table 3 below.

Table 3. The 6M categories for immediate consequence for the MRO service provider with description and example.

6M Category	Consequence Description and Example
1. Machine-related consequences	Damage to machinery, e.g., damaged borescope
2. Mother Nature-related consequences	Adverse effect on MRO environment, e.g., damage of test cell or facility
3. Man-related consequences	Consequences for employees, e.g., additional training or certification needed
4. Method-related consequences	Changes of methods required, e.g., revision of standard work protocols and subsequent re-training of staff
5. Material-related consequences	Additional part preparation, e.g., water jet wash
6. Management consequences	Reputational consequences, e.g., degradation of engine shop status

Subsequent Consequences for the Airline

In the airline situation, the effect of a defective part in an engine has a different set of consequences, as shown in Table 4. These can reach from less critical gate returns and flight delays, to engine failure during flight operation with the potential to cause accidents or harm to passengers and cabin crew.

Table 4. The 6M categories for subsequent consequences for the airline with description and example.

6M Category	Consequence Description and Example
1. Machine-related consequences	Damage to the engine or aircraft, e.g., uncontained engine failure
2. Mother Nature-related consequences	Contamination of airport or nature after engine failure, e.g., debris from engine falls from aircraft
3. Man-related consequences	Harm to passengers and cabin crew, e.g., fatality
4. Method-related consequences	New procedures, e.g., additional checks before flight operation
5. Material-related consequences	Material failure, e.g., propagation of a defect leads to part separation (FOD)
6. Management consequences	Reputational or financial consequences for airline, e.g., compensation for causing harm

4.4.3. Combined Threat and Consequence Structure Using 6M

The combined threat and consequence structure of the Bowtie diagram applying 6M is presented in Figure 6 below. The diagram illustrates the structure of the concept and that there is no limit for the number of threats in each category. It provides a systematic guide to identify threats and consequences. Furthermore, a threat of an M category does not necessarily result in a consequence of the same category.

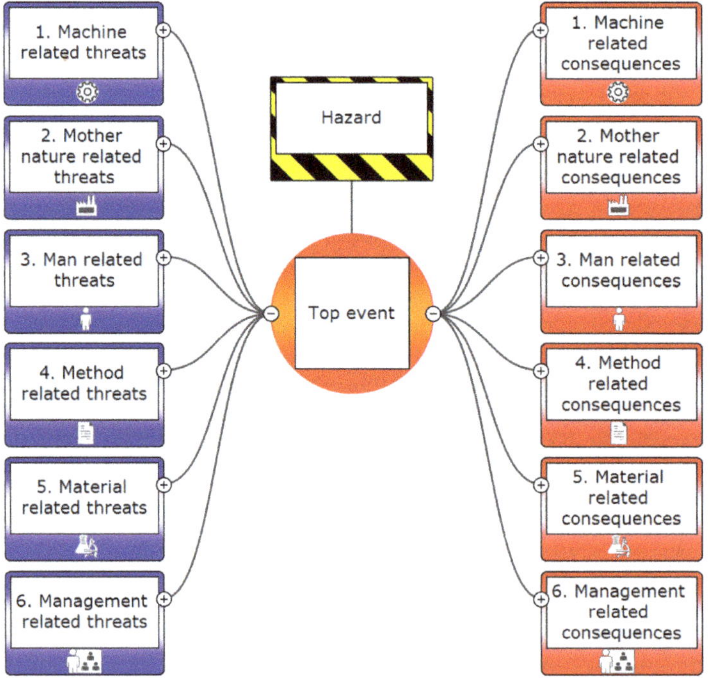

Figure 6. Bowtie with threats and consequences structured based on the 6M approach.

4.5. Barrier Structures Using 6M

4.5.1. Generic 6M Barrier Structure

One limitation of Bowtie is that barriers are not presented in a time or process following manner [7]. This limitation, however, allows grouping the barriers based on their nature and following the 6M categorisation, without changing the overall Bowtie structure. Providing this 6M structure for barriers supports and structures the brainstorming sessions, which will remain an essential part of the element identification process. The framework is presented below—see Figures 7 and 8—and shows one barrier per category. It should be noted that each of these barriers is a representation for all barriers of its type. There may be threat or consequence paths that have no barriers of one or more 6M categories, whereas they may have multiple barriers of another category. A description and example of each barrier category can be found in Table 5. For more barrier samples please refer to the case study below.

Table 5. The 6M categories for prevention and mitigation barriers with example.

6M Category	Barrier Description and Example
1. Machine-related barriers	Machinery and inspection tool-related barriers, e.g., backup tools availability
2. Mother Nature-related barriers	Work environmental barriers (external and internal environment, e.g., appropriate work place design
3. Man-related barriers	Operator or inspector-related barriers, e.g., airmanship, self-awareness, and experience
4. Method-related barriers	Prevention and mitigation processes and procedures, e.g., standard working procedures
5. Material-related barriers	Material-related barriers
6. Management barriers	Operational management-based barriers, e.g., provision of appropriate training

4.5.2. Colour Coding of Barriers

To support the core function of Bowtie, being a communication tool that is easy to understand, we propose colour coding the barriers by 6M category. The colour assignment was to some extent random and did not follow a particular scheme. In addition, each barrier should have a colour that has not been used before for any other element of the Bowtie diagram. The colour assignment of each M category is shown in Figures 7 and 8. If greater emphasis is necessary, either to draw attention to specific categories, or to provide greater clarity for the visually impaired, additional demarcation could be added in the form of a symbol. Some suggested symbols are illustrated in Figures 7 and 8 below, though it should be noted that these have been added manually as this is not currently a feature of the BowtieXP software used here. Alternatively, hatching may be added to the Bowtie elements.

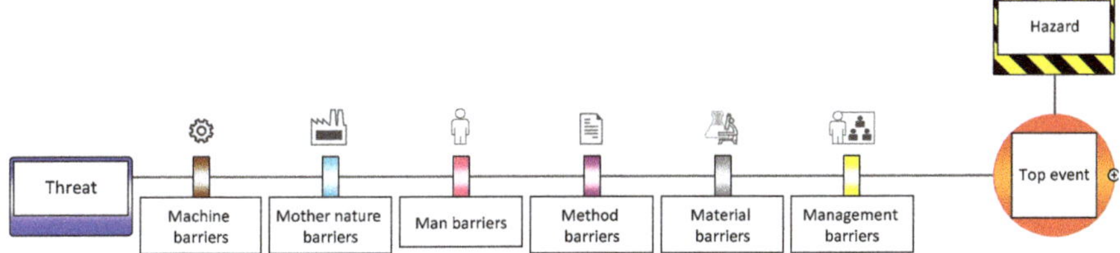

Figure 7. Threat path with coloured barrier categories.

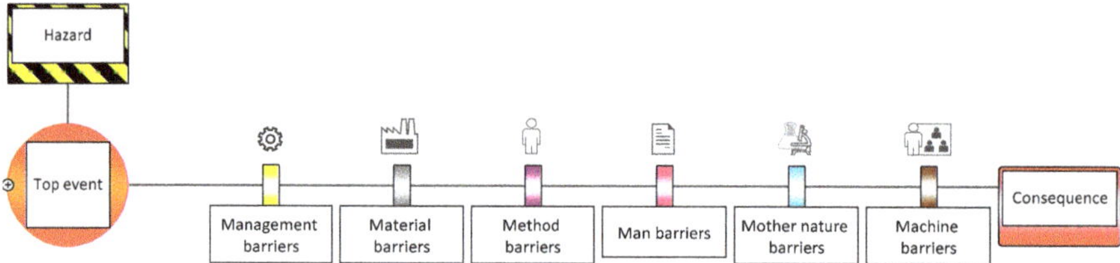

Figure 8. Consequence path with coloured barrier categories.

4.5.3. Escalation Factor Paths with a 6M Structure

In principle, the escalation factor path on the prevention and mitigation side of the Bowtie diagram could follow the same 6M structure as demonstrated for the threat and consequence path in Figures 7 and 8. The result would look like Figure 9. After the higher-level structure of the Bowtie has been completed, it may in some situations be necessary to extend the analysis to the escalation factors, and in this case, the methodology proposed here offers a way this may be approached.

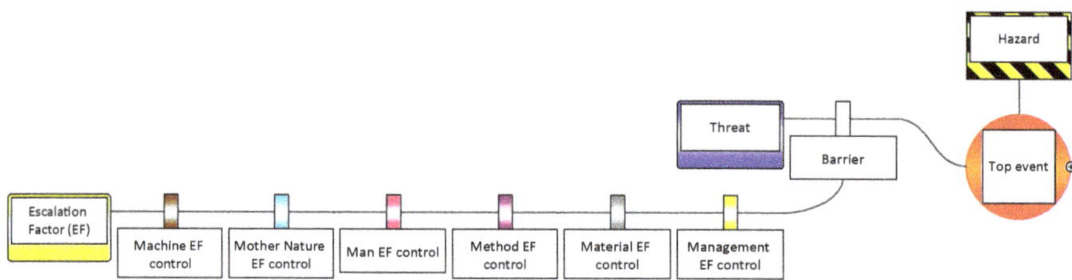

Figure 9. Escalation factor path with coloured barrier categories.

4.5.4. Barrier Modules

Since barriers repeat themselves multiple times along the Bowtie, we propose defining "barrier modules". This has the potential of making the Bowtie development process more time efficient since not every barrier with its entire escalation factor and escalation factor control path has to be repeated. Furthermore, it tidies up the diagram without diminishing comprehensiveness. A visualisation of the barrier module is presented in Figure 10.

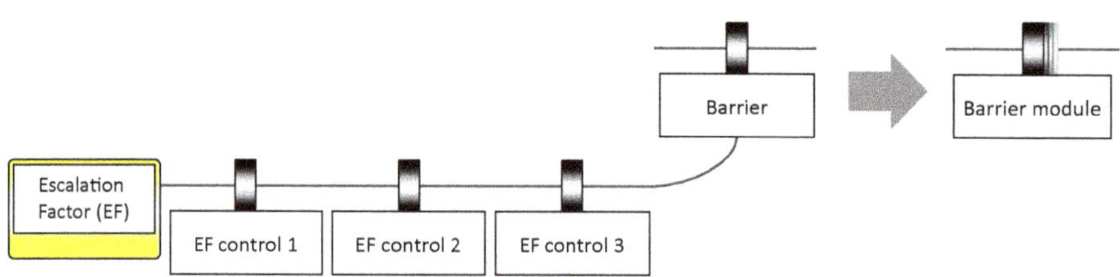

Figure 10. Potential visualisation of barrier modules encapsulating escalation factors and its controls.

When introducing barrier modules, one limitation might be that all barriers are expected to have the same efficiency. Some barriers (modules) might be repeated because they are of the same nature; their effectiveness in regards to the threat, however, might be different. For example, when considering 'fire' being the threat, the extent of the fire might vary significantly, e.g., a burning candleholder, a house fire or a wildland fire. In each case, a fire-extinguishing agent would be a barrier. However, for the small fire, a handheld fire extinguisher might be sufficient, while for a house fire, a fire truck is the right level of prevention, and for a wildland fire, an extinguishing plane may be required. This limitation must be considered when using barrier modules.

4.6. Full Bowtie with a 6M Structure

Combining the 6M structure for threats and consequences from Section 4.4.3, the 6M structure for barriers from Section 4.5.1, and the colour-coding scheme from Section 4.5.2, results in a 6M × 6M matrix structure on both sides of the Bowtie diagram. The full Bowtie structure is shown in Appendix A. For better legibility, both sides of the diagram are presented individually in Figures 11 and 12.

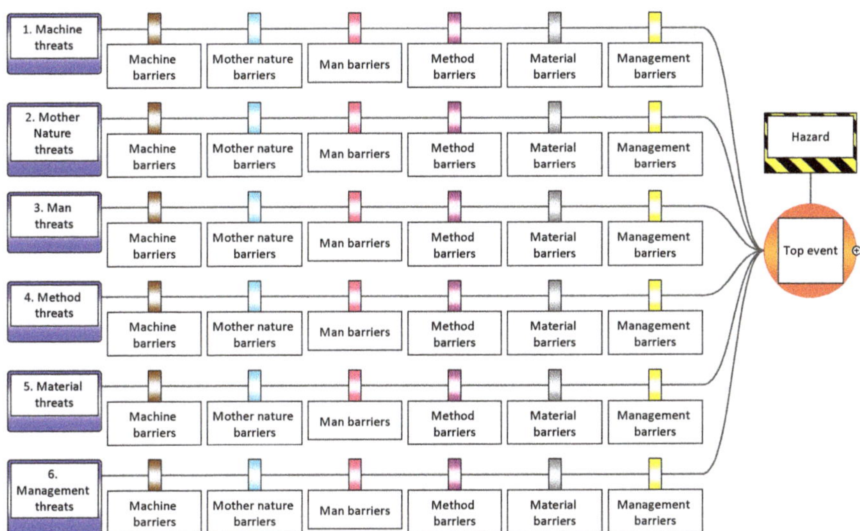

Figure 11. Threat side of the Bowtie diagram with 6M prevention barrier structure.

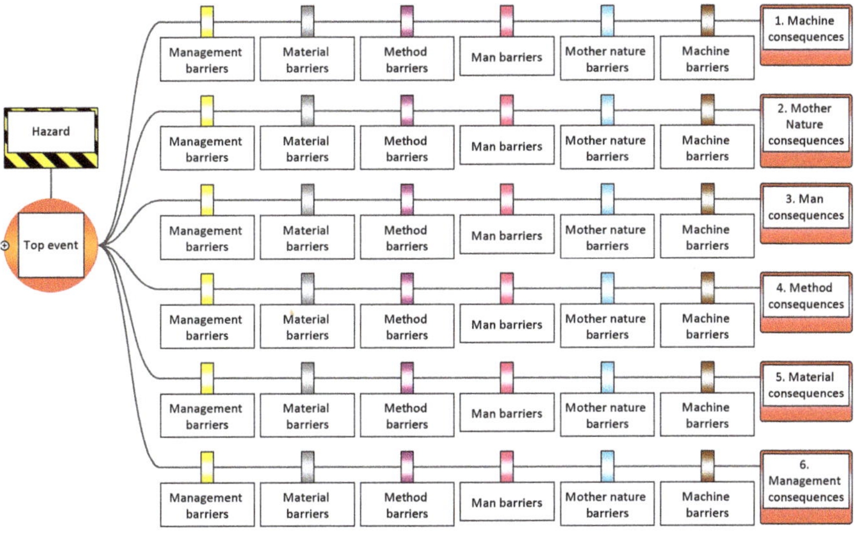

Figure 12. Consequence side of the Bowtie diagram with 6M mitigation barrier structure.

4.7. Application to a Case Study

The conceptual framework was applied to the specific case of visual inspection of aero engine parts in an MRO environment. Borescope inspection plays a crucial part in engine maintenance, since it allows inspecting parts inside the engine for defects, such as nicks, dents, cracks, tears, and fractures, without the need for a costly teardown. Missing such a defect during visual inspection is highly critical for the airworthiness of the engine and passenger safety. Hence, we defined the top event as being the risk of a 'Defect missed during inspection'. The next step was the identification of the threats, consequences and barriers. We asked each specialist from the maintenance and inspection domain to identify risks inherent in the process of borescope inspection, and what means of prevention and mitigation are or could be in place. The insights were extracted from field notes taken during the observation and the Bowtie diagrams were drawn. It shall be noted that the main emphasis was put on the prevention side, which is common practice in Bowtie application [4,26]. The reason behind this approach is that prevention efforts are more cost-effective and hence more attractive from a management perspective [77].

A general limitation of Bowtie and other root cause diagrams is the scalability and legibility when analysing complex systems, as the diagrams tend get quite large. When using BowtieXP software, there is a function to show and hide different layers of the Bowtie diagram. The layers include: (a) only hazard and top event; (b) hazard, top event, threats and consequence; (c) hazard, top event, threats and consequences with all barriers; (d) hazard, top event, threats, consequences, barriers and escalation factors; (e) threats, consequences, barriers, escalation factors and escalation factor controls. This is helpful when presenting the diagram to an audience who does not need all the details, but without losing any of the data in the background. Unfortunately, this is a manual task and there is no automatism for expanding or condensing Bowtie diagrams, or hiding individual threat or consequence paths. It would be helpful if this feature could be enhanced in future versions of BowtieXP software.

Possibly, multiple threats of the same category can be merged into one 'higher-level' threat path (similar to the barrier modules introduced in Section 4.5.4), e.g., summarising fatigue, distraction, and complacency, into a single human factors threat path instead of listing all twelve human factors (HFs) individually. However, this requires that all threats have the same barriers, which often is not the case.

In order to represent the Bowtie diagram of this research in a legible and receptive way, it was divided into six Sub-Bowties based on the M categories. The six Sub-Bowties for the threat side are shown in Figures 13–18. The consequence side of the diagram is presented in Figure 19. Additionally, for purposes of illustration, the size of all Bowtie diagrams was somewhat artificially limited to a maximum six barriers per threat and consequence path. This is solely a limitation to provide legible diagrams within the journal constraints.

The full-sized Bowtie diagrams with all barriers can be found in Appendix A Figures A2–A8. For a higher resolution version of the developed diagrams, refer to Supplementary Materials Figures S1–S8.

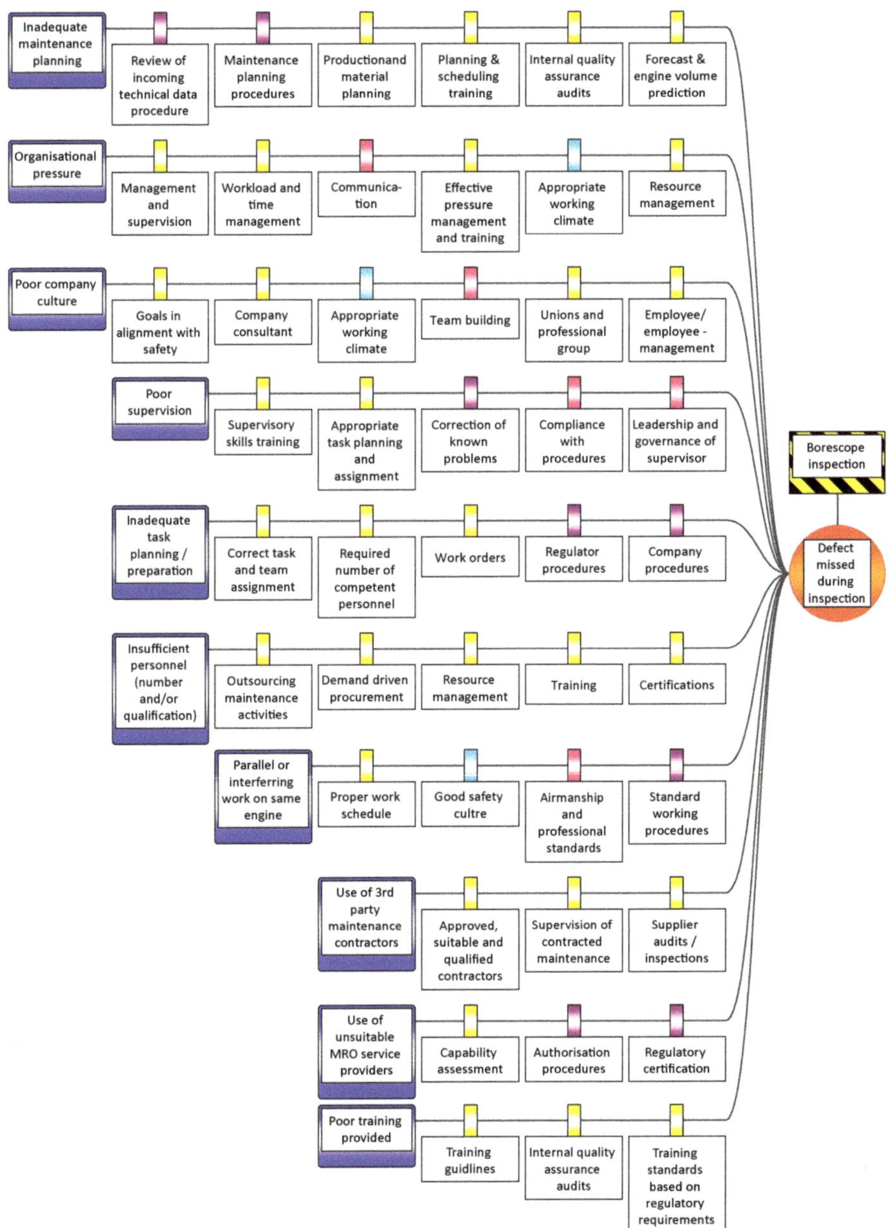

Figure 13. Management-related threat paths with barriers.

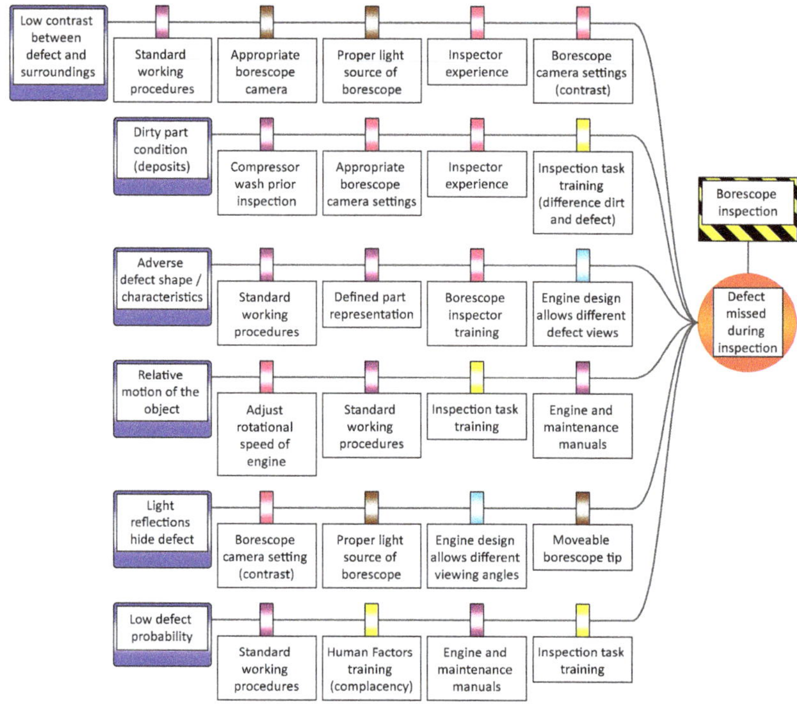

Figure 14. Material-related threat paths with barriers.

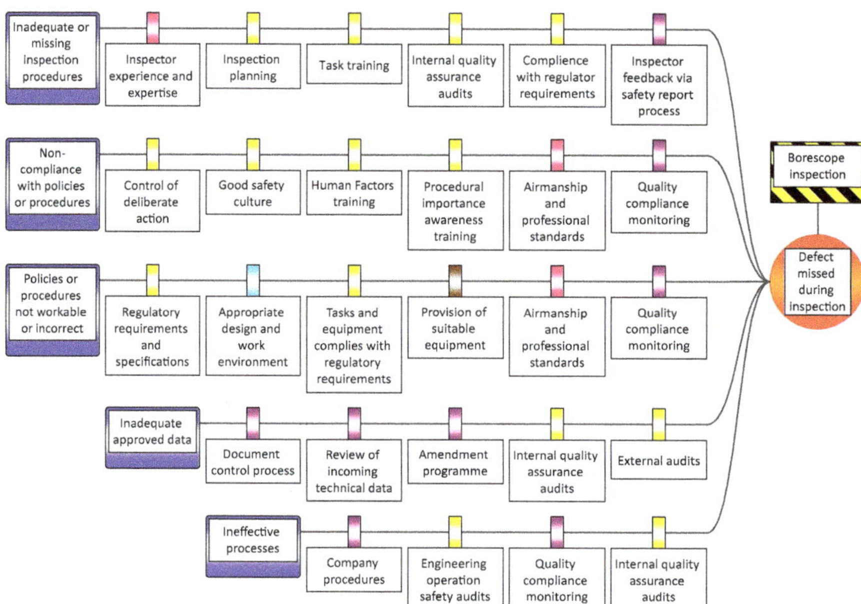

Figure 15. Method-related threat paths with barriers.

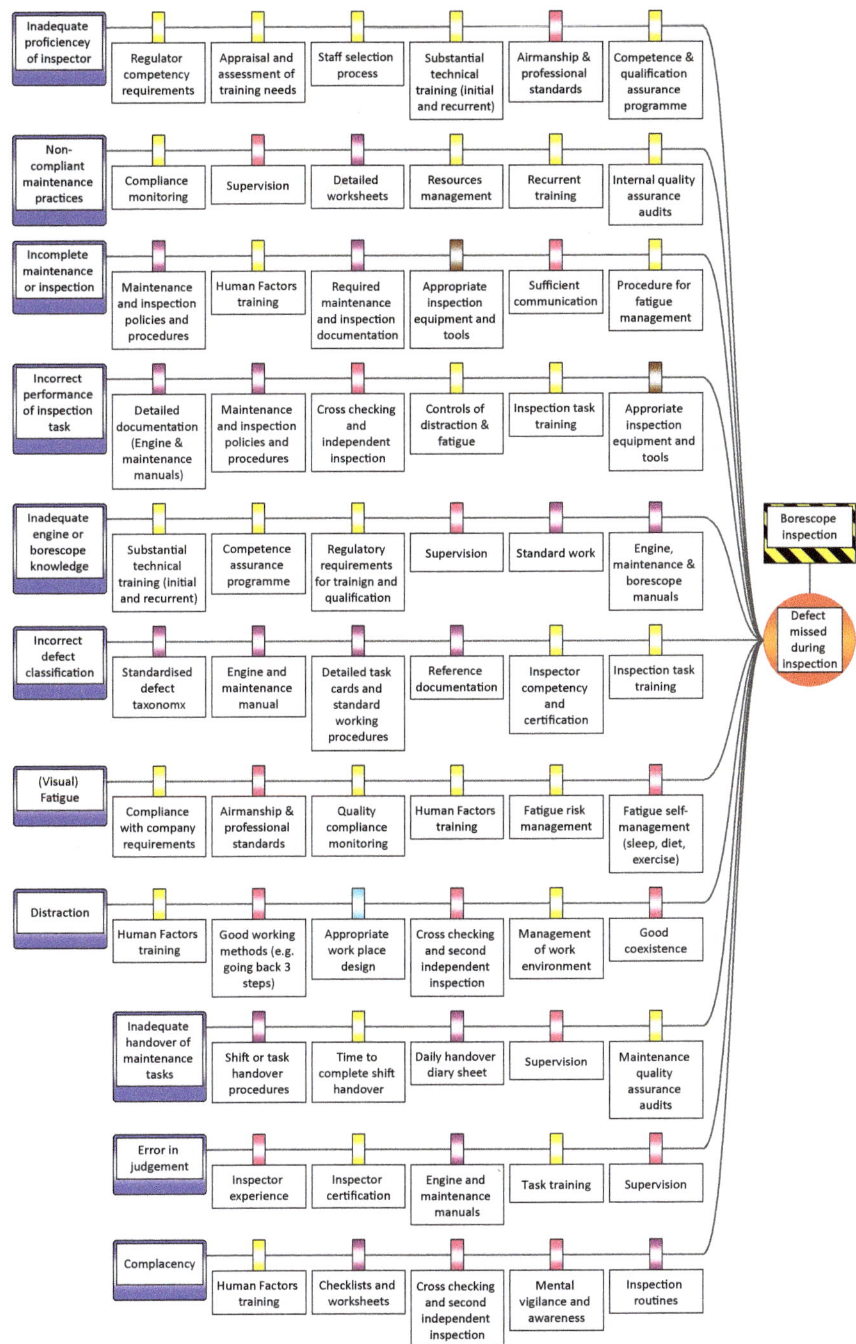

Figure 16. Man-related threat paths with barriers.

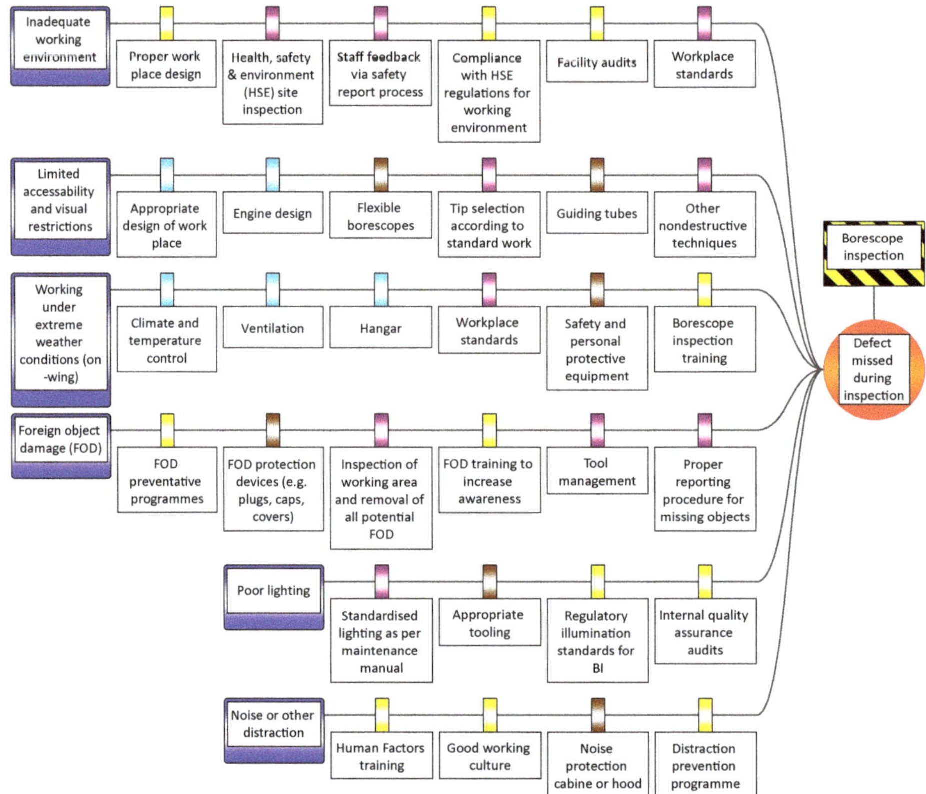

Figure 17. Mother Nature-related threat paths with barriers.

The categorisation and barrier colour coding were made in collaboration with the risk and management team of our industry partner. The categorisation was based on the responsibility and exerting agency of the threat or barrier. This decision was made after a discussion with the industry experts, about the Bowtie elements that could be placed in more than one category, such as task-related threats. In this particular example, the threat could be placed in the man category since the inspection personnel performs the task. On the other hand, it could also be a method-related threat, since a task is part of a process or procedure. Based on the decision above, we categorised it as a man-related threat, since the human performs the task and human performance is always critical in this industry. It is generally accepted that human errors cause over 70% of all aircraft accidents [78]. Furthermore, 80% of all maintenance errors involve human factors [79]. Hence, it can be expected that the threat paths in the man category will make up the majority of all threats in the Bowtie diagram.

The diagrams produced as part of this research should not be considered comprehensive. We limited the consequence side of the Bowtie to the immediate consequence rather than the full consequence chain.

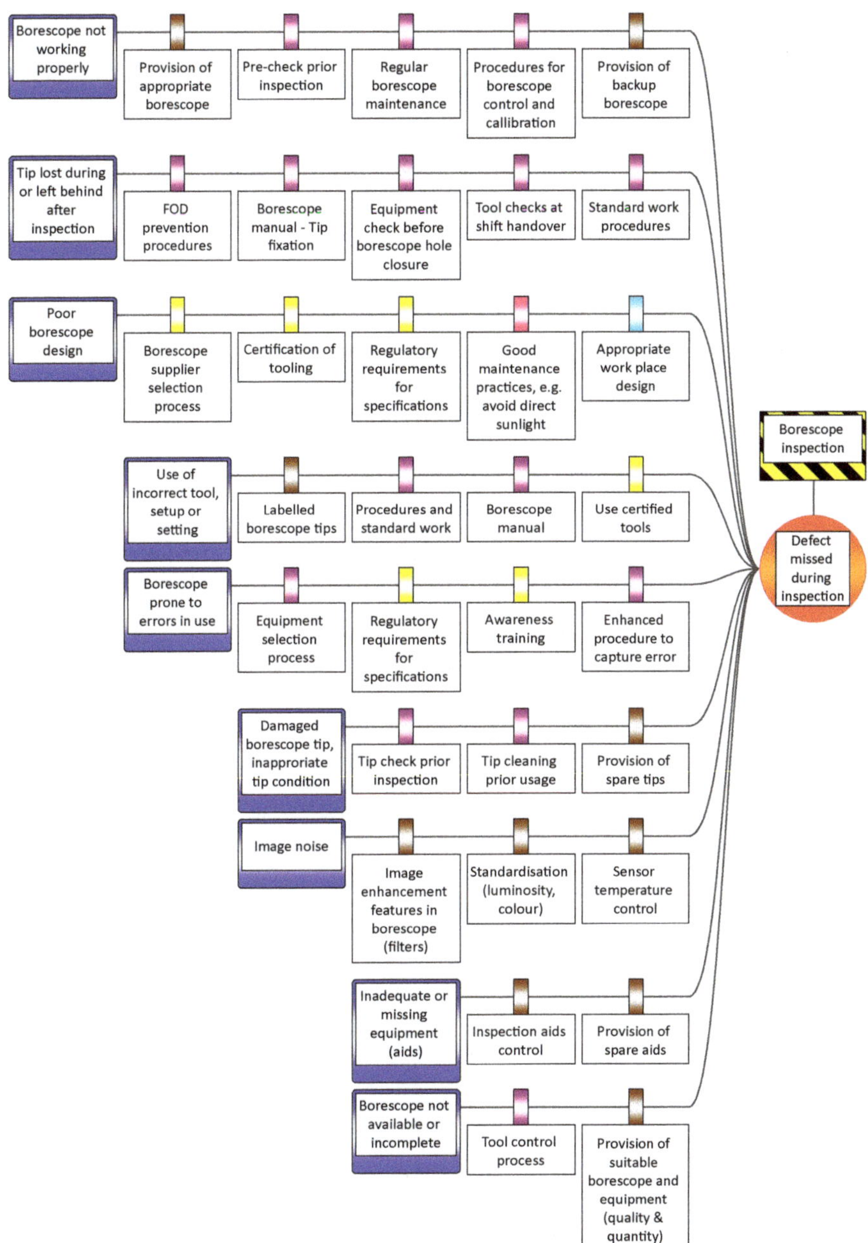

Figure 18. Machine-related threat paths with barriers.

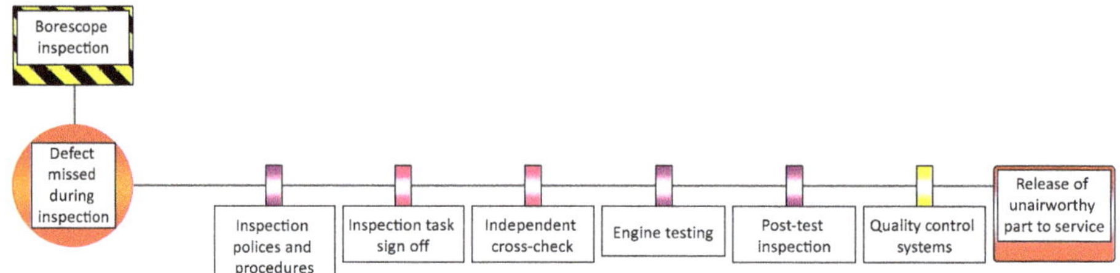

Figure 19. Consequence path with barriers.

5. Discussion

5.1. Summary of Outcomes

This work proposes a new methodology for integrating structured frameworks to the Bowtie method. Ishikawa's 6M approach was chosen to structure the Bowties and the accompanying brainstorming sessions. We showed that a contextualisation of the 6M categories was required for application to a maintenance environment. While constructing the Bowtie diagram, it was found that there is inconsistency and confusion about the hazard and top event relationship. A consistent interpretation was provided to overcome this problem. Furthermore, it was found that there are cascading consequences for different stakeholders. Depending on the focus of the risk analysis and the target audience, there are different consequences, which we demonstrated in this work. Moreover, the visualisation and receptivity of the diagram was improved by assigning different colours to each barrier category, which supports the main purpose of the Bowtie method, i.e., functioning as communication tool. Finally, the proposed conceptual framework was tested by applying it to the specific case of visual borescope inspection of aero engine parts.

5.2. Implications for Practitioners

The proposed structured approach was tested in the aviation maintenance area. However, the method could be applied in other areas within or outside the aviation industry. It might be of particular interest to other high-reliability organisations (HROs), such as oil and gas, nuclear power generation, health care, or wildland firefighting [80,81]. This is supported by the fact that the structured approach could make the development of Bowtie somewhat simpler and hence promote the broader application of Bowtie.

The framework provides non-risk experts with a tool to perform risk assessment. Operators, who may have limited risk management skills but a better knowledge of the system and processes than a risk analyst, might use the tool to identify threats, consequences, and means of prevention and mitigation. The framework encapsulates a wide range of previous known areas relevant to risk assessment and ensures that most common and obvious threats, consequences and barriers are not missed. Furthermore, it enables the analyst to put the emphasis on the threat, consequence, and barrier identification, rather than on the construction of the diagram itself. The conceptual work could also be used to structure the accompanying brainstorming sessions of the Bowtie development process, which has the potential to overcome some of the limitations mentioned in Section 2.4.

The categorisation of threats and consequences may help to better address them by appropriate means. Categorising barriers in turn may help practitioners to gain a better overview of the types of barriers in place and how diverse a threat or consequence path is in terms of barrier types. Furthermore, it may help to identify appropriate and efficient barriers that prevent more than one threat or consequence path, i.e., barriers that occur on multiple paths. This could be beneficial when identifying and eliminating ineffective barriers or barriers that only prevent one path, and rather improve barriers that prevent multiple threats. This brings in a management perspective of barrier prioritisation and investment strategies.

The idea of cascading consequence could allow an organisation to break the complex MRO process down into smaller blocks and perform a risk assessment for each of these process steps with the relevant process experts. This goes along with the previously mentioned practicability of the proposed method by non-risk analysts and may improve the quality of the Bowtie diagrams.

5.3. Limitations of the Work

The proposed methodology may be of limited use when analysing novel systems outside the manufacturing and maintenance industry, where 6M originated. It is important to accept that there is not only one right solution. Every model needs to provide a certain extent of flexibility that enables it to be applicable to the broader industry. The categories need to be tailored to suit the different needs and concerns of the specific industry and organisation that is applying it [82]. This limitation was already addressed and we showed that the categories can be contextualised and adjusted to the area under investigation. It was found that the level of risk assessment plays an important role when contextualising the categories.

In other industries, different categorisations have already been applied such as the '8Ps marketing mix' or the '4S cause categories' in the service industry. Each of these categorisations could theoretically be applied to Bowtie following the principles presented in this paper. It is recommended to use a common approach to avoid arbitrary structures, which would act adversely on the attempt to provide consistency.

As mentioned in the previous section, the approach covers the most common risk areas based on previous experience. However, this involves the risk of missing Bowtie elements that have not previously occurred. The use of strictly defined categories may limit the imagination when identifying threats, consequences, or barriers. Analysts will need to ensure they are not so fixated on the method that they fail to anticipate new threats.

In some cases, the classification is not explicit as threats or barriers may fit into two of the proposed categories. From a risk point of view, it is not essential where and under which category an element is listed, as long as it is listed and brought to attention, so that it can be further analysed. In the case study, we made the decision to categorise the elements based on their nature and exerting agent.

The process of developing a Bowtie diagram following the proposed structure can be time consuming and people may focus too much on trying to fill in all gaps, although it is realistic and acceptable that there is not a threat, consequence, or barrier in each category type for every case.

There is a caveat regarding the Management category. The intent is to represent the operations management, as opposed to management-theory, leadership and vision. Consequently, the management threats shown here are aimed for an audience of operators, who have the operational knowledge to know how the integrity of the work may be compromised. Business executives normally do not know every process in detail and are not risk experts, and hence tend not to create Bowtie diagrams.

While there are many risk assessment methods (e.g., Bowtie, FTA, FMEA, Zonal analysis, and Ishikawa), and they all cope with single threats, they often struggle to represent multiple simultaneous failures. Reason [83] stated that often multiple barriers fail at the same time, which then releases the top event, and ultimately has the potential to cause severe damage or result in a catastrophe. Consequently, any type of method that fixates on identifying root causes has the intrinsic detriment of under-emphasising the temporal relationships of causality between the contributory factors. It is particularly difficult to represent how organisational factors (such as work culture) affect physical failure, since the causal mechanisms are indistinct and perhaps easier to obfuscate [84]. Many enquiries into major disasters focus on the physical root causes and the accident sequence: the organisational root causes are treated differently, are termed 'contributory factors', and are not easily representable with some diagrammatic methods. Bowtie analysis is not particularly efficient at representing complex relationships of causality, neither natively nor with the changes proposed in this paper. This is evident in the need to repeatedly represent causal chains on the diagram, hence our suggestion to use modules. It does not readily capture the more abstract organisational factors such as organisational culture and perverse agency [85]. Nonetheless, Bowtie does excel at representing the failings of the operational systems alongside the physical faults. This plus its simple depiction make it an effective communication tool by which operators can build a shared understanding (and hence a local work culture) of how their tasks contribute to a larger good. Hence, we propose that the purpose of any risk analysis tool is to capture *sufficient* complexity of the real system behaviour as to direct improvement efforts and consolidate work-culture around actions that improve safety outcomes.

5.4. Implications for Future Research

We identify the potential for future research in the following areas. Now that there is a more systematic approach for developing Bowtie, this means that there can be different representations of it, similar to a Gantt chart and a network diagram, which are complementary representations of the same project plan. While the Gantt chart is a visual representation, it can also be expressed as a table. It is conceivable that there could be a similar spreadsheet representation of Bowtie. If so, this may provide a mechanism to add additional information about the likelihoods and frequencies of the threats and the effectiveness of the barriers, and include other application critical factors. In the presented case study, these factors could include defect detectability, engine history, and other influence factors. Furthermore, the spreadsheet has the potential to calculate the risk of each threat and the overall hazard considering these factors.

A user interface could be developed for automated query of the values for the Bowtie elements, i.e., hazard, top event, threats, consequences, barriers, escalation factors and escalation factor barriers, following the proposed structure. These values might be used to automatically generate a starting Bowtie. This has the potential to generate Bowtie diagrams quicker and more efficiently. Moreover, the automation of Bowtie would allow selecting different levels of detail and presenting the most relevant elements for a target audience, or based on the likelihood and impact. This might be carried out by applying different filters in the Bowtie interface and retrieving the data from the spreadsheet accordingly. The generated Bowtie could then be limited to (say) the most important ten threats (highest risk) and the five most effective barriers of each threat and consequence path. This has not only the potential to significantly reduce the size and complexity of the Bowtie diagram, but also to further support a standardised presentation and to highlight the critical elements, where most emphasis should be put on improvement efforts.

6. Conclusions

The purpose of this research was to overcome the arbitrariness of the Bowtie methodology. This work makes several novel contributions by addressing the research purpose. Firstly, it provides a structured way of performing Bowtie analysis and constructing the diagram accordingly by following the 6M approach. This required contextualisation of the 6M categories for application in a maintenance area, which differs to a production environment. Secondly, it was applied to borescope inspection of aero engine parts and extended the risk analysis beyond the tool (borescope device), and included other relevant risks related to methods, management, material, work environment and human factors.

Supplementary Materials: The following are available online at http://www.mdpi.com/2226-4310/7/7/86/s1, Figure S1: Full Bowtie diagram with 6M prevention and mitigation barriers; Figure S2: Management-related threat paths with barriers; Figure S3: Material-related threat paths with barriers; Figure S4: Method-related threat paths with barriers; Figure S5: Man-related threat paths with barriers; Figure S6: Mother Nature-related threat paths with barriers; Figure S7: Machine-related threat paths with barriers; Figure S8: Consequence path with a 6M barrier structure.

Author Contributions: Conceptualization, J.A. and D.P.; methodology, J.A. and D.P.; software, J.A.; formal analysis, J.A. and D.P.; investigation, J.A.; resources, D.P.; data curation, J.A.; writing—original draft preparation, J.A.; writing—review and editing, J.A. and D.P.; visualisation, J.A.; supervision, D.P.; project administration, D.P.; funding acquisition, J.A. and D.P. All authors have read and agreed to the published version of the manuscript.

Funding: This research project was funded by the Christchurch Engine Centre (CHCEC), a maintenance, repair and overhaul (MRO) facility based in Christchurch and a joint venture between the Pratt and Whitney (PW) division of United Technologies Corporation (UTC) and Air New Zealand (ANZ).

Acknowledgments: We sincerely thank staff at the Christchurch Engine Centre for their support and providing insights into visual inspection and risk management. In particular, we thank Tim Coslett, Marcus Wade, Tim Fowler and Allan Moulai.

Conflicts of Interest: J.A. was funded by a PhD scholarship through this research project. The authors declare no other conflicts of interest.

Appendix A

Figure A1 shows the full Bowtie diagram with the 6M structure for threats and consequences, as well as for the colour-coded prevention and mitigation barriers.

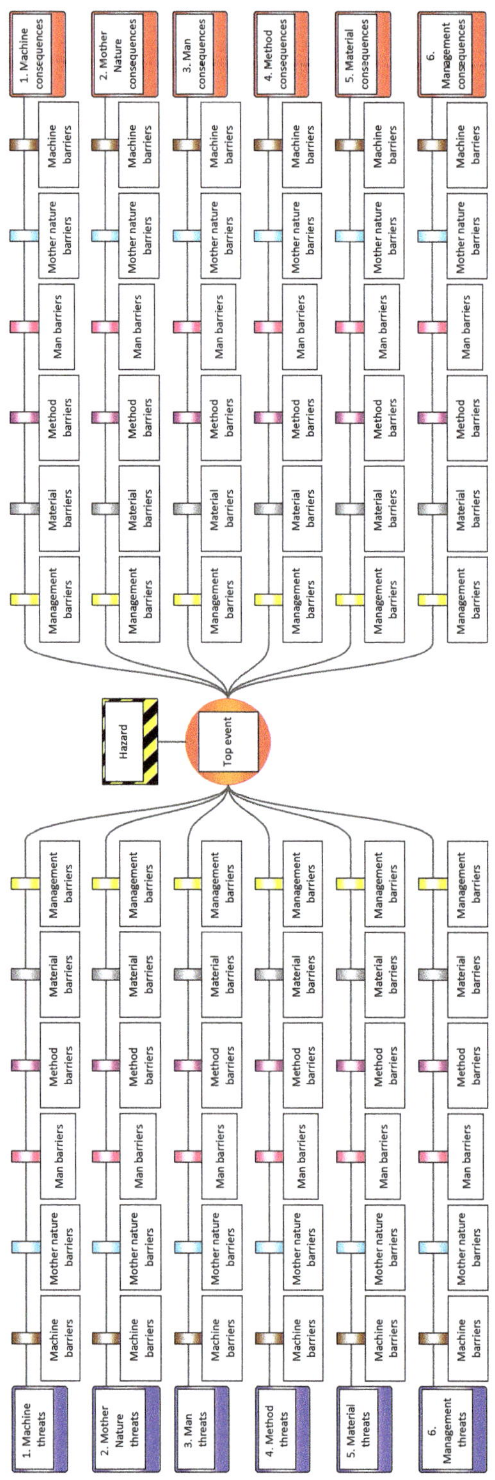

Figure A1. Full Bowtie structure with a 6M × 6M matrix on both sides of the diagram.

Figures A2–A7 show the Sub-Bowties of the threat side of the Bowtie diagram including all prevention barriers. Figure A8 presents the consequences and the prevention barriers in place.

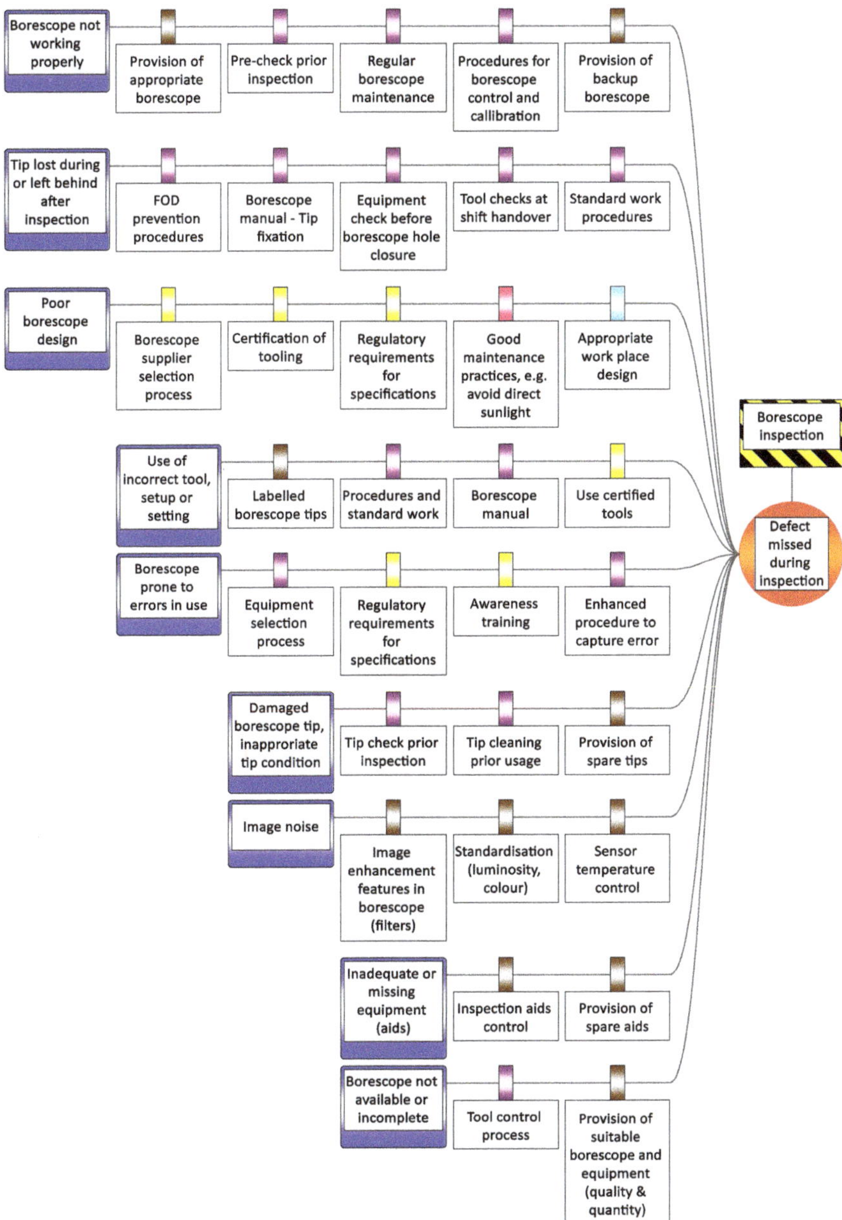

Figure A2. Machine-related threat paths with barriers.

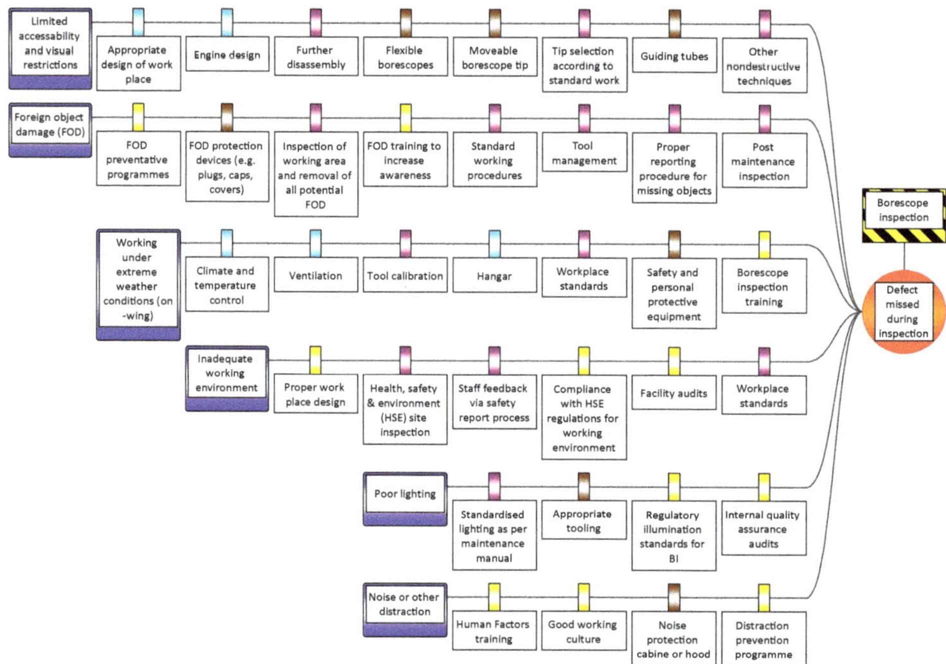

Figure A3. Mother Nature-related threat paths with barriers.

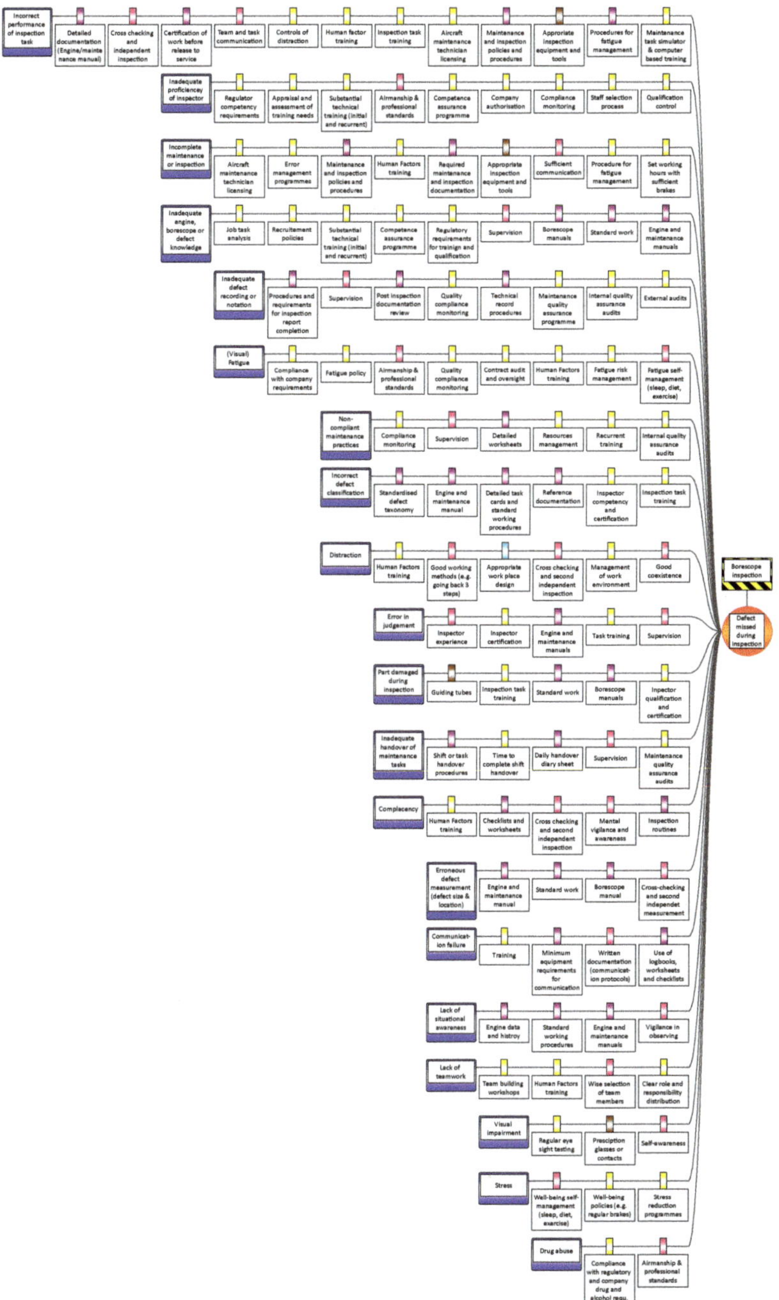

Figure A4. Man-related threat paths with barriers.

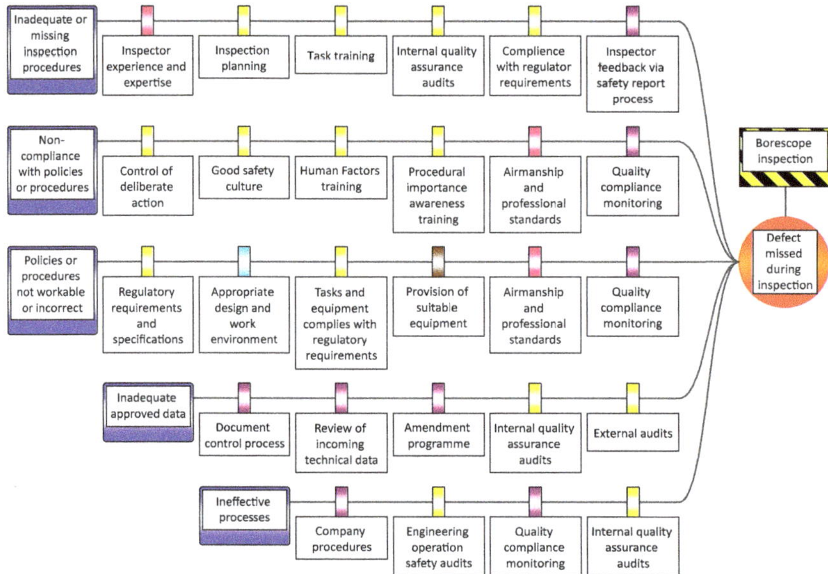

Figure A5. Method-related threat paths with barriers.

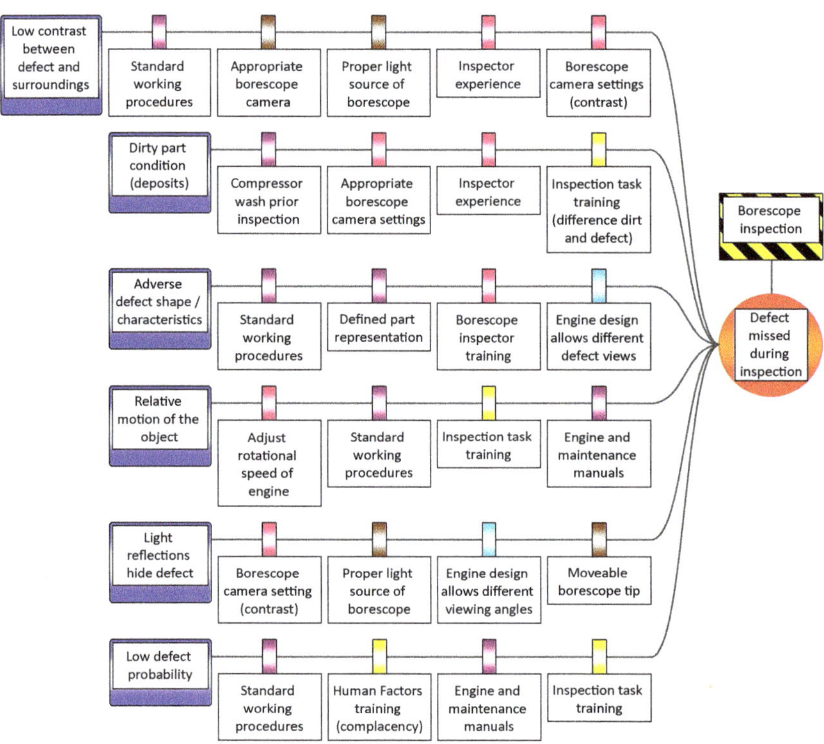

Figure A6. Material-related threat paths with barriers.

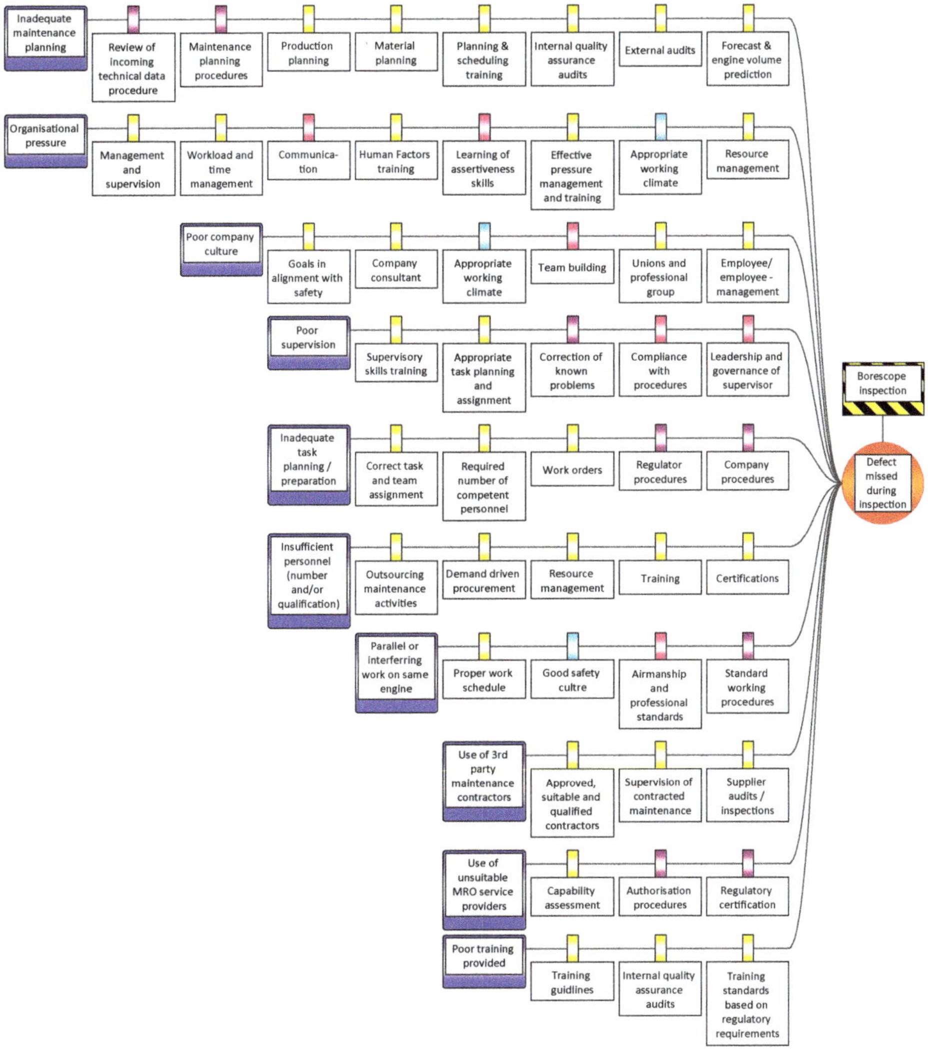

Inadequate maintenance planning
Review of incoming technical data procedure
Maintenance planning procedures
Production planning
Material planning
Planning & scheduling training
Internal quality assurance audits
External audits
Forecast & engine volume prediction
Organisational pressure
Management and supervision
Workload and time management
Communica-tion
Human Factors training
Learning of assertiveness skills
Effective pressure management and training
Appropriate working climate
Resource management
Poor company culture
Goals in alignment with safety
Company consultant
Appropriate working climate
Team building
Unions and professional group
Employee/ employee - management
Poor supervision
Supervisory skills training
Appropriate task planning and assignment
Correction of known problems
Compliance with procedures
Leadership and governance of supervisor
Borescope inspection
Inadequate task planning / preparation
Correct task and team assignment
Required number of competent personnel
Work orders
Regulator procedures
Company procedures
Defect missed during inspection
Insufficient personnel (number and/or qualification)
Outsourcing maintenance activities
Demand driven procurement
Resource management
Training
Certifications
Parallel or interferring work on same engine
Proper work schedule
Good safety cultre
Airmanship and professional standards
Standard working procedures
Use of 3rd party maintenance contractors
Approved, suitable and qualified contractors
Supervision of contracted maintenance
Supplier audits / inspections
Use of unsuitable MRO service providers
Capability assessment
Authorisation procedures
Regulatory certification
Poor training provided
Training guidlines
Internal quality assurance audits
Training standards based on regulatory requirements

Figure A7. Management-related threat paths with barriers.

Borescope inspection
Defect missed during inspection
Inspection polices and procedures
Inspection task sign off
Independent cross-check
Piece part inspection
Other NDI methods
Installer checks
Engine testing
Post-test inspection
Quality control systems
Release of unairworthy part to service

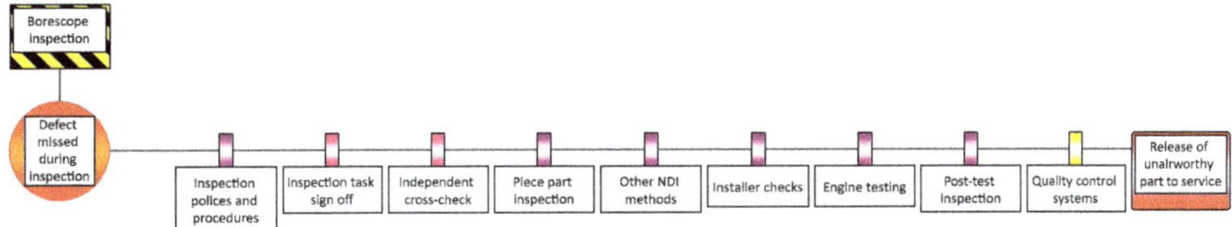

Figure A8. Consequence path with mitigation barriers.

References

1. Shahriar, A.; Sadiq, R.; Tesfamariam, S. *Risk Analysis for Oil & Gas Pipelines: A Sustainability Assessment Approach Using Fuzzy based Bow-Tie Analysis*; Elsevier: Amsterdam, The Netherlands, 2012; Volume 25, pp. 505–523. [CrossRef]
2. De Ruijter, A.; Guldenmund, F. The bowtie method: A review. *Saf. Sci.* **2016**, *88*, 211–218. [CrossRef]
3. Federal Aviation Administration (FAA). Bow-Tie Analysis. Available online: http://www.hf.faa.gov/workbenchtools/default.aspx?rPage=Tooldetails&subCatId=43&toolID=21 (accessed on 1 October 2018).
4. Civil Aviation Authority (CAA). Bowtie Risk Assessment Models. Available online: https://www.caa.co.uk/Safety-Initiatives-and-Resources/Working-with-industry/Bowtie/ (accessed on 25 September 2018).
5. European Aviation Safety Agency (EASA). *The European Plan for Aviation Safety (EBAS) 2018–2022*; European Aviation Safety Agency: Cologne, Germany, 2018.
6. Civil Aviation Authority (CAA) of New Zealand. Making Safe Aviation Even Safer. In *Civil Aviation Authority Sector Risk profile of Medium and Large Aircraft Air Transport*; CAA: Wellington, New Zealand, 2017.
7. Aust, J.; Pons, D. Bowtie Methodology for Risk Analysis of Visual Borescope Inspection during Aircraft Engine Maintenance. *Aerospace* **2019**, *6*, 110. [CrossRef]
8. Delmotte, F. A Sociotechnical Framework for the Integration of Human and Organizational Factors in Project Management and Risk Analysis. Master's Thesis, Virginia Tech, Blacksburg, VA, USA, 2003.
9. Brown, K. Review of the South Island Rail Coal Route—MET351-X-REP-001. Available online: www.ltsa.govt.nz/rail/coal-route (accessed on 27 September 2018).
10. Van Scyoc, K.; Hughes, G. Rail ruminations for process safety improvement. *J. Loss Prev. Process Ind.* **2009**, *22*, 689–694. [CrossRef]
11. Papazoglou, I.A.; Bellamy, L.J.; Hale, A.R.; Aneziris, O.N.; Ale, B.J.M.; Post, J.G.; Oh, J.I.H. I-Risk: Development of an integrated technical and management risk methodology for chemical installations. *J. Loss Prev. Process Ind.* **2003**, *16*, 575–591. [CrossRef]
12. De Dianous, V.; Fievez, C. ARAMIS project: A more explicit demonstration of risk control through the use of bow–tie diagrams and the evaluation of safety barrier performance. *J. Hazard. Mater.* **2006**, *130*, 220–233. [CrossRef]
13. Burgess-Limerick, R.; Horberry, T.; Steiner, L. Bow-tie analysis of a fatal underground coal mine collision. *Ergon. Aust.* **2014**, *10*, 1–5.
14. Abdo, H.; Kaouk, M.; Flaus, J.M.; Masse, F. A safety/security risk analysis approach of Industrial Control Systems: A cyber bowtie combining new version of attack tree with bowtie analysis. *Comput. Secur.* **2018**, *72*, 175–195. [CrossRef]
15. Bernsmed, K.; Frøystad, C.; Meland, P.H.; Nesheim, D.A.; Rødseth, Ø.J. Visualizing cyber security risks with bow-tie diagrams. In *International Workshop on Graphical Models for Security*; Springer: Berlin/Heidelberg, Germany, 2017; pp. 38–56.
16. Faulkner, A.; Nicholson, M. *Data-Centric Safety: Challenges, Approaches, and Incident Investigation*; Elsevier: Amsterdam, The Netherlands, 2020.
17. Culwick, M.D.; Merry, A.F.; Clarke, D.M.; Taraporewalla, K.J.; Gibbs, N.M. Bow-Tie Diagrams for Risk Management in Anaesthesia. *Anaesth. Intensive Care* **2016**, *44*, 712–718. [CrossRef]
18. Abdi, Z.; Ravaghi, H.; Abbasi, M.; Delgoshaei, B.; Esfandiari, S. Application of Bow-tie methodology to improve patient safety. *Int. J. Health Care Qual. Assur.* **2016**, *29*, 425–440. [CrossRef]
19. Janssen, E. Patient Safety BowTies. Available online: http://www.patientsafetybowties.com/knowledge-base/5-why-bowties-in-healthcare (accessed on 13 February 2020).
20. Badreddine, A.; Amor, N.B. A Bayesian approach to construct bow tie diagrams for risk evaluation. *Process Saf. Environ. Prot.* **2013**, *91*, 159–171. [CrossRef]
21. Maragakis, I.; Clark, S.; Piers, M.; Prior, D.; Tripaldi, C.; Masson, M.; Audard, C. Guidance on Hazard Identification. In *Safety Management System and Safety Culture Working Group (SMSWG)*; European Commercial Aviation Safety Team (ECAST): Cologne, Germany, 2009; pp. 6–8.
22. Acquisition Safety & Environmental Management System (ASEMS). Bow-Tie Diagram. Available online: https://www.asems.mod.uk/content/bow-tie-diagram (accessed on 4 April 2020).
23. Ishikawa, K. *Introduction to Quality Control*; Productivity Press: New York, NY, USA, 1990.

24. Lewis, S. Lessons Learned from Real World Application of the Bow-tie Method. In Proceedings of the 6th Global Congress on Process Safety, Antonio, TX, USA, 22–24 March 2010; Available online: https://www.aiche.org/academy/videos/conference-presentations/lessons-learned-real-world-application-bow-tie-method (accessed on 27 October 2018).
25. Civil Aviation Authority (CAA). CAA 'Significant Seven' Task Force Reports. In *CAA PAPER 2011/03*; CAA: London, UK, 2011. Available online: https://publicapps.caa.co.uk/docs/33/2011_03.pdf (accessed on 25 October 2018).
26. Joint Industry Program. The BowTie Examples Library. In *Joint Industries Project*; CGE Risk Management Solutions: Leidschendam, The Netherlands, 2016.
27. International Civil Aviation Organization (ICAO). *Safety Management Manual (SMM)*, 3rd ed.; ICAO: Montreal, QC, Canada, 2012.
28. Sklet, S. Safety barriers: Definition, classification, and performance. *J. Loss Prev. Process Ind.* **2006**, *19*, 494–506. [CrossRef]
29. Badreddine, A.; Romdhane, T.B.; HajKacem, M.A.B.; Amor, N.B. A new multi-objectives approach to implement preventive and protective barriers in bow tie diagram. *J. Loss Prev. Process Ind.* **2014**, *32*, 238–253. [CrossRef]
30. Jacinto, C.; Silva, C. A semi-quantitative assessment of occupational risks using bow-tie representation. *Saf. Sci.* **2010**, *48*, 973–979. [CrossRef]
31. Visser, J.P. Developments in HSE management in oil and gas exploration and production. *Saf. Manag. Chall. Chang.* **1998**, *1*, 43–66.
32. Manton, M.; Moat, A.; Ali, W.; Johnson, M.; Cowley, C. Representing Human Factors in Bowties as per the new CCPS/EI Book. In Proceedings of the CCPS Middle East Conference on Process Safety, Sanabis, Bahrain, 10 October 2017.
33. Delvosalle, C.; Fiévez, C.; Pipart, A. ARAMIS Project: Reference Accident Scenarios Definition in SEVESO Establishment. *J. Risk Res.* **2006**, *9*, 583–600. [CrossRef]
34. Hamzah, S. Use bow tie tool for easy hazard identification. In Proceedings of the 14th Asia Pacific Confederation of Chemical Engineering Congress, Singapore, 21–24 February 2012.
35. Flight Safety Foundation. Basic Aviation Risk Standard. In *Offshore Helicopter Operations*; Flight Safety Foundation: Alexandria, VA, USA, 2015.
36. Kang, J.; Zhang, J.; Gao, J. Analysis of the safety barrier function: Accidents caused by the failure of safety barriers and quantitative evaluation of their performance. *J. Loss Prev. Process Ind.* **2016**, *43*, 361–371. [CrossRef]
37. Neogy, P.; Hanson, A.; Davis, P.; Fenstermacher, T. *Hazard and barrier analysis guidance document. Department of Energy, Office of Operating Experience Analysis and Feedback, Report No. EH-33*; Springer: Berlin/Heidelberg, Germany, 1996.
38. Chevreau, F.R.; Wybo, J.L.; Cauchois, D. Organizing learning processes on risks by using the bow-tie representation. *J. Hazard. Mater.* **2006**, *130*, 276–283. [CrossRef]
39. Wilson, P.F. *Root Cause Analysis: A Tool for Total Quality Management*; Quality Press: Milwaukee, WI, USA, 1993.
40. Hall, D.; Hulett, D.; Graves, R. *Universal Risk Project—Final Report*; PMI Risk SIG: Newtown Township, PA, USA, 2002.
41. Rose, K.H. *A Guide to the Project Management Body of Knowledge (PMBOK® Guide)*, 5th ed.; Project Management Institute: Hampton, VA, USA, 2013; Volume 44.
42. Hillson, D. Use a risk breakdown structure (RBS) to understand your risks. In Proceedings of the project Management Institute Annual Seminars & Symposium, San Antonio, TX, USA, 3–10 October 2002.
43. Tanim, M.M.Z. *Risk Management in International Business Handbook*; 2019.
44. Lester, A. *Project Management, Planning and Control: Managing Engineering, Construction and Manufacturing Projects to PMI, APM and BSI Standards*; Elsevier Science & Technology: Saint Louis, UK, 2013.
45. Chung, W.; Zhu, M. Risk Assessment Based on News Articles: An Experiment on IT Companies. In Proceedings of the International Conference on Information Systems (ICIS 2012), Orlando, FL, USA, 16–19 December 2012.
46. Deshpande, P. Study of Construction Risk Assessment Methodology for Risk Ranking. In Proceedings of the International Conference on Education, E Learning and Life Long Learning, Kuala Lumpur, Maleysia, 16–17 November 2015.

47. Sandle, T. *Approaching Risk Assessment: Tools and Methods; Newsletter*; Global Biopharmaceutical Resources Inc.: Clarksburg, MD, USA, 2012; Volume 1, pp. 1–23.
48. Kurian, G.T. *The AMA Dictionary of Business and Management*; AMACOM: Nashville, TN, USA, 2013.
49. Pons, D. Strategic Risk Management: Application to Manufacturing. *Open Ind. Manuf. Eng. J.* **2010**, *3*, 13–29. [CrossRef]
50. International Organization for Standardization (ISO). Information Technology. In *Security Techniques—Information Security Management Systems Requirements*; Joint Technical Committee ISO/IEC JTC1. Subcommittee SC 27; International Organization for Standardization (ISO): Geneva, Switzerland, 2013.
51. Jouini, M.; Ben, L.; Ben Arfa Rabai, L.; Aissa, A. Classification of security threats in information systems. *Procedia Comput. Sci.* **2014**, *32*, 489–496. [CrossRef]
52. Paradies, M. *TapRoot—Root Cause Tree Dictionary*; System Improvements, Inc.: Knoxville, TS, USA, 2015; Volume 9.
53. International Air Transport Association (IATA). *Safety Report*; International Air Transport Association (IATA): Geneva, Switzerland; Montreal, QC, Canada, 2006.
54. Liliana, L. A new model of Ishikawa diagram for quality assessment. In *IOP Conference Series: Materials Science and Engineering*; IOP Publishing: Bristol, UK, 2016; Volume 161, p. 012099.
55. Radziwill, N. Creating Ishikawa (Fishbone) Diagrams With R. *Softw. Qual. Prof.* **2017**, *20*, 47–48.
56. Hristoski, I.; Kostoska, O.; Kotevski, Z.; Dimovski, T. Causality of Factors Reducing Competitiveness of e-Commerce Firms. *Balk. Near East. J. Soc. Sci.* **2017**, *3*, 109–127.
57. Eckes, G. *Six Sigma for Everyone*; John Wiley & Sons: Hoboken, NJ, USA, 2003.
58. Burch, R.F.; Strawderman, L.; Bullington, S.F. Global corporation rollout of ruggedised handheld devices: A Lean Six Sigma case study. *Total Qual. Manag. Bus. Excell.* **2016**, *27*, 1–16. [CrossRef]
59. Bradley, E. *Reliability Engineering: A Life Cycle Approach*; CRC Press: Boca Raton, FL, USA, 2016.
60. See, J.E. *Visual Inspection: A Review of the Literature*; Sandia National Laboratories: Albuquerque, NM, USA, 2012.
61. Wiegmann, D.A.; Shappell, S.A. *A Human Error Approach to Aviation Accident Analysis: The Human Factors Analysis and Classification System*; Routledge: London, UK, 2017.
62. U.S. Department of Transportation. FAA-H-8083-32A, Federal Aviation Administration (FAA) Session. In *Aviation Maintenance Technician Handbook—Powerplant*; FAA: Washington, DC, USA, 2018; Volume 2.
63. Wiener, E.; Nagel, D. *Human Factors in Aviation*; Academic Press Limited: London, UK, 1988.
64. Gong, L.; Zhang, S.; Tang, P.; Lu, Y. An integrated graphic–taxonomic–associative approach to analyze human factors in aviation accidents. *Chin. J. Aeronaut.* **2014**, *27*, 226–240. [CrossRef]
65. Said, M.; Mokhtar, A. Significant human risk factors in aviation maintenance. *Sains Hum.* **2014**, *2*, 31–34.
66. Chang, Y.H.; Wang, Y.C. Significant human risk factors in aircraft maintenance technicians. *Saf. Sci.* **2010**, *48*, 54–62. [CrossRef]
67. Johnson, W.; Maddox, M. A PEAR shaped model for better human factors. *Cat Mag.* **2007**, *2*, 20–21.
68. CGE Risk. *BowtieXP*; CGE Risk: Leidschendam, The Netherlands, 2019.
69. Wilken, M.; Hüske-Kraus, D.; Klausen, A.; Koch, C.; Schlauch, W.; Röhrig, R. Alarm Fatigue: Causes and Effects. In Proceedings of the GMDS, Oldenburg, Germany, 17–21 September 2017; pp. 107–111.
70. Wong, K.C. Using an Ishikawa diagram as a tool to assist memory and retrieval of relevant medical cases from the medical literature. *J. Med. Case Rep.* **2011**, *5*, 120. [CrossRef]
71. Hessing, T. 6M's in Six Sigma (Six Ms or 5Ms and one P or 5M1P). Available online: https://sixsigmastudyguide.com/six-ms-6ms-or-5ms-and-one-p-5m1p/ (accessed on 15 February 2020).
72. Boca, G.D. 6M in Management Education. *Procedia Soc. Behav. Sci.* **2015**, *182*, 4–9. [CrossRef]
73. Vaanila, T. Process Development Using the Lean Six Sigma Methodology: Case: Oy AGA Ab, Linde Healthcare. Bachelor's Thesis, HAMK Häme University of Applied Sciences, Hämeenlinna, Finland, 2015.
74. Gwiazda, A. Quality tools in a process of technical project management. *J. Achiev. Mater. Manuf. Eng.* **2006**, *18*, 439–442.
75. Levine, M.E. Alternatives to regulation: Competition in air transportation and the Aviation Act of 1975. *J. Air Law Commer.* **1975**, *41*, 703.
76. Eceral, T.Ö.; Köroğlu, B.A. Incentive mechanisms in industrial development: An evaluation through defense and aviation industry of Ankara. *ProcediaSoc. Behav. Sci.* **2015**, *195*, 1563–1572. [CrossRef]

77. Fraser, J.; Simkins, B. *Enterprise Risk Management: Today's Leading Research and Best Practices for Tomorrow's Executives*; John Wiley & Sons: Hoboken, NJ, USA, 2010; Volume 3.
78. Rankin, W. MEDA Investigation Process. In *Boeing Commercial Aero*; Boeing: Chicago, IL, USA, 2007.
79. Federal Aviation Administration (FAA). Addendum: Chapter 14—Human Factors (PDF). In *Published Separately from Aviation Maintenance Technician Handbook FAA-H-8083-30*; FAA: Washington, DC, USA, 2018.
80. Christianson, M.K.; Sutcliffe, K.M.; Miller, M.A.; Iwashyna, T.J. Becoming a high reliability organization. *Crit Care* **2011**, *15*, 314. [CrossRef] [PubMed]
81. Sutcliffe, K.M. High reliability organizations (HROs). *Best Pract. Res. Clin. Anaesthesiol.* **2011**, *25*, 133–144. [CrossRef]
82. The Institute of Operational Risk (IOR). Risk Categorisation. Available online: https://www.ior-institute.org/sound-practice-guidance/risk-categorisation (accessed on 2 May 2020).
83. Reason, J. *Human Error*; Cambridge university press: Cambridge, UK, 1990.
84. Pons, D. Pike river mine disaster: Systems-engineering and organisational contributions. *Safety* **2016**, *2*, 21. [CrossRef]
85. Ji, Z.; Pons, D.; Pearse, J. Why do workers take safety risks?—A conceptual model for the motivation underpinning perverse agency. *Safety* **2018**, *4*, 24. [CrossRef]

Article

Integration-In-Totality: The 7th System Safety Principle Based on Systems Thinking in Aerospace Safety

Johney Thomas [1,2,*,†], Antonio Davis [2,†] and Mathews P. Samuel [3,†]

1 Hindustan Aeronautics Limited, LCA-Tejas Division, Bengaluru 560037, India
2 International Institute for Aerospace Engineering & Management, Jain (Deemed-to-be University), Bengaluru, Karnataka 562112, India; antonio.davis@jainuniversity.ac.in
3 Regional Centre for Military Airworthiness (Engines), CEMILAC, DRDO, Bengaluru 560093, India; drmatmail@gmail.com
* Correspondence: johney.thomas@hal-india.co.in
† These authors contributed equally to this work.

Received: 1 September 2020; Accepted: 9 October 2020; Published: 14 October 2020

Abstract: Safety is of paramount concern in aerospace and aviation. Safety has evolved over the years, from the technical era to the human-factors era and organizational era, and finally to the present era of systems-thinking. Building upon three foundational concepts of systems-thinking, a new safety concept called "integration-in-totality principle" is propounded in this article as part of a "seven-principles-framework of system safety", to act as an integrated framework to visualize and model system safety. The integration-in-totality principle concept addresses the need to have a holistic 'vertical and horizontal integration', which is a key tenet of systems thinking. The integration-in-totality principle is illustrated and elucidated with the help of a simple "Rubik's cube model of integration-in-totality principle" with three orthogonal axes, the 'axis of perspective' of vertical integration, and the two 'axes of perception and performance' of horizontal integration. Safety analysis along the three axes with a 'bidirectional synthesis' and 'continuum approach' is further elaborated with relevant case studies, one among them related to the Boeing 737 MAX aircraft twin disasters. Safety is directly linked to quality, reliability and risk, through a self-reinforcing reflexive paradigm, and airworthiness assurance is the process through which safety concepts are embedded in a multidisciplinary aviation environment where the system of systems is seamlessly operating. The article explains how the system safety principle of integration-in-totality is related to reliability and airworthiness of an aerospace system with the help of the 'V-model of systems engineering'. The article also establishes the linkage between integration-in-totality principle and strategic quality management, thus bridging the gap between two parallel fields of knowledge.

Keywords: integration-in-totality principle; seven-principles-framework of system safety; systems thinking; systems engineering; system safety principles; strategic quality management; risk management; reliability; airworthiness

1. Introduction

Accidents and serious incidents continue to occur in the field of aviation and no further emphasis is required on the requirement to abate potential hazards in aviation systems. Though the probability of accidents has come down over the years, the severity of the consequences of an aviation accident can be catastrophic, as seen in the case of the two Boeing 737 MAX aircraft disasters at Indonesia and Ethiopia in October 2018 and March 2019, respectively, that resulted in the tragic deaths of 346 people. This has brought about a renewed focus on safety as the paramount cause of concern in both civil and military aviation, and an important knowledge field for study and action.

The concept of safety has evolved over the years, from the technical era to the human-factors era and the organizational era, and finally to the present era of systems-thinking [1–9]. In order to account for the nuances of safety concepts in the context of the modern complex aerospace systems, a "seven-principles-framework of system safety principles" has been developed by the authors. This proposed framework is built upon the five system safety principles (comprising of fail-safe, safety-margin, ungraduated-response, defence-in-depth and observability-in-depth principles) conceptualized by Saleh et al. [6], with the addition of the 'human-factors principle' as the 6th system safety principle, and a newly developed concept called "integration-in-totality principle" as the 7th system safety principle. Thus, in the remaining sections of this article, the authors will discuss the new safety concepts of the *'integration-in-totality principle'*, as well as the *'seven-principles-framework of system safety principles'* to which it belongs. These new concepts are meant to enhance the understanding on safety-critical socio-technical systems in their entirety, incorporating the key tenets of systems thinking.

Before getting into the details of the *'integration-in-totality principle'*, it is worthwhile to present a brief background, and the need to have a fresh outlook and an augmentation of the existing concepts. It is a well-established fact that there exist multiple root causes and failure modes in real life complex systems that could complement each other. According to Latino [4], the three basic types of causes are: (i) Technical/physical: the actual physical mechanism of the failure; (ii) Human: the human practices that allowed the physical root causes to exist; and (iii) Latent: the way a facility is managed and/or designed that creates the human root causes. Often the physical roots lead to the multiple human and latent roots, and hence it is important to truly understand the physical roots of a failure to find the larger causes. This has been pointed out by the authors [1,2] among other researchers [3,4]. According to Hulme et al. [5], "there is a need to update our understanding of the different viewpoints of the systems-thinking approach, upgrade the accident analysis methodologies to a unified one, and further explore the opportunities towards development of a novel comprehensive accident analysis approach". The development of the 'integration-in-totality principle' is a forward step in that direction.

Even though the five system safety principles suggested by Saleh et al. [6,7] are effective in illuminating the technical/technological/physical aspects of accident causation and understanding the preventive measures thereof, they are not addressing the human-factors and organizational aspects of system safety to capture the human and latent root causes. Hence the authors have included the 'human-factors principle' popularized by International Civil Aviation Organization (ICAO) [8,9] as the 6th system safety principle, in addition to the five basic/technical system safety principles of Saleh et al. The importance of the human factors principle as one of the cardinal principles of system safety needs no further emphasis, and especially in aviation activities one should adopt and train their personnel in the human factors principle. Furthermore, the authors felt the need to include one more system safety principle to suitably address the latent root causes based on systems thinking and systems engineering, in order to make the principles more comprehensive. Hence the new system safety principle of 'integration-in-totality principle' has been propounded as the 7th system safety principle, to take care of the systems-theoretic aspects of accident analysis and prevention. Before proceeding to the details of the new concepts, an overview of the organization of this paper is presented below to give a broader perspective of the discussions.

At the outset, in Section 2, an elaborate discussion on the proposed framework called the *'Seven-Principles-Framework of System Safety Principles'* is provided. Having discussed the broad framework, the *'Integration-in-Totality Principle'*, the 7th system safety principle newly introduced by the authors, is explained in detail in Section 3. The 'integration-in-totality principle' is illustrated and elucidated in this Section with the help of the simile of a "Rubik's cube model" having three orthogonal axes, viz. "axis of perspective", "axis of perception", and "axis of performance". The two properties of this model called the "continuum approach" and the "bidirectional synthesis" along the three axes are also discussed in Section 3. Now it is required to discuss the connection and linkage of the integration-in-totality principle with the system thinking domain, and the same has been taken up in Section 4. The authors identified "five key tenets of systems-thinking" and mapped them against the present set of system safety principles, which revealed and amplified the need to add not only the 'human-factors principle' to take care of the human aspects, but also the 'integration-in-totality principle' to take care of the key systems thinking tenet of "vertical and horizontal integration".

The 'theoretical foundation' of the 'integration-in-totality principle' is presented in Section 5. It is comprised of three foundational concepts of systems thinking, viz. the 'abstraction hierarchy' proposed by Rasmussen [10–14], the 'design-control-practice (DCP) diagram' of Stoop [15–18], and the 'mental models in systems-theoretic framework' described by Leveson [19–21], which are related to the 'axis of perspective', 'axis of performance' and 'axis of perception', respectively.

Having elaborated the integration-in-totality principle from a theoretical angle, it is pertinent to present a few case studies chosen to demonstrate how it could be implemented, and the same has been taken up in subsequent sections. Section 6 further elaborates on the "macro-meso-micro levels of vertical integration" along the 'axis of perspective' and provides a case study on the application of the concept in defect investigation and failure analysis of an aero-engine component. Section 7 is devoted to elucidating the significance of perception and mental models in aviation safety, which has not been explored in the safety literature to the fullest extent. Analysis and understanding of an accident or a safety event along the 'axis of perception' of the 'integration-in-totality principle' can remove the distortions in perceptions, and thus help find the truth in any given situation. The recent aviation twin disasters of Boeing 737 MAX aircraft have been analyzed as a case study to illustrate the application of the understanding along the path of "intent-execution-manifestation" in the 'axis of perception'.

Section 8 is meant for explaining the usefulness of 'bi-directional synthesis' along the "design-manufacture-operation" life-cycle continuum. A case study based on the analysis of the test data of 200 aero-engines along the reverse path of test-assembly-manufacture helped improving the engine performance and safety, by working back on the assembly procedures of the compressor modules and manufacturing practices of the compressor blades.

A model on 'quality-reliability-risk-safety paradigm' is presented in Section 9 to highlight the relationship between these four aspects so fundamental to the aerospace and aviation field. Section 10 presents a very interesting analysis of the 'V-model of systems engineering' mapped with respect to the axes of perspective, perception and performance of the 'integration-in-totality principle', thus establishing the applicability of the 'integration-in-totality principle' in the field of reliability analysis and airworthiness certification. Section 11 narrates the suitability of the 'integration-in-totality principle' in risk management.

Finally, Section 12 establishes the linkage between two emerging and parallel fields of knowledge, viz. 'systems thinking and system safety principles', and 'strategic quality management'. It is diagrammatically shown how the 'integration-in-totality principle', developed by the authors as the 7th system safety principle based on 'systems thinking in safety', can be used as a pivotal concept in 'strategic quality management'. The technical discussion is concluded in Section 13.

2. System Safety Principles and the Seven-Principles-Framework

System safety principles are general, high-level, domain-independent and technologically-agnostic principles, adoptable as detailed safety measures for dealing with various safety hazards. Once incorporated, the system safety principles are expected to vastly improve the safety of socio-technical systems. The five basic/technical system safety principles, originally formalized by Saleh et al., and built upon the notion of the level of hazard and its escalation along the path of accident causation [6,7], are described below:

(1) The fail-safe principle [22] mandates that the system design should prevent or mitigate the unsafe consequences of the failure of a system;

(2) The safety margin principle [23] requires that features be put in place to maintain the operational conditions and the associated hazard level at some "distance" away from the estimated critical hazard threshold or accident-triggering threshold;

(3) The ungraduated response principle [24] posits that the first course of action to explore for accident prevention and mitigation is the possibility of eliminating a hazard altogether, regardless of the extent of its belligerence, using creativity and technical ingenuity
(4) The defence-in-depth principle [25–27] calls for safety protection by means of multiple lines of defences or safety barriers along the potential accident sequences.
(5) The observability-in-depth principle [26,27] requires that various features be put in place to observe and monitor for the system state and breaches of any safety barrier, and reliably provide this feedback to the operators, so that all safety-degrading events or states (that the safety barriers are meant to protect against) are observable.

In order to have a comprehensive set of safety principles, a "seven-principles-framework of system safety principles" has been developed by the authors, which is shown in Figure 1.

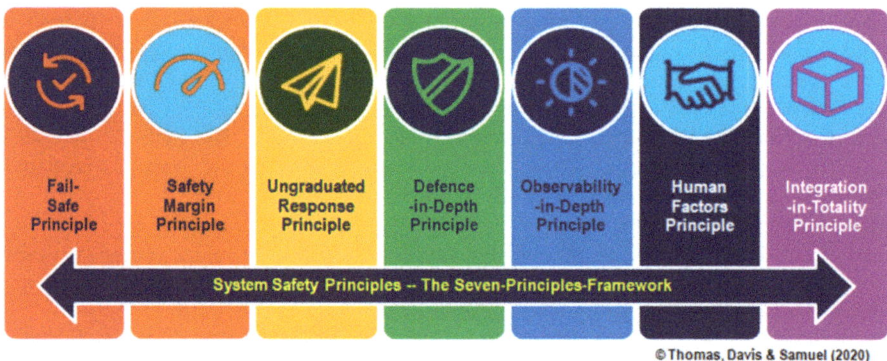

Figure 1. The Seven-Principles-Framework of System Safety Principles.

Here, two additional system safety principles have been added over and above the aforementioned "basic/technical five", covering the human and systemic aspects of system safety. These "additional two" included in the 'seven-principles-framework of system safety principles' are:

(1) The human factors principle [8,9,28] which calls for due consideration of the pivotal resource of human personnel in a production system, and their interaction with the other resources or factors of production including the other human beings, for smooth and effective functioning of the system.
(2) The integration-in-totality principle, which the authors expound in this article, requires that every aspect in a socio-technical system be integrated vertically and horizontally. Furthermore, it views, analyzes and understands the system bi-directionally along the continuum of three axes of perspective, perception, and performance, to have necessary cohesiveness in operations with convergence of purpose in safety.

There are different ways to comprehend and appreciate integration as a systems requirement. In general, one can select any one of the three basic approaches or their combinations towards achieving integration in a system. The first approach is the "interface approach" in accordance with the 'SHELL model' [8], which endeavors perfect interface and smooth interaction between the 'liveware' (meaning human-beings) and the remaining workplace elements/components of software, hardware, environment, and other liveware. The second approach is the "resource approach" as per the '5M model' [9], based on the interplay between various resources viz. man, machine, medium, mission and management. These two approaches form part of the 'human factors principle', which is propounded by the International Civil Aviation Organization (ICAO) as part of the safety management system, as elaborated in the ICAO Safety Management Manual [8].

However, in the current context, the authors are focusing on a third and perhaps the most important approach, which can be called the "continuum approach", which has not been adequately captured in the safety literature. The details of the 'integration-in-totality principle', developed based on the continuum approach of systems thinking in aerospace Safety, is further elaborated in the next section.

3. Integration-In-Totality Principle and the Rubik's Cube Model

The "integration-in-totality principle" proposed by the authors calls for viewing, analyzing and understanding socio-technical systems bi-directionally along three axes, viz. (i) the axis of performance, (ii) the axis of perception, and (iii) the axis of perspective. Though conceptually appealing, the integration of these diverse dimensions needs further illustration and elucidation.

In order to illustrate and illuminate the 'integration-in-totality principle' wherein three dimensions of organizational continuum along three axes have been integrated together, the authors have developed a "Rubik's cube model of integration-in-totality principle", as shown in Figure 2.

The first dimension of continuum, the "axis of perspective", represent the "macro-meso-micro" levels of systems thinking in the conventional 'vertical integration' approach, which can have many different interpretations depending upon the context. They could include the continuum permeating the echelons of regulatory command to management control to operator compliance (command-control-compliance), the purpose-function-equipment comprehension, or a system-subsystems-components level understanding. It allows one to migrate from, and bi-directionally navigate between, a bird's eye-view of wider and general understanding to a worm's eye-view of closer and detailed look.

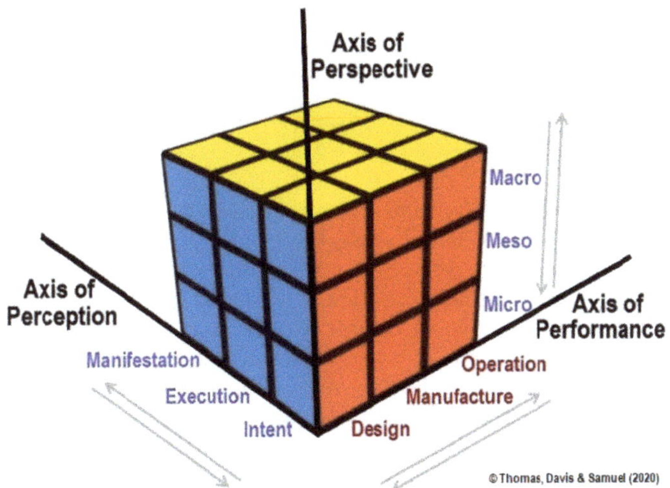

Figure 2. The Rubik's Cube Model of Integration-in-Totality Principle.

The second dimension of continuum, the "axis of perception", consists of the "intent-execution-manifestation" pathway which the authors propose here in this article as a novel concept of "horizontal integration" in systems thinking, in addition to the 'vertical integration'. The axis of perception reflects the perceptions and mental models being maintained by different participants in the system.

Finally, the third dimension of continuum, the "axis of performance", comprises of the major stages in the product life cycle, viz. design, manufacture and operation. This can further be telescopically expanded, as the need arises, into a design-development-manufacture-assembly-testing-operation-modification continuum. The 'axis of performance' provides an additional orthogonal element of "horizontal integration" in systems thinking.

Thus, the 'integration-in-totality principle' captures the essence of an integrated "continuum approach" along the three axes of perspective (macro-meso-micro), perception (intent-execution-manifestation), and performance (design-manufacture-operation). The integration-in-totality principle is proposed as a stand-alone principle, along with the five basic/technical system safety principles proposed by Saleh et al. [6], and the human factors principle popularized by ICAO [8], within the 'seven-principles-framework of system safety principles'.

The traversal from the highest level to the lowest level and then back to the highest level, like that from the bird's eye-view to the worm's eye-view and vice-versa, can be called "bi-directional synthesis", which is in fact applicable along each of the three axes, viz. 'axes of perspective, perception, and performance'. This property reinforces the dynamics of the 'continuum approach'. The 'bi-directional synthesis' is represented by the bi-directional arrows shown along each of the continuum axes in the 'Rubik's cube model of integration-in-totality principle'.

The bidirectional interplay between the three axes of continuum, viz. the 'axis of perspective' providing the vertical integration, and the two orthogonal 'axes of perception and performance' giving the horizontal integration, is at the core of the dynamics of the 'integration-in-totality principle'. 'Integration-in-totality principle' can be particularly useful in the realm of safety investigations, since the analysis along the 'axis of perspective' of vertical integration can take care of the factors that are typically found at the higher echelons of a socio-technical system, like the command and policies of the regulatory agencies, and the control and practices of the company management, which are not fully captured by the present set of accident analysis models. Furthermore, the analysis along the 'axis of perception' and 'axis of performance', the two orthogonal axes of horizontal integration, can provide a more comprehensive and insightful analysis with a lot of flexibility, for understanding and analyzing a safety-critical socio-technical system in its entirety and instituting necessary preventive interventions early on.

4. Vertical and Horizontal Integration—A Key Tenet of Systems-Thinking

4.1. The Five Key Tenets of Systems Thinking

Grant et al. [29] tried to capture the spirit of systems thinking by synthesizing the core features of contemporary accident causation models, as a basis to develop a formal methodology for anticipating and preventing accident causation and occurrence. They identified a set of 15 basic systems thinking tenets across the different accident causation models. It was found that, despite considerable variation in the different philosophical approaches towards accident causation, these tenets are universally supported. The authors analyzed the 15 basic systems thinking tenets suggested by Grant et al. It was found that the 15 tenets can further be consolidated into the "five key tenets of systems thinking", in order to have a simplified and focused understanding. This effort in consolidation helped in correlating the 'systems thinking tenets' to the 'system safety principles'. It also revealed the inadequacy of the present set of the five basic/technical system safety principles in covering the complete set of systems thinking tenets. A comparative matrix prepared by the authors showing the 'five key tenets of systems thinking' mapped against the relevant 'system safety principles' is presented in Table 1.

Table 1. The Five Key Tenets of Systems Thinking and the Correlated System Safety Principles.

S/N	The Fifteen Basic Systems-Thinking Tenets Identified by Grant et al. (2018), with their Description		Consolidated Set of "Five Key Tenets of Systems-Thinking"	"System Safety Principles" Corresponding to the Key Systems Thinking Tenets
1	Unruly Technologies	Unforeseen and unpredictable behaviours of new technologies that are introduced into the system	Complex and Unruly Technologies	Fail-Safe Principle Margin-of-Safety Principle Ungraduated-Response Principle Defence-in-Depth Principle Observability-in-Depth Principle
	Constraints	System elements that impose limits on, or influence, the behaviour of other system elements to ensure safe operation		
2	Non-linear Interactions	Complex interactions that produce dynamic unpredictable sequences and outcomes	Non-linear Interactions and Emergence	Fail-Safe Principle Margin-of-Safety Principle Ungraduated-Response Principle Defence-in-Depth Principle Observability-in-Depth Principle Human-Factors Principle
	Dependence on Initial conditions	Characteristics of the original state of the system that are amplified throughout and alters the way the system operates at a later point in time		
	Emergence	Outcomes that result from the interactions between elements in the system that cannot be fully explained by examining the elements alone		
	Linear Interactions	Direct and predictable cause and effect relationships between system elements and production sequences		
3	Performance Variability	System elements change performance and behaviour to meet the conditions in the world and environment in which the system works	Performance Variability and Functional Resonance	Fail-Safe Principle Margin-of-Safety Principle Ungraduated-Response Principle Defence-in-Depth Principle Observability-in-Depth Principle Human-Factors Principle
	Contribution of the Protective Structure	The formal and organized structure intended to protect and optimize system safety, but instead competes for resources with negative effects *[ETTO Principle]*		
	Decrementalism	Minor modifications to system elements and/or normal performances that gradually create a significant change with safety risks *[Normalization of Deviance]*		
	Normal Performance	The way that activities are actually performed within a system *[Work-as-Done]*, regardless of formal rules and procedures *[Work-as-Imagined]*		
4	Functional Dependencies	Necessary relationships and path dependence between tightly coupled system elements (i.e., components that serve a functional purpose)	Functional Dependencies and Control-Feedback	Fail-Safe Principle Margin-of-Safety Principle Ungraduated-Response Principle Defence-in-Depth Principle Observability-in-Depth Principle Human-Factors Principle
	Coupling	The degree or 'tightness' and interconnectivity of the interactions that exist between system elements		
	Modularity	Sub-systems and elements that interact but are designed and operated independently of each other		
	Feedback loops	Communication structure and information flow to evaluate control requirements of hazardous processes		
5	Vertical Integration	Interaction between elements across levels of the system hierarchy	Vertical and Horizontal Integration	Integration-in-Totality Principle (Newly introduced)

It was found from the analysis that the 'human factors principle' should be added as a 6th system safety principle to the set of five basic/technical system safety principles, since all the key tenets (except probably for the 'complex and unruly technologies' tenet) are directly influenced by human factors. The analysis also revealed that the 'integration-in-totality principle' is required to be introduced as the 7th system safety principle to completely take care of the need for embracing the conventional systems thinking tenet of 'vertical integration', which in fact requires further integration with the two dimensions of 'horizontal integration' presented in this article.

4.2. Need for Both Vertical and Horizontal Integration—The Case for Integration-In-Totality

"Systems thinking is all about relationships and integration", said Sydney Dekker in his seminal works on Systems thinking concepts and tenets [30,31]. 'Vertical integration' is only one part of the totality of integration. Even though Grant et al. listed 'vertical integration' as one among the fifteen basic systems thinking tenets, the authors felt that 'integration-in-totality' is achieved only through a holistic "vertical and horizontal integration". Hence the authors, in their compilation of the "five key tenets of systems thinking", substituted the tenet of 'vertical integration' with 'vertical and horizontal integration' to reflect the need of complete integration in the true spirit of systems thinking. The next section is devoted to narrate how a combination of vertical integration and horizontal integration is created to generate the "integration-in-totality principle", with strong theoretical foundation from three important foundational concepts from the field of "systems thinking".

5. Integration-In-Totality Principle—Three Concepts Constituting the Theoretical Foundation

5.1. The Axis of Perspective—Abstraction Hierarchy and the Macro-Meso-Micro Levels of Vertical Integration

In his pioneering Systems thinking concept of "abstraction hierarchy", Rasmussen [10–14] proposed five top-down hierarchical levels of abstraction, viz. functional purpose, abstract function, generalized functions, physical functions, and physical form, shown in Figure 3.

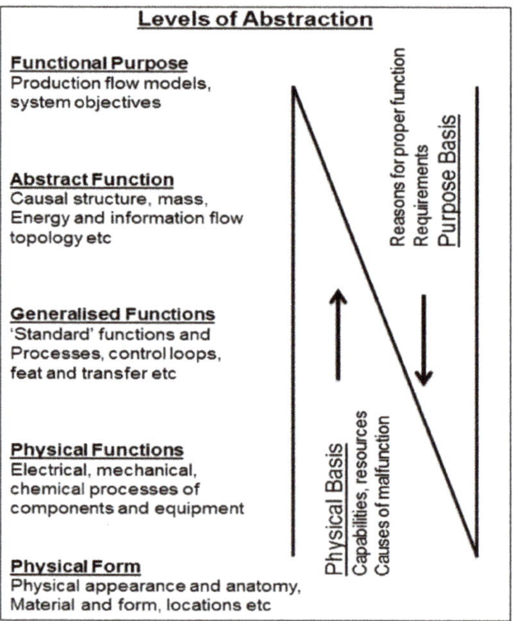

Figure 3. The Levels of Abstraction Hierarchy by Rasmussen (source: [10]).

The concept of abstraction hierarchy can nevertheless be simplified into the three levels of purpose, function, and physical-form. These levels of abstraction hierarchy are the basis for the "macro-meso-micro" levels of 'vertical integration' in the integration-in-totality principle. In the systems analogy, these three levels could be related to the system (having a purpose), sub-systems (having their own functions), and components/equipment (having the physical-form). This understanding calls for a vertical integration of the system, the sub-systems and the equipment so as to capture the entirety of the system. However, the macro-meso-micro levels have different connotations in different system contexts, as explained in subsequent sections. The 'bidirectional synthesis' with 'continuum approach' along the different levels of vertical integration is ingrained in the abstraction hierarchy, as evidenced by the bi-directional arrows shown in the diagram.

5.2. The Axis of Performance—The Design-Control-Practice (DCP) Diagram

The "design-control-practice (DCP) diagram", shown in Figure 4, was proposed by Stoop [15–18]. The DCP diagram is constructed of three sets of bi-directional arrows representing three axes. The macro-meso-micro levels of the vertical axis here represent the control levels of governance-oversight, management-control and operator-compliance respectively. The diagonal axis indicates the engineering design cycle of goal-function-form. It can be seen that both the vertical and diagonal axes of the DCP diagram have a one-to-one correspondence with the macro-meso-micro levels of 'vertical integration' derived from the concept of abstraction hierarchy (which in fact have different connotations in different system contexts), and represented by the 'axis of perspective' in the 'Rubik's cube model of integration-in-totality Principle'.

The horizontal axis of the DCP diagram represents a 'design-develop-construct-operate-adapt' bi-directional continuum which additionally provides 'horizontal integration', which has been taken as the basis for the 'axis of performance' of 'design-manufacture-operation' continuum in the 'Rubik's cube model of integration-in-totality principle'. The need for adopting the concepts of 'continuum approach' and the 'bidirectional synthesis' in safety-related analyses is evident from the three bi-directional arrows used in the construction of the DCP diagram.

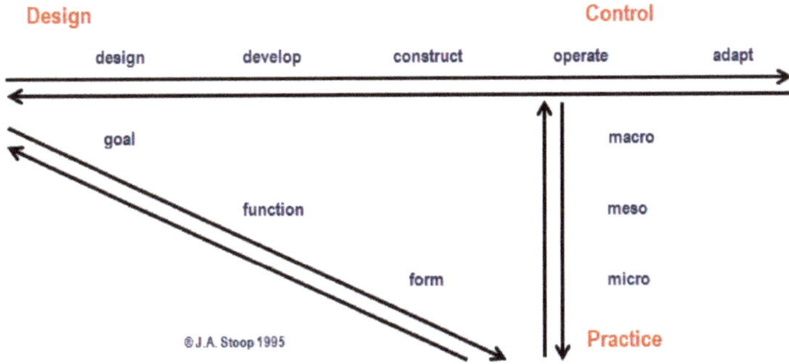

Figure 4. The Design-Control-Practice (DCP) Diagram by Stoop (Source: [16]).

5.3. The Axis of Perception—The Role of Mental Models in Systems-Theoretic

The 'horizontal integration' cannot be limited to the life-cycle continuum of design-manufacture-operation. Perception and mental models play an important role in understanding a socio-technical system in its entirety. That is the reason why one more horizontal axis orthogonal to the other two axes is provided in the form of 'axis of perception' in the 'Rubik's cube model of integration-in-totality principle', having an 'intent-execution-manifestation' pathway along its length. The 'axis of perception' is meant to capture the possible variances in the realms of design-manufacture-operation in the life-cycle continuum, and also between the macro-meso-micro levels.

The 'axis of perception' has been conceived in accordance with the 'role of mental models in systems-theoretic framework', suggested by Leveson [19–21] who opined that the human behavior within a system-theoretic framework is based on the three elements of (i) the designer's model, (ii) the actual system model, and (iii) the operator's mental model, as shown in Figure 5. The bi-hexagonal arrows in the figure have been added by the authors to indicate the need and scope for the 'bi-hexagonal synthesis' with the 'continuum approach' along the path.

The designer deals with idealized description which is generally known as the "intent". The actual system is a result of the "execution" as per the specifications. The operators continually test their mental model of the process against the reality, which results in the "manifestation". Thus, the authors have defined the "axis of perception" of the integration-in-totality principle as an 'intent-execution-manifestation' continuum, deriving from the aforementioned concept of 'mental models of system-theoretic framework' from Leveson [19–21].

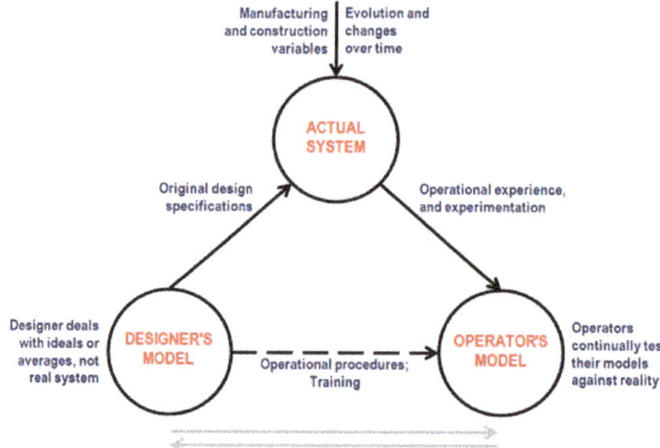

Figure 5. The Role of Mental Models in Systems-theoretic Framework (Adapted from [19]).

The 'axis of perception' of the integration-in-totality principle, with its horizontal integration along the orthogonal axis of 'intent-execution-manifestation', takes care of the perceptive mental models involved in understanding and analyzing a socio-technical system. This is adding up to the horizontal integration provided by the 'axis of performance' along the 'design-manufacture-operation' continuum. Hence it can be seen that the systems safety principle of 'integration-in-totality' is perfectly in alignment with the key systems thinking tenet of 'vertical and horizontal integration', with the three axes of performance, perception and perspective providing the pathways for analyzing any socio-technical system by applying the concepts of the 'continuum approach' to ensure the necessary system integration, and the 'bidirectional synthesis' for comprehensive analysis along the integrated pathways.

Thus we can see that three important concepts of "systems thinking" by three prominent thinkers in the field of safety have been combined by the authors in this article to conceptualize the 'Rubik's cube model' having the 'continuum approach' and the 'bi-directional synthesis', in order to develop the 'integration-in-totality principle' as the "7th system safety principle".

6. The Axis of Perspective in Integration-In-Totality Principle, and the Macro-Meso-Micro Levels

6.1. Skill-Rule-Knowledge Framework and Macro-Meso-Micro Perspective Levels

The "skill-rule-knowledge (SRK) framework" developed by Rasmussen in 1983 has been a pioneering work on systems thinking, along with the abstraction hierarchy proposed by him the same year [10–12]. The SRK framework posits that the human behavior is a reflection of complexity of the environment; and is basically 'teleological', i.e., driven by purposive goals; and is shaped by signals, signs and symbols in the environment. It gives a description of the abstraction hierarchy, explaining the operational aspect of the functional properties of a system, relating it to the various levels of the operator's cognitive processing at three levels based on skills, rules and knowledge. It provides an integrated approach to the design of human-machine systems, combining the concepts of control engineering and psychology [13,14].

The authors further innovated and reframed the 'SRK framework' in the form of a "FRAMED-IN-FRAM® diagram" to bring in better clarity on how it is a reflection of the 'axis of perspective' of 'vertical integration' which is fundamental to the 'integration-in-totality principle'. The FRAMED-IN-FRAM® diagram is an improved version of the functional resonance analysis method (FRAM) diagram [32], developed by the authors. Interested readers are referred to Thomas, et al. [1,2] for further information on the FRAMED-IN-FRAM® diagram. The FRAMED-IN-FRAM® diagram for the SRK framework, presented in Figure 6, shows how the behaviour and control, based on the three levels of skill, rule and knowledge, works through signals, signs and symbols of perceptual, conceptual and explicit nature respectively. It also illuminates how they work at the three organizational levels, viz. strategic, tactical, and operational levels, which correspond to the 'macro-meso-macro' levels respectively of the 'Rubik's cube model of integration-in-totality principle'.

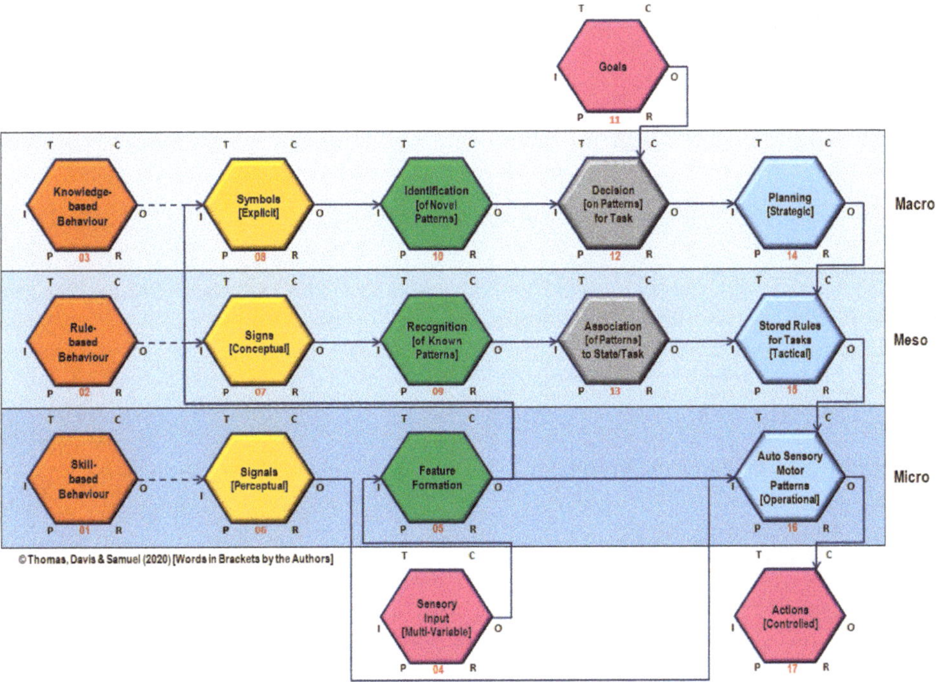

Figure 6. Skill-Rule-Knowledge Framework of Rasmussen, Interpreted using the FRAMED-IN-FRAM® Diagram.

6.2. Macro-Meso-Micro Perspectives in Different Contexts

The macro-meso-micro levels of vertical integration in the axis of perspective of the integration-in-totality principle can be understood/interpreted in many different ways depending upon the context in which they exist. From a systems-theoretic point of view, it could be the system-subsystem-component levels of understanding and analyzing the entity being examined.

The vertical integration achieved along the axis of perspective in integration-in-totality principle at the macro-meso-micro levels in different contexts is presented in Figure 7.

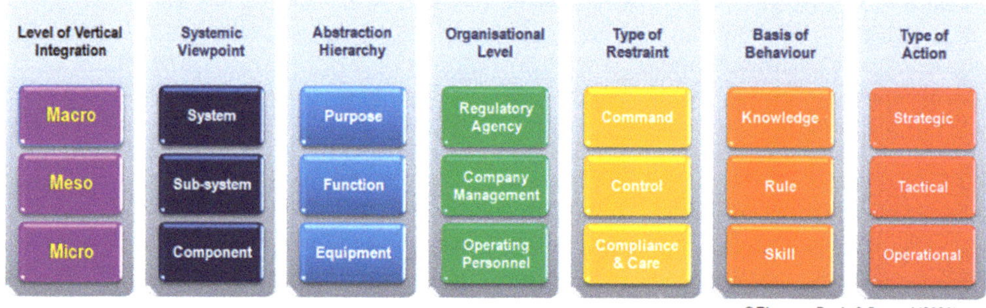

Figure 7. Macro, Meso and Micro Levels of Vertical Integration in Different Contexts.

In terms of the abstraction hierarchy, the macro-meso-micro levels correspond to the purpose-function-equipment levels, as explained in the previous section. In an organizational situation, the macro-meso-micro levels could be the echelons of regulatory agency, company management and operating personnel, with the corresponding restraint actions of command, control, and compliance & care, respectively, as envisaged by Stoop in the DCP diagram [15–18].

As per the SRK framework proposed by Rasmussen, explained earlier with the help of a FRAMED-IN-FRAM® diagram, the macro-meso-micro levels have knowledge, rule and skill as the basis of behavior, with corresponding actions being strategic, tactical and operational, respectively.

6.3. The Micro-Meso-Macro Levels of the Axis of Perspective in a Typical Case Study

The case study presented in an earlier technical article by the authors [1] can be shown as an example of the application of the concept of macro-meso-micro levels for detailed analysis. The case study pertains to the crack developed at the shear neck of the drive shaft of the oil cooling system (OCS shaft) of a turbo-shaft engine. Three major influencing sources were identified for occurrence of the crack (which happened because of excitation of 'backward whirl' phenomenon in the OCS shaft as shown alongside). Interestingly, the three influencing sources were at the three macro-meso-micro levels from the systemic viewpoint, viz. the aircraft (system), aero engine (sub-system) and the OCS shaft (component), as shown in Figure 8.

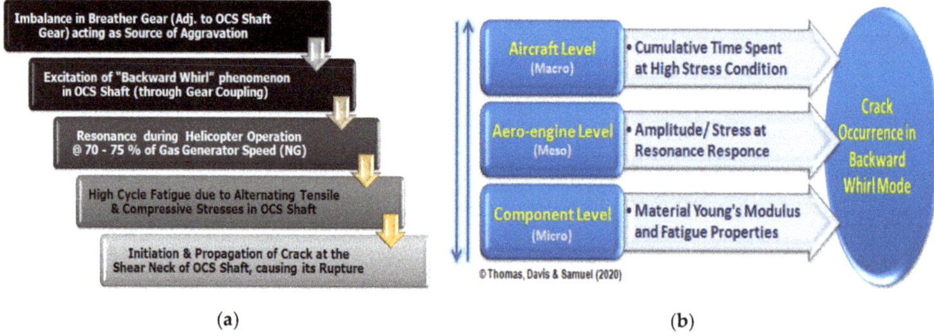

(a) (b)

Figure 8. The Micro-Meso-Macro Levels of Component, Aero-engine and Aircraft; (**a**) Progression of Events Leading to Crack and Rupture, and (**b**) the Influencing Sources of Crack and Rupture. (Source: Thomas, Davis & Samuel [1]).

7. Axis of Perception—The Intent-Execution-Manifestation Pathway

7.1. The World of Perspectives and Perceptions

A good starting point for further discussions on the need for integration-in-totality could be the illustrations by the authors on the different facets of perspective, based on the ideas from a Deloitte pamphlet [33], given in Figure 9. The illustrations show that perceptions vary depending upon the perspective or the viewpoint.

The 'big picture', shifting from a worm's eye-view to a man's eye-view to a bird's eye-view and vice versa, is in fact the 'macro-meso-micro' level viewpoints along the 'axis of perspective' of the 'vertical integration' concept of integration-in-totality principle. As we go higher up in the ladder, things become smaller, but the field of vision become larger and wider to have a totally different perspective. The 'flip side' calls for looking from the exactly opposite direction to get a totally different understanding of the same thing, just as the rotation of an object understood to be clockwise when looking from above is perceived as an anti-clockwise rotation when looked from below, as illustrated. This is in fact the property of 'bi-directional synthesis' ingrained in the integration-in-totality principle. 'Looking through others eyes' and 'view from the future' provide entirely new perspectives and perceptions. The 'analogous angle' and the 'unexpected answer' provide new options to be considered and selected from in any given situation. Other than the facet of 'big picture' which belong to the 'axis of perspective', all the other facets are captured by the 'axis of perception' of integration-in-totality principle.

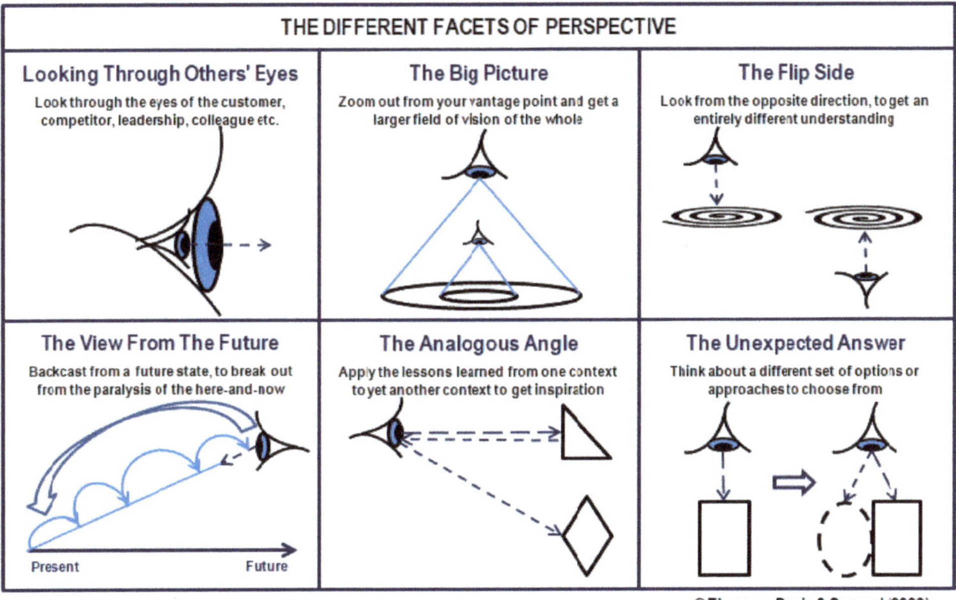

Figure 9. The Different Facets of Perspective, Illustrated.

7.2. The Axis of Perception—Perceptions Vary

The role of perception in understanding the truth and reality is best exemplified by the story of "The Blind Men and the Elephant" from Indian folklore, wherein the same elephant was variedly interpreted to be a snake, spear, fan, tree, wall, and rope by the six blind men who touched the trunk, tusk, ear, leg, side and tail respectively. "Our perception of truth depends on our point of view", writes Losmilzo [34] as a caption to the illustration shown in Figure 10, wherein "truth" is shown as a three-dimensional object which has shadows of square, circular and triangular shapes when projected in the three orthogonal directions, all of which are perceived as "true".

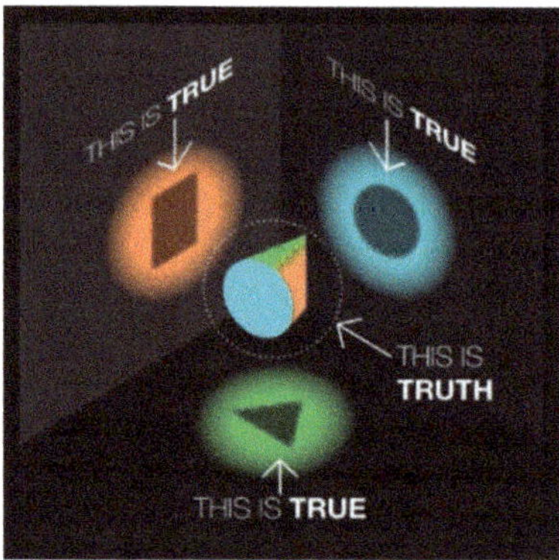

Figure 10. "Our perception of Truth depends on our viewpoint" (Adapted from [34]).

The 'axis of perception' in the integration-in-totality principle captures the variance in perception due to the difference in viewpoint by different stakeholders at different levels depending upon their own field of endeavour like design, manufacture, or operation. This variance can be clarified through the 'bi-directional synthesis' with 'continuum approach' along the 'axis of perception' of 'intent-execution-manifestation' in the integration-in-totality principle.

7.3. The Intent-Execution-Manifestation Continuum of the Axis of Perception in a Typical Case Study

The inadequacy of the five basic/technical system safety principles [6] in facilitating complete understanding, analysis and interpretation of aviation accidents and safety events was earlier highlighted by the authors. The human factors principle, and the human-factor-focused accident analysis methods like human factors analysis and classification system (HFACS) also fail to fully achieve this objective, due to a disconnect with the technical aspects of the present-day aerospace systems which are basically complex, software-driven and automated. In such a situation, integration-in-totality principle with its ability to provide multi-dimensional interpretations can provide multifarious insights into the specific problem.

It would be interesting to see how the integration-in-totality principle could be applied to analyze the twin disasters of Boeing 737 MAX aircraft [35,36] mentioned in the Introduction. During the upgrade to Boeing 737 MAX aircraft with bigger engines, the engines were moved up the wing to get sufficient ground clearance, causing the aircraft nose to lift up higher during take-off. This increased the possibility of aircraft stall due to a higher angle of attack (AoA). The maneuvering characteristics augmentation system (MCAS) was introduced by the designers as a software solution to overcome the problem. The 'design intent' was to achieve an automatic "aircraft nose down (AND)" by means of a stabilizer trim input actuated by the MCAS when the 'critical angle of attack' is reached or exceeded.

However, in both the disaster cases, one of the two AoA sensors installed on the aircraft became faulty, indicating an AoA value higher than the actual value. The feedback from the sensor on the higher angle of attack (~20° in the Indonesian aircraft case, and ~57° in the Ethiopian aircraft one) resulted in the stabilizer trim input actuation by the MCAS, making the aircraft automatically and uncontrollably pitch down. The pilot applied the manual "aircraft nose up (ANU)" electric trim to counter the 'AND' as and when it was encountered, but the faulty AoA sensor kept sending the wrong signal triggering the MCAS to cause automatic aircraft nose down repeatedly. The erroneous reading by the faulty AoA sensor threw up multiple and confusing signals to the aircrew in the cockpit, and the pilots were not trained to handle such an automation surprise.

This vicious cycle of the automatic 'AND' by the MCAS and the manual 'ANU' by the pilot continued many times, and finally the pilot had to give up the control to the MCAS automation under duress, causing the aircraft to plunge downwards and crash in both the disaster cases, as shown in the "FRAMED-IN-FRAM® diagram" (Thomas et al. [1,2]), given in Figure 11.

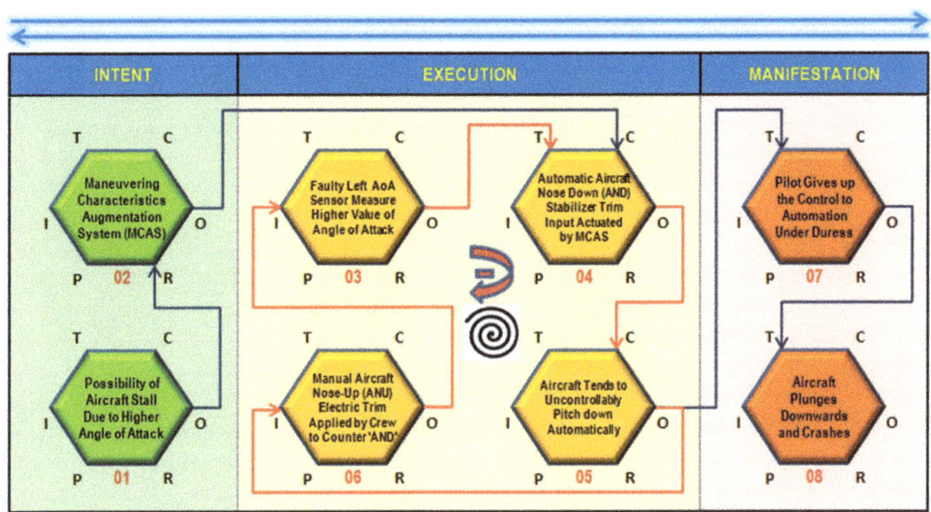

Figure 11. Intent-Execution-Manifestation Continuum—Case Study on Boeing 737 MAX Disasters.

The case study shows the disconnect between the 'intent' of the designers, the 'execution' by the MCAS and the pilot, and finally the 'manifestation' of the disasters due to the disconnect. Had such possibilities been anticipated as a mental model, necessary checks and controls could have been instituted in the design stage itself so as to obviate the fatal disasters.

The traditional accident analysis methods like AcciMap [13] would have tried to understand the event along the macro-meso-micro levels of vertical integration, which can be captured by the "axis of perspective' of the integration-in-totality principle. The design-related aspects of the MCAS and its integration into the aircraft system and its testing and certification could be captured by the 'axis of performance'. However, the 'axis of perception' provides a powerful tool for understanding and analysis in the form of an intuitional mental model along the 'intent-execution-manifestation' continuum as shown in the case study, highlighting the applicability of integration-in-totality principle in general and the axis of perception in particular in safety investigations.

8. Axis of Performance—The Design-Manufacture-Operation Continuum

8.1. The Axis of Performance—The Pathway for Improvement Processes

The continuum of design-development-manufacturing-assembly-test-operation along the axis of performance in the integration-in-totality principle is the real pathway for improvement processes in a system, applying the intent-execution-manifestation mental models of the axis of perception, and the macro-meso-micro levels of the axis of perspective simultaneously, and hence the Rubik's cube simile for the integration-in-totality principle.

The analysis along the continuum of the 'axis of performance' has to happen bi-directionally. Normally, the flow of information and the consequent action, if any, happen uni-directionally along the forward direction only. But there is a need to have a bi-directional flow of information and action in the value chain of production/overhaul of an aircraft or aero engine between all the stages and sub-stages.

For example, the expected acceptance test parameters of an aero-engine are made available with the assembly personnel and the expected assembly acceptance criteria of the manufactured components are taken care by the people involved in manufacture/overhaul of the aero-engine, as shown in Figure 12.

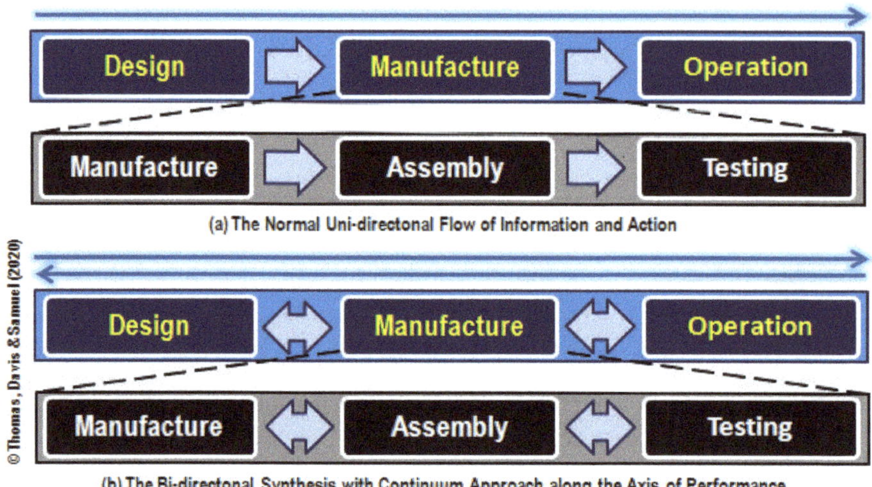

Figure 12. Bi-directional Synthesis along the Axis of Performance.

8.2. The Design-Manufacure-Operation Continuum of the Axis of Performance—A Case Study

Quantitative and qualitative analysis bi-directionally along the life-cycle continuum of design-manufacture-operation of an airborne system can help improving performance and safety of the system. As a case study, the authors would like to present a glimpse into a research done by them on performance enhancement of a turbofan aero-engine. The engine type used to have pre-mature withdrawals before completion of the specified time between overhaul (TBO) due to performance deterioration, manifesting in the form of higher turbine entry temperature (TET), consequent upon the higher fuel burning requirement to get the required engine thrust. An analysis of the engine test data of 200 engines for various engine performance parameters revealed very interesting results. Two of the typical trend graphs (for the TET and the compressor pressure ratio, with respect to the compressor mass flow rate) are shown in Figure 13.

The graphs show that the more the compressor mass flow rate, the lesser is the turbine entry temperature, and the higher is the compressor pressure ratio. Working backward along the axis of performance, the analysis of the assembly procedures revealed the various reasons for a reduction in the compressor pressure ratio, and in turn the compressor mass flow rate, leading to higher turbine entry temperature, thus making the engine susceptible to early withdrawal due to performance deterioration, like a higher blade tip run-out.

Working further backward along the axis of performance, the contributing factors at the component manufacturing stage which eventually led to the higher blade tip run-out could be found out. Improvement actions taken in the manufacturing stage on the blade realization processes and in the assembly stage on the assembly procedures, and establishing the best practice rules accordingly, helped in getting a lower turbine entry temperature and thus higher thrust at the testing stage. This could substantially reduce the susceptibility of the aero-engine for pre-mature withdrawals from the operating unit due to performance deterioration, since sufficient margin of safety was provided in the engine pass-out stage itself by aiming for an engine with lesser TET.

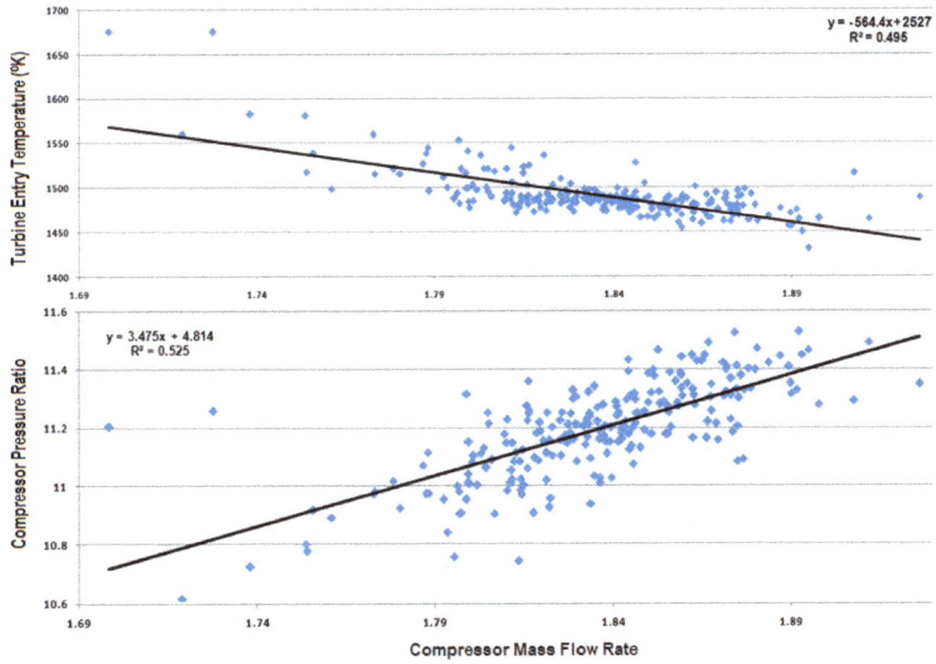

Figure 13. The Case study on Bi-directional Synthesis along the Axis of Performance.

9. The Quality-Reliability-Risk-Safety Paradigm

The concepts of quality, reliability, risk and safety are correlated, as shown in the FRAMED-IN-FRAM® diagram of quality-reliability-risk-safety paradigm (Thomas et al. [2]) in Figure 14.

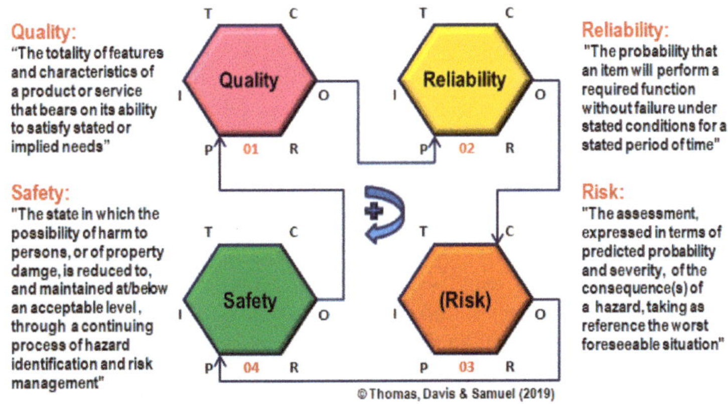

Figure 14. The Quality-Reliability-Risk-Safety Paradigm.

10. Integration-In-Totality Principle—Linkages to Systems Engineering and Airworthiness

In this section, the 'integration-in-totality principle' is explained in the context of systems engineering concepts applicable to reliability and airworthiness, and the linkage between the integration-in-totality principle and the "V-model of systems engineering" is established.

Systems engineering is the structured approach towards definition, implementation, integration and operation of a system to meet its functional, physical and operational performance requirements, in the given environment over the planned life cycle. The V-model captures the essence of the systems engineering process [37].

10.1. The Integration-In-Totality Principle Represented in the V-Model of Systems Engineering

It is interesting to note that the system safety principle of integration-in-totality, with its axes of perspective, perception and performance, can be depicted in the V-model of systems engineering, as shown in Figure 15.

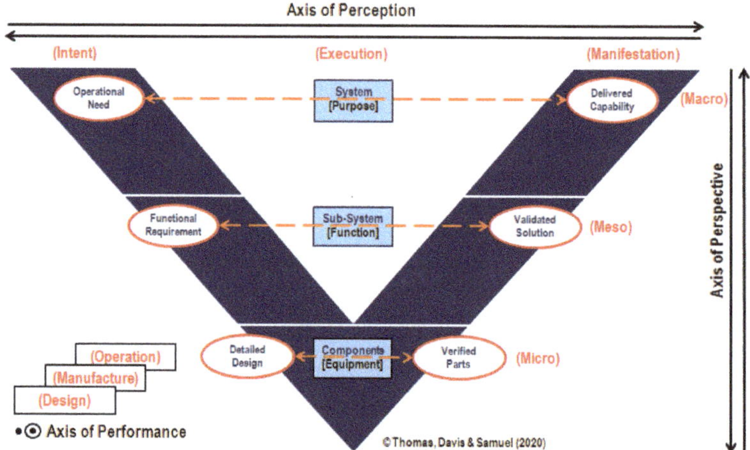

Figure 15. Integration-in-Totality Principle, Depicted in the V-Model of Systems Engineering.

10.2. The Systems Engineering Process and the Macro-Meso-Micro Levels of 'Axis of Perspective'

The "axis of perspective", comprising of the macro-meso-micro levels of vertical integration in the integration-in-totality principle, can be viewed in two different ways in the V-model of systems engineering. The creation of a "system" in systems engineering is meant to meet the mission objective or the "purpose" [37]. This is achieved by means of various design teams (applying concurrent engineering concepts) working on multiple "subsystems" having their own "function". At a lower level, specialized design groups (applying the engineering design process) design the "components", forming part of the "equipment". Thus it can be seen that systems engineering follows the macro-meso-micro levels of system-subsystem-component bi-directionally, which in turn corresponds to the abstraction hierarchy levels of purpose-function-equipment, fundamental to the axis of perspective of the integration-in-totality principle.

The left leg of the systems engineering V-model represents the 'Formulation phases of decomposition and definition', wherein 'tearing down' of the system is done to reveal the complete system architectural design. The right leg of the V-model, on the other hand, represents the 'Implementation phases of integration and verification' that are effectively 'building up' the system from the component level to the functional sub-systems to the complete system. This traversal from the highest level to the lowest level and then back to the highest level, like the traversal from the bird's eye-view to the worm's eye-view and back, is in accordance with the "bi-directional synthesis" with "continuum approach" along the 'axis of perspective' of the integration-in-totality principle, as shown alongside the figure of V-model by bi-directional arrows. The same 'bi-directional synthesis' and 'continuum approach' are applicable along the 'axis of perception' and the 'axis of performance' as well.

10.3. The Systems Engineering Process and the Intent-Execution-Manifestation of 'Axis of Perception'

The V-model of systems engineering, which is basically a process model, calls for moving down along the left leg by completing each phase sequentially and then moving up the right leg, applying the 'eleven systems engineering functions' at each stage to achieve the objectives [37]. This process traverses along the mental model path of intent-execution-manifestation of the "axis of perception" of the integration-in-totality principle. It can be seen that there is a one-to-one correspondence between the Intent and the manifestation at each level of execution (viz. the operational need of the system and the delivered capability; the functional requirement of the subsystem and the validated solution; and the detailed design of the equipment and the verified parts).

10.4. The Systems Engineering Process and the Design-Manufacture-Operation Path of 'Axis of Performance'

The engineering design process (EDP) in the V-model follows the "axis of performance" of design-manufacture-operation. As illustrated in the representative V-model of systems engineering in Figure 15 linking it to the integration-in-totality principle, the axis of performance can also be shown perpendicular to the plane of the page bi-directionally, since the same V-model having the axes of perspective and perception is applicable not only for design in the plane of the diagram, but also for the parallel planes for manufacture and operation as well.

10.5. Integration-In-Totality Principle in Airworthiness Certification

The operational requirements of an aircraft or an aero engine are specified by the customer and designed, manufactured and maintained by the contractor firm having the 'system design responsibility (SDR)'. The design organization holds the 'type approval' which is obtained through an elaborate type certification process undertaken by a dedicated airworthiness certification agency. The 'military type qualification process' regulates the procedures concerning the military aircraft 'type qualification' for performance and 'certification' for airworthiness, and the qualification and suitability for installation of pertinent systems.

Typically, the verification process of a 'type design' for airworthiness is done in a three-stage process, viz. (i) definition of the type in accordance with approved documentation or design standard, (ii) definition of the 'means of compliance' to demonstrate each requirement as per the qualification programme plan, and (iii) demonstration of compliance with the safety requirements. It can be seen from the foregoing discussions that the integration-in-totality principle, with its three axes of perspective, perception and performance, can be used as a valuable theoretical foundation for airworthiness certification, including for continuing and continued airworthiness, since it takes care of all the related aspects of reliability, risk, safety and quality.

11. Integration-In-Totality Principle—Linkage to Risk Management

11.1. Risk Management and System Safety

"Safety is the state in which risk (of personal harm or property damage) is reduced to and maintained at or below an acceptable level, through a continuing process of hazard identification and risk management", according to the ICAO definition [8]. Quantitative risk management is done based on the assessment of 'probability' of occurrence of safety hazards/events and 'severity' of their consequences. The integration-in-totality principle, being the system safety principle based on systems thinking in safety, has got major relevance in the process of identifying the hazards and managing the associated risks. This is done by way of mitigating the risk through necessary corrective actions in the short term and eliminating the risk altogether through effective preventive actions for the long term.

11.2. Risk Management along the Axes of Perspective, Perception and Performance

Risk management in an organization is carried out at different levels. 'Organizational risk management' is concerned with the threats and opportunities external to the organization, and hence is 'strategic' in nature. 'Operational risk management', on the other hand, deals with the weaknesses and strengths within the organization and are therefore 'tactical' and 'operational' in practice. Hence it can be seen that risk management has a strategic-tactical-operational continuum of vertical integration as shown in Figure 7, along the 'axis of perspective' of the integration-in-totality principle.

Risk management also works along the 'axis of perception'. The disconnect between the design intent, manufacturing execution and the operational manifestation are to be captured by applying forward-looking and backward-looking logics respectively between the safety event and the cause/consequence using the various inductive and deductive techniques of system safety analysis. This requires 'bidirectional synthesis' along the intent-execution-manifestation continuum in the axis of perception of integration-in-totality principle.

Analyzing the system along the 'axis of performance' of the design-development-manufacturing-testing-operation continuum also is equally important for risk management, to understand the system deficiencies and vulnerabilities along the path. Bi-directional synthesis along the chain of adjacent operations, treating the personnel dealing with the next phase or process or operation as the external/internal customer is very important for achieving risk mitigation at each stage, bringing down the probability of occurrence of safety events and severity of their consequences. In order to mitigate risk, and enhance quality, reliability and safety, it is necessary to act upon the accident precursors, pathogens and latent defects in a near-miss management framework early on along the axis of performance of the integration-in-totality principle.

12. Integration-In-Totality Principle—Linkage to Strategic Quality Management

12.1. Strategic Quality Management—A Convergence Concept

Quality as an organizational function has evolved over the years, from inspection to quality-control to quality-assurance to company-wide-quality-control to total-quality-management to strategic-quality-management. In the process, the tenets of quality also got enlarged with a snowballing effect, encompassing and subsuming the product, process, system, people, improvement-cycle and risk [38]. Strategic quality management (SQM) is a convergent concept, combining the basic concepts of total quality management and corporate strategy management [39].

12.2. Integration-In-Totality Principle and Strategic Quality Management

The integration-in-totality principle is the pivotal concept which can bridge the gap between the two parallel knowledge fields of safety and quality, by integrating the concepts of systems thinking in aerospace safety and strategic quality management, as shown in Figure 16.

Strategic quality management and system safety principles represent the latest developments in the fields of quality and safety, respectively. Quality and safety are linked through the quality-reliability-risk-safety paradigm presented in an earlier section [2], and strategic quality management has risk-based thinking as one of the cornerstones [39]. Hence application of strategic quality management and integration-in-totality principle together in the systems thinking framework can help understand aerospace systems like aircraft, aero engines, etc. better and achieve performance enhancement of the system, applying quantitative analysis using predictive analytics, and also employing qualitative analysis techniques like functional resonance analysis method (FRAM).

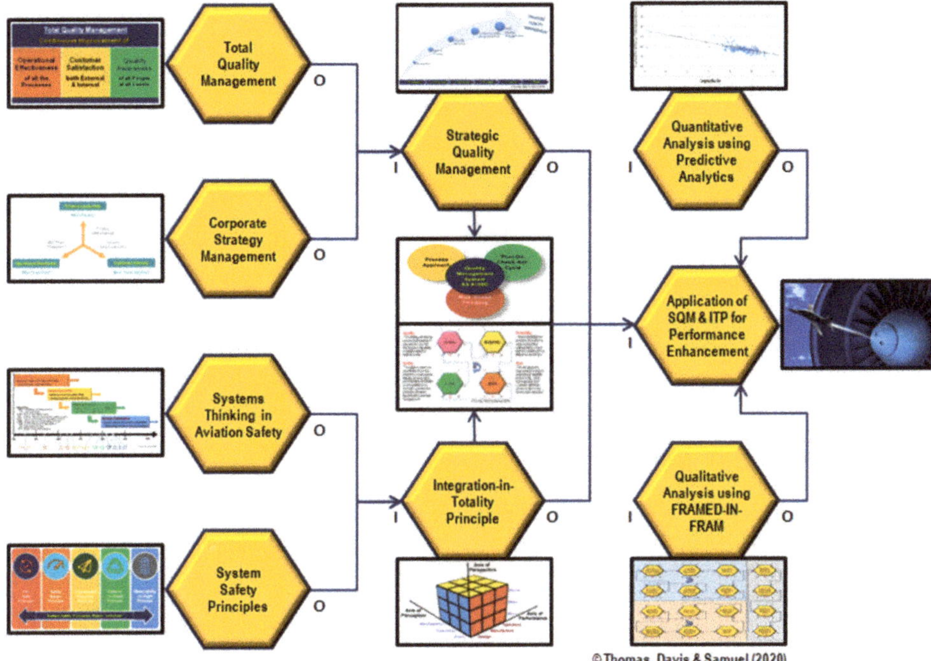

Figure 16. Theoretical Foundation of Applying Integration-in-Totality Principle in Strategic Quality Management.

13. Conclusions

A new safety concept called "integration-in-totality principle" has been introduced in this article as the 7th system safety principle. A "seven-principles-framework of system safety principles" is proposed, adding two more principles to the five basic/technical system safety principles conceptualized by Saleh et al. The "integration-in-totality principle" is illustrated with the simile of a "Rubik's cube model of integration-in-totality principle" having three axes, viz. the axis of performance, the axis of perception, and the axis of perspective, reinforcing the key systems thinking tenet of "vertical and horizontal integration". The relevance of 'bidirectional synthesis' of a socio-technical system with a 'continuum approach' along these three axes to facilitate systems thinking is articulated, drawing upon the 'abstraction hierarchy' and the 'SRK framework' by Rasmussen, 'DCP diagram' by Stoop and 'mental models in systems-theoretic framework' by Leveson. The article also explores the linkage of integration-in-totality principle to strategic quality management and risk management, bridging the gap between two parallel fields of knowledge. The integration-in-totality principle is interpreted in terms of the 'V-model of systems engineering', to establish its linkage to reliability and airworthiness of an aerospace system. It is expected that the new safety concepts shall augment the existing body of knowledge and trigger further research in the field of systems thinking and strategic quality management.

Author Contributions: All the authors contributed equally to this work. All authors have read and agreed to the published version of the manuscript.

Funding: This research received no external funding.

Acknowledgments: This research has been a part of the academic programme at International Institute for Aerospace Engineering & Management, Jain (Deemed-to-be University), Bengaluru, India. The theme discussed herein is independently developed by the authors for the academic work. The authors would like to express their deep sense of gratitude to all the colleagues who have associated directly or indirectly towards creation of knowledge in the field, and thank the managements of their organizations for providing all support for the study and the permission to publish this technical article.

Conflicts of Interest: There is no conflict of interest whatsoever.

References

1. Thomas, J.; Davis, A.; Samuel, M.P. Strategic Quality Management of Aero Gas Turbine Engines, Applying Functional Resonance Analysis Method. In *Proceedings of the National Aerospace Propulsion Conference*; Mistry, C.S., Kumar, S.K., Raghunandan, B.N., Sivaramakrishna, G., Eds.; Lecture Notes in Mechanical Engineering; Springer: Singapore, 2021; pp. 65–91. Available online: https://doi.org/10.1007/978-981-15-5039-3_4 (accessed on 31 August 2020).
2. Thomas, J.; Davis, A.; Samuel, M.P. Quality–Reliability–Risk–Safety Paradigm—Analyzing Fatigue Failure of Aeronautical Components in Light of System Safety Principles. In *Fatigue, Durability, and Fracture Mechanics*; Seetharamu, S., Jagadish, T., Malagi, R.R., Eds.; Lecture Notes in Mechanical Engineering; Springer: Singapore, 2021; pp. 267–304. Available online: https://doi.org/10.1007/978-981-15-4779-9_18 (accessed on 8 October 2020).
3. Sachs, N.W.; Beckman, M. Figuring out why Things Breakdown. In *Tribology & Lubrication Technology*; STLE, Society of Tribologists and Lubrication Engineers: Park Ridge, IL, USA, 2019; pp. 38–45.
4. Latino, M.A.; Latino, R.J.; Latino, K. *Root Cause Analysis: Improving Performance for Bottom-Line Results*, 4th ed.; CRC Press: Boca Raton, FL, USA, 2011; ISBN 978143950923.
5. Hulme, A.; Stanton, N.A.; Walker, G.H.; Waterson, P.; Salmon, P.M. What do applications of systems thinking accident analysis methods tell us about accident causation? A systematic review of applications between 1990 and 2018. *Saf. Sci.* **2019**, *117*, 164–183. [CrossRef]
6. Saleh, J.H.; Marais, K.B.; Favarò, F.M. System safety principles: A multidisciplinary engineering perspective. *J. Loss Prev. Process. Ind.* **2014**, *29*, 283–294. [CrossRef]
7. Gnoni, M.G.; Saleh, J.H. Near-Miss Management Systems and Observabiliy-in-Depth: Handling Safety Incidents and Accident Precursors in Light of Safety Principles. *Saf. Sci.* **2017**, *91*, 154–167. [CrossRef]
8. International Civil Aviation Organisation (ICAO). *Safety Management Manual (SMM), Doc. 9859*, 4th ed.; ICAO Headquarters: Montreal, QC, Canada, 2018.
9. FAA Air Traffic Organisation. *Safety Management System Manual April 2019*; Federal Aviation Administration: Richmond, VA, USA, 2019.
10. Waterson, P.; le Coze, J.-C.; Andersen, H.B. Recurring themes in the legacy of Jens Rasmussen. *Appl. Ergon.* **2017**, *59*, 471–482. [CrossRef] [PubMed]
11. Rasmussen, J. Skills, rules, and knowledge; signals, signs, and symbols, and other distinctions in human performance models. *IEEE Trans. Syst. Man Cybern.* **1983**, 257–266. [CrossRef]
12. Rasmussen, J.; Vicente, K.J. Coping with human errors through system design: Implications for ecological interface design. *Int. J. Man-Mach. Stud.* **1989**, *31*, 517–534. [CrossRef]
13. Rasmussen, J.; Svedung, I. *Proactive Risk Management in a Dynamic Society*; Swedish Rescue Services Agency: Karlstad, Sweden, 2000; ISBN 91-7253-084-7.
14. Le Coze, J.C. Reflecting on Jens Rasmussen's legacy. A strong program for a hard problem. *Saf. Sci.* **2015**, *71*, 123–141. [CrossRef]
15. Stoop, J.; de Kroes, J.; Hale, A. Safety science, a founding fathers' retrospection. *Saf. Sci.* **2017**, *94*, 103–115. [CrossRef]
16. Stoop, J.A. Safety: A system state or property? *J. Saf. Stud.* **2016**, *2*. [CrossRef]
17. Stoop, J.A.; Dechy, N.; Dien, Y.; Tulonen, T. Past and Future in Accident Prevention and Learning: Single Case or Big Data? In Proceedings of the ESReDA 50th Seminar, Sevilla, Spain, 18–19 May 2016.
18. Stoop, J.A.; van der Burg, R. From Factor to Vector, a System Engineering Design Perspective on Safety. Ph.D. Thesis, Delft University of Technology, Delft, The Netherlands, January 2014.
19. Leveson, N.G. Applyng systems thinking to analyze and learn from events. *Saf. Sci.* **2011**, *49*, 55–64. [CrossRef]
20. Leveson, N.G. *Engineering a Safer World: Systems Thinking Applied to Safety*; MIT Press: Cambridge, MA, USA, 2011; ISBN 978-0-262-01662–9.

21. Leveson, N.G.; Stephanopoulos, G. A system-theoretic, control-inspired view and approach to process safety. *AIChE J.* **2013**, *60*, 2–14. [CrossRef]
22. Saleh, J.; Marais, K.; Bakolas, E.; Cowlagi, R. Highlights from the literature on accident causation and system safety: Review of major ideas, recent contributions, and challenges. *Reliab. Eng. Syst. Saf.* **2010**, *95*, 1105–1116. [CrossRef]
23. Favarò, F.M.; Saleh, J.H. Toward risk assessment 2.0: Safety supervisory control and model-based hazard monitoring for risk-informed safety interventions. *Reliab. Eng. Syst. Saf.* **2016**, *152*, 316–330. [CrossRef]
24. Saleh, J.H.; Geng, F.; Ku, M.; Walker, M.L. Electric propulsion reliability: Statistical analysis of on-orbit anomalies and comparative analysis of electric versus chemical propulsion failure rates. *Acta Astronaut.* **2017**, *139*, 141–156. [CrossRef]
25. Cowlagi, R.V.; Saleh, J.H. Co-Ordinability and Consistency in Accident Causation and Prevention: Formal System Theoretic Concepts for Safety in Multilevel Systems. *Risk Anal.* **2013**, *33*, 420–433. [CrossRef]
26. Bakolas, E.; Saleh, J.H. Augmenting defense-in-depth with the concepts of observability and diagnosability from Control Theory and Discrete Event Systems. *Reliab. Eng. Syst. Saf.* **2011**, *96*, 184–193. [CrossRef]
27. Favaro, F.M.; Saleh, J.H. Observabilit-in-Depth: An Essential Complement to the Defence-in-Depth Safety Strategy in the Nuclear Industry. *Nuclear Eng. Technol.* **2014**, *46*, 1–14. [CrossRef]
28. Shanmugam, A.; Robert, T.P. Human factors engineering in aircraft maintenance: A review. *J. Qual. Maint. Eng.* **2015**, *21*, 478–505. [CrossRef]
29. Grant, E.; Salmon, P.M.; Stevens, N.J.; Goode, N.; Read, G.J. Back to the future: What do accident causation models tell us about accident prediction? *Saf. Sci.* **2018**, *104*, 99–109. [CrossRef]
30. Dekker, S.W.A. *Why We Need New Accident Models*; Technical Report 2005-02; Lund University School of Aviation: Lund, Sweden, 2015.
31. Dekker, S.W.; Pruchnicki, S. Drifting into failure: Theorising the dynamics of disaster incubation. *Theor. Issues Ergon. Sci.* **2013**, *15*, 534–544. [CrossRef]
32. Hollnagel, E. *FRAM: The Functional Resonance Analysis Method: Modelling Complex. Socio-Technical Systems*; Ashgate Publishing Limited: Surrey, UK, 2012; ISBN 978-1-4094-4551-7.
33. Deloitte. *10 Moves to Make Moments Matter*; Deloitte Development LLC: London, UK, 2017.
34. Our Perception of Truth Depends on Our Viewpoint 2.0. 2016. Available online: https://imgur.com/gallery/obWzGjY (accessed on 31 August 2020).
35. Comittee on Transportation and Infrastructure. *The Design, Development & Certification of the Boeing 737 MAX*; Final Committee Report; Comittee on Transportation and Infrastructure: Washington, DC, USA, September 2020.
36. National Transportation Safety Board. *Assumptions Used in the Safety Assessment Process and the Effects of Multiple Alerts and Indications on Pilot Performance*; Safety Commission Report; National Transportation Safety Board: Washington, DC, USA, 2019.
37. NASA. Chapter 2: The Systems Engineering (SE) Process. National Aeronautics and Space Administration. Available online: https://www.nasa.gov/pdf/598887main_Auburn_PowerPoints_SE.pdf (accessed on 31 August 2020).
38. Thomas, J.; Davis, A.; Samuel, M.P. Aerospace Organizational Excellence: Quality System Standards and Global Best Practices. In Proceedings of the CSDO Golden Jubilee Seminar on Excellence through Maintainability in Aviation, Bengaluru, India, 13–14 December 2018.
39. Thomas, J.; Davis, A.; Samuel, M.P. Strategic Quality Management and Risk-Based Thinking. *J. Aerospace Qual. Reliabil.* **2019**, *7*, 1–6.

Publisher's Note: MDPI stays neutral with regard to jurisdictional claims in published maps and institutional affiliations.

Article

Analysis of Continuing Airworthiness Occurrences under the Prism of a Learning Framework

James Clare [1] and Kyriakos I. Kourousis [1,2,*]

1 School of Engineering, University of Limerick, V94 T9PX Limerick, Ireland; james.clare@ul.ie
2 School of Engineering, RMIT University, Melbourne, VIC 3001, Australia
* Correspondence: kyriakos.kourousis@ul.ie

Received: 15 December 2020; Accepted: 2 February 2021; Published: 5 February 2021

Abstract: In this research paper fifteen mandatory occurrence reports are analysed. The purpose of this is to highlight the learning potential incidents such as these may possess for organisations involved in aircraft maintenance and continuing airworthiness management activities. The outputs from the mandatory occurrence reports are aligned in tabular form for ease of inclusion in human factors' continuation training material. A new incident learning archetype is also introduced, which intends to represent how reported incidents can be managed and translated into lessons in support of preventing event recurrence. This 'learning product' centric model visually articulates activities such as capturing the reported information, establishing causation and the iterative nature of developing a learning product.

Keywords: aircraft maintenance; airworthiness; learning from incidents; aviation safety; learning taxonomy

---

1. Introduction

Structured and continuous safety management actions, such as collection of data, analysis and intervention can be enabled with the support of the necessary safety intelligence. High quality maintenance and management tasks are some of the essential inputs for safe operations. Continuous information 'harvested' from incident reporting arising from these tasks, is another major part of learning and preserving acceptable levels of safety [1]. Thankfully, serious incidents are becoming less frequent [2] but often because of environmental, cognitive and human centric demands, reportable and unreportable events do occur. The main underpinning aviation regulation in Europe, European Union (EU) regulation 2018/1139 [3] refers to 'management system' and mandates an operator to implement and maintain a management system to ensure compliance with these essential requirements for safe operations; it also aims for continuous improvement of the safety system through learning from incidents.

In the area of continuing airworthiness, the fundamentals of management systems are also extended to incident and occurrence reporting through the implementing conduit of EU regulation 1321/2014 [4]. It is common for incidents to be discovered within organisations and reported with the assistance of such 'systems of systems' [5]. On an operational level, initial human factors training, and company procedures are intended to specify and re-affirm the class and type of occurrence and incident that should be reported. Recent developments in Europe in the guise of EU regulation 376/2014 [6] empower voluntary and confidential reporting and are independent of all other individual obligations. The paper recounts an analysis of 15 occurrences drawn from a repository of reportable incidents. Each incident was assessed, and the report data interpreted to support potential primary and secondary causation factors. To translate these learning points into tangible lessons, causation factors are harmonised with a taxonomy for learning. This taxonomy is based upon the Transport Canada 'Dirty Dozen' [7] human factors terms which feature common aviation human error preconditions. Additionally, a framework is presented in the paper to demonstrate how learning from incidents can be leveraged with best effect in the industry segment. Mandatory reportable incidents are notified through the formal mechanism of reporting. Once the incident enters its lifecycle, it ideally transverses a process that transforms the information gathered into knowledge. This knowledge is intended to assist with the prevention of similar future events.

2. Safety Reporting Background

2.1. International and European Regulatory Context

Safety information databases containing appropriate details of events with potential and latent ancillary contributors are available and can be considered with the assistance of continuous analysis. In the United States a combined effort by the aviation industry, organisations and individuals, known as the Aviation Safety Reporting System (ASRS) [8] collect reports that are submitted on a voluntary basis. The outputs from this initiative set out to identify system deficiencies and raises correspondence directly with the responsible people. The intention is to affect learning and improvements that correlate with corrective actions that avert event recurrence.

On a wider scale through the diligent offices of the International Civil Aviation Organization (ICAO), standards and recommended practices that define contracting state reporting and analysis obligations, have been developed as a result of the collective efforts of participating states. For example, Annex 13 Aircraft Accident Investigation [9] to the Chicago Convention [10] defines the standards that require states to report accidents involving aircraft with a maximum take-off weight (MTOW) of 2250 kg and above. The document also contains details of reportable incidents (MTOW 5700 kg) that are considered important in terms of safety and accident prevention. An accident/incident data reporting system (ADREP) is operated and managed by ICAO. Safety data from (ICAO) member states are received, verified and retained in the ADREP system. The repository contains an aggregate of occurrences/incidents/accidents reported by the contracting states. The Accident/Incident Reporting Manual [11] document defines the report content, its composition and means of transmittal to ICAO. A common group of general codes known as a taxonomy is used to standardise the inputs for reporting. In an effort to improve harmonisation and exchange of information, most European aviation competent authorities have already migrated to the ICAO common ADREP taxonomy.

The EU, in recognition of its duty of care to the travelling public acknowledges that it must continue to improve levels of aviation safety. Based on a global expectation [12] of the imminent increase in aviation activity, significant challenges are evident if EU is to only preserve current levels of safety. Presently, air passengers enjoy the benefits of a safe industry based on the technological advancements, recognition of human performance and limitations, compliance primarily with prescriptive regulations and the learning potential arising from past accidents and incidents. The EU regulation 376/2014 [6] was developed to enable the collection, analysis, and follow-up of reportable incidents and occurrences. It mandates provisions for reporters to submit mandatory occurrence reports (MOR's) and voluntary occurrence reports (VOR's). There are discriminating conditions that must be met in order to determine which 'conduit' is required to report a hazard or incident. The regulation also defines reporting timelines for initial reporting (within 72 h of discovery) and for reporting to the competent authority (within a further 72 h). Organisations are also required to have a process in place to implement timely follow-up and notification of their analysis to their competent authority.

In Europe, reporting entities are encouraged to submit reports through a reporting portal moderated by the European aviation safety agency (EASA). Civil aviation competent authorities have access to the portal and the incidents and accidents are categorised in accordance with a standard aviation data reporting program (ADREP) taxonomy. They are then uploaded to a European coordination for accident and incident reporting systems database (ECCAIRS). This multi-modal European transport database can facilitate the collection, analysis and sharing of transport safety data.

2.2. Learning from Incidents: Underpinning Theory

According to Leveson [13], a holistic view of an organisation's capability in terms of learning from incidents can be enhanced by shifting the focus from the individual to what is happening across the system. In the world of 'operational aviation' the concept of Safety Management Systems (SMS) has been for the most part successfully embraced and applied where mandated. Deming [14] the respected purveyor of quality assurance methodologies asks the question, 'what is a system?' He continues to answer, 'a system is a network of interdependent components that work together to try to accomplish the aim of the system'. This description of the system suggests that the process (in safety management parlance) is 'a network of interdependent components'. Safety management philosophy requires specific points to be formally addressed so that the safety management process of operational risk can be explicitly expressed and therefore effectively managed. One of these points is preventing the recurrence of incidents and occurrences through learning from past events to achieve an acceptable level of safety.

Today, in many jurisdictions it is a requirement for aircraft maintenance and continuing airworthiness management organisations to maintain an occurrence-reporting system. European regulatory requirements [6] and organisation procedures [4] normally require the event to be investigated, documented and the causal factors considered. Additionally, corrective and/or immediate actions are often necessary to prevent re-occurrence. Learning from these incidents can often provide potential solutions to preventing safety crises in the future by looking back at what has happened and deriving lessons learned and predicting probable future challenges, [15].

'Learning from incidents' (LFI) is a valuable tool in many domains. Much research has been devoted to understanding how this process can be expressed and measured, how worthwhile lessons can be learned through more efficient and effective learning, as proffered by Drupsteen and Guldenmund [16], Hovden et al. [17] Jacobsson et al. [18]. A main tenet of this reporting system is the ability to report any error or potential error in a 'free and frank' way. This philosophy is intended to be supported by what is termed a just culture, where the outcome for the individual is not based on punitive measures or being inappropriately punished for reporting or co-operating with occurrence investigations. The occurrence reporting system is also intended to be a 'closed-loop' system where feedback is given to the originator and effective actions are implemented within the organisation to address the embryonic or evident safety hazards. The concept is progressive in terms of its potential for contribution to identifying and addressing less than optimal performance of human, organisational and technical systems. Understanding that adverse and unwelcome events can be minimised through diligent reporting, event analysis and learning and subsequent necessary intervention is a positive trait with respect to improving acceptable levels of safety.

Argyris and Schön [19] (pp. 20–21) highlight the importance of learning to detect and address effective responses to errors. Their 'theory in action' concept is the focal point for this determination. The first of its two components, 'theory in use' is one that guides a person's behaviour. This is often only expressed in tacit form and is how people behave routinely. Very often these observed habits are unknown to the individual. The second element is known as 'espoused theory', namely what people say or think they do. Drupsteen and Guldenmund [16] mention that espoused theory comprises of 'the words we use to convey what we do, or what we like others to think we do'.

Enabling this learning channel, ICAO Doc 9859 [19] defines a template for aviation operators and regulators to support the application of a variety of proactive, predictive and reactive oversight methodologies. In addition to routine monitoring schemes, voluntary and mandatory reporting, post incident follow-up; there are regular safety oversight audits. These audits and inspections often set out to establish if there is a difference between espoused theory and the theory in use, e.g., is the task being correctly performed in accordance with the documented procedure/work instruction or is there a deviation from approved data and practice? However, Drupsteen and Guldenmund [16] caution auditors not to 'focus too much on the documentation of procedures' alone. In such cases the audit oversight may be ineffective because of its sole focus on espoused theories of the organisation only and not the theory-in-use. They progress to translate this idea of poor focus on theory in action and recommend a solution by suggesting a valid learning component arising from the incidents. They also highlight the 'espoused' aspect where those attempting to learn from incidents often fail to experience the desired learning because outcomes are not fully aligned with the practical objectives of an LFI initiative. For learning to be most effective, espoused theory and theory in use should be reasonably well aligned.

Aircraft maintenance and continuing airworthiness management activities that are performed in European member states are moderated by rules that mandate reporting of defined incidents and occurrences. Repositories of reported data tend to be populated only from sources predominantly aligned with mandatory incident/occurrence reporting requirements. Conventional safety oversight models only verify the presence of reporting media and repositories in this segment of the industry. Traditionally there has been a focus amongst organisations to ensure details of reports are submitted in line with state's mandatory reporting obligations. However, it is possible such a narrow focus on a single element (i.e., reporting alone) of an incident in its lifecycle could negate the potential learning benefits that might accrue from considering other likely related sources. As a result, the absence of clear regulatory requirements capable of augmenting learning from incidents could be considered an impediment to effective learning in the domains affected by EU regulation 1321/2014 [4]. The featured industry sector is regulated by the application and upkeep of numerous requirements in each jurisdictions of operation. In general, oversight duties tend to be carried by regulating states and operators in support of safe and profitable activity. However, a growing tendency to just increase some regulatory requirements across the segments may not always offer the same safety returns necessary for states in the future.

Up until some years ago, basic risk mitigation methods had remained unchanged. The previously reactive initiatives had largely been based on post-event analysis of accidents and incidents. At present, learning from past incidents, occurrences and accidents must be credited with playing a major part in helping evolve the paradigm to the more proactive means of risk management in many aviation segments we know today. Accident models (Heinrich and Reason) can sometimes inadvertently contribute to an over-simplification of how accident and incident contributing factors are perceived. This can result in striving to establish a singular root cause. Understandably the propensity for those tasked with accident and incident investigation is sometimes to establish a linear view based only on apparent causal factors. Proactively identifying precursors to events or potential conditions can greatly assist in averting latent or undiscovered conditions. Since the early 1990s, the potential for organisations to learn from incident precursors and conditions has been worthy of attention. Cooke [20] endorses a suggestion that increased reporting of incidents enhances continuous improvements in high reliability industries. In the continuing airworthiness segment of the industry, here is often a regulatory driven focus on establishing a single root cause. The importance of adequate resources and efforts to determine accurate incident causation and the measures to prevent reoccurrence should be a primary concern. Until ED 2020/002/R [21] is fully implemented, it is possible that the custodians of current regulatory requirements are satisfied once a root cause is established. Could it be that the current popular practice of pursuing (singular) root cause focus can be a lost opportunity when additional related sources exist?

The harvesting of information from incident reporting systems is a necessary input to continuously develop appropriate and effective recurrent training material. The inclusion of basic qualification criteria for human factor trainers in the regulatory requirements should also be addressed. However, it is questionable if the perpetuation of these measures alone would support more effective delivery and application of lessons learned throughout the segment. One means of addressing this impending issue is to remodel regulatory, operational and training requirements to consider a new approach in the segment. Reflecting a combination of actions, events and conditions in a new basic model supporting human factor continuation training, may lay the foundations to better elucidate event causation and yield improved and sustainable safety recommendations in the featured segment.

3. A Model Supporting Learning from Incidents

3.1. Model Design and Description

Currently European measured levels of aviation safety are generally considered as acceptable. As domain activity is expected to increase in the coming decades, further steps to improve or at least preserve contemporaneous levels of safety will have to continue to be developed. One of the main facets of safety management is the reporting, collection, analysis and follow-up to incidents according to Annex 19 [22]. This is also highlighted in an EU communication COM/2011/0670 [23] and (EU) 376/2014 [6]. A primary reason for the emphasis on reporting and subsequent learning from incidents (LFI) is to enable and support a shift from prescribed safety oversight to a risk-based programme. This is seen as the best fit to enable and effect improvements in areas that will present the most risk [24]. Figure 1 presents one view of a generic incident lifecycle [25] integrated with an interactive framework arising from the researchers work. This 'proposed enhancement' could augment a learning dimension in the cycle of an incident.

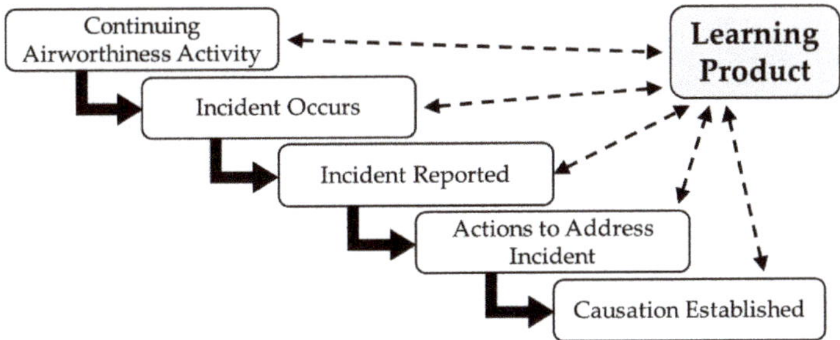

Figure 1. Incident learning product and process (Broken line denotes iterative learning feedback).

Figure 1 also illustrates a view of the overall process employed to acquire, process and store incident data. The 'broken line' arrows signify an iterative action at each stage of processing the incident. The purpose of this is to ask and record what can be learned at each point? The motif of how a learning product originates from the regulatory perspective is also featured. The effectiveness of the learning from the event is considered in terms of how it can be gauged. This is evident from feedback originating from the actions in the cycle when the learning product is being developed. Closing the learning loop is also necessary and reflected in graphic form. In addition to this, assessing actions at each incident stage is intended to support an analysis of how effective resulting actions are in terms of preventing recurrence of the incident. Actions to prevent the recurrence of the same or similar events can be embodied as a result of how effective the learning was. As such the novelty of this framework exists in its clear visual representation rather than the actual arrangement of the specific stages recorded. Traditionally the industry focus on incidents and occurrences has pivoted solely around the reporting requirements. These obligations are the backdrop against which mandatory reporting activity takes place. The establishment of causation is required by regulatory process but little or no suitability of same is mandated by requirement in support of any potential for learning. The featured framework serves to present the main elements of an incident during its lifecycle and highlight the aspects to be considered when incidents are being used in support of developing effective safety lesson delivery.

3.2. Model Implementation

The area of focus for this paper is aircraft maintenance and continuing airworthiness management [4] activities. It was decided to establish contact with an Irish Aviation Authority (IAA) European central repository for aviation accident and incident reports (ECCAIRS) focal point. Following a briefing, a specific permission was granted to review a data set of deidentified mandatory occurrence reports (MOR's) for the purpose of academic analysis. The operational theatre of activity involved licensed air carriers operating large aircraft on the Irish civil aircraft register. The permission allowed an initial physical database search to be performed from June 2019 to November 2019 using 'Part 145 (maintenance) and Part M (continuing airworthiness management)' as the search terms for de-identified report content. Approximately 200 results came back. The narrative and content of each report was reviewed by the researchers for applicability to the analysis. This exercise refined the reports under review to a data set of 85. Figure 2 presents an overview of the analysis framework, described in the sequel.

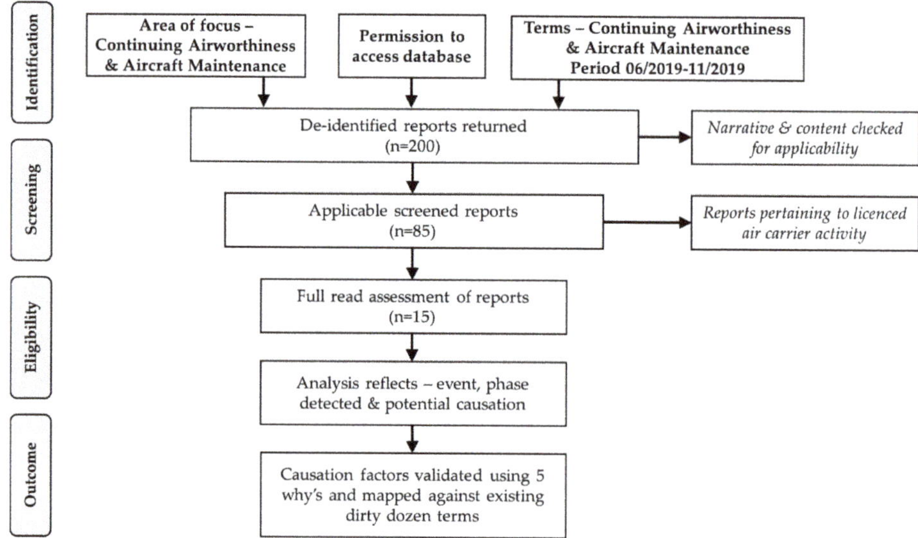

Figure 2. Overview of the analysis framework applied. The systematic review phases (identification-screening-eligibility-outcome) follow the methodology of Liberati et al. [26].

3.3. Model Validation: Report Causal Elements

A third round of full read screening of the set yielded 15 deidentified reports applicable to the exercise topic. Each featured event was considered under the following elements: the actual event, maintenance phase detected and likely potential causation factors. Table 1 contains an overview of this analysis output. Each of the 15 analysed occurrence reports provided a description of the featured event and some were helpfully contextualised with a chronological timeline when included in the report body. This later assisted with appreciating all the potential causation elements for each event. However, the reported verbiage tended to terminate mostly with a focus on consequential impact rather than causal information. For the sake of consistency across the analysis, the authors decided to apply a systematic approach to elicit and validate causal factors from the data. The process was based on a clear definition of root cause as proffered by Paradies and Busch [27] as: '*the most basic cause that can be reasonably identified and the management has control to fix*'.

Table 1. Results of the analysis of 15 incidents and mapping against the 'Dirty Dozen'.

	Potential Causation Factors for Each Incident	Dirty Dozen 'Taxonomy'	
		Primary	Secondary
1	Incorrect tooling Competence for task Task sequencing	Lack of resources Lack of knowledge (Knowledge, skills, ability) Complacency Lack of awareness	Norms established Lack of communication Lack of assertiveness
2	Continuing Airworthiness Management Organisation (CAMO) management system competence Production pressure	Lack of knowledge (procedures & knowledge, skills, ability) Lack of communication Pressure	Lack of resources Stress Fatigue/Stress/Distraction
3	Inadequate aircraft maintenance programme (AMP) inspection task Product design	Norms established Lack of communication	Lack of awareness Lack of resources
4	Inadvertent damage Incorrect gauge of locking wire & locking technique	Distraction Lack of knowledge (Knowledge, skills, ability) Fatigue Lack of teamwork	Complacency Lack of awareness
5	CAMO work request incorrect Maintenance procedure not followed	Lack of knowledge (Knowledge, skills, ability) Pressure Lack of assertiveness	Norms established Lack of resources
6	Procedure design Production pressure Competence for task CAMO management system competence	Lack of knowledge Pressure Lack of awareness Lack of communication	Lack of resources Fatigue/Stress/Distraction Lack of supervision
7	Production pressure Competence for maintenance task	Pressure Lack of knowledge (Knowledge, skills, ability)	Fatigue/Stress/Distraction Lack of resources
8	Maintenance data availability Production pressure Competence for task	Lack of resources Lack of knowledge Pressure	Norms established Fatigue Lack of awareness
9	Procedure design Production pressure Supervision Competence for maintenance task	Lack of knowledge (Procedures) Pressure Lack of knowledge (Knowledge, skills, ability)	Lack of awareness Lack of communication Complacency Lack of assertiveness
10	Incorrect tooling Competence for maintenance task Task sequencing	Lack of knowledge Lack of awareness Lack of communication	Norms Lack of resources
11	Maintenance data Procedure design Production pressure Competence for maintenance task Post task leak-check	Lack of knowledge (Knowledge, skills, ability) Pressure Lack of awareness	Lack of teamwork Complacency Fatigue/Stress/Distraction Lack of resources
12	Production pressure Competence for maintenance task Maintenance data availability Supervision	Pressure Lack of knowledge Lack of teamwork	Fatigue/Stress/Distraction Lack of resources

Table 1. *Cont.*

	Potential Causation Factors for Each Incident	Dirty Dozen 'Taxonomy'	
		Primary	Secondary
13	Competence for maintenance task Production pressure	Lack of knowledge Pressure	Fatigue/Stress/Distraction
14	Culture Risk taking Competence Supervision	Norms established Lack of resources Lack of knowledge (Knowledge, skills, ability)	Lack of awareness Complacency Lack of communication
15	CAMO procedure competence Culture Supervision	Lack of knowledge (Knowledge, skills, ability) Norms Stress	Lack of awareness Lack of resources Pressure

Many analysis tools [e.g., Fault tree analysis (FTA), functional resonance analysis model (FRAM), systems theoretic accident model and process (STAMP), sequentially timed events plotting (STEP)] are available and can be applied in support of a systematic review aimed at establishing causal factors. However, each of the aforementioned is generally applied in support of more voluminous operational applications and a degree of familiarity and adequate resources are usually required to ensure an efficacious outcome. As the incident reports (n = 15) under review already had causal factors ascribed, the authors deemed a simple analysis tool to be appropriate. According to Card [28], the '5 Why's technique' is a widely used technique applied in support of root cause analysis and is used by many statutory organisations globally. Ohno [29] (p. 123) highlights that by repeating why five times, the nature of the problem as well as its solution becomes clear. As the authors of this paper were aware, sole reliance on a tool like the 5 Whys has limitations. In particular, exclusive operational reliance on its prowess as a revealing panacea could inveigle its users in to over-simplifying an event and thereby be seduced into pursuing an inappropriate singular cause. As a result, the tool was applied solely as a mechanism to validate the already operator ascribed event categorisations and causal factors.

4. Results

Each mandatory occurrence report (MOR) was thoroughly reviewed, and the content of the event and related actions carefully assessed. However, without an intimate knowledge of the operational environment, history of the aircraft reliability and related operational dynamic and contextual influences for example, it was not possible to definitively establish if the recorded causation and related factors were indisputably accurate for each event. Notwithstanding the foregoing, based on the authors experience and judgement the recorded causation factors were harmonised with a taxonomy derived from the elements of the Transport Canada [7] 'dirty dozen' terms associated with common error preconditions. The elements are generally identified as, Lack of communication, Distraction, Lack of resources, Stress, Complacency, Lack of teamwork, Pressure, Lack of awareness, Lack of knowledge, Fatigue, Lack of assertiveness, Norms.

The purpose of aligning the 'potential incident causation factors' with a known taxonomy is to assist with developing clear learning product content and learning objectives. Regulatory code or guidelines for the continuing airworthiness domain do not require a formal approach to learning such as those defined by Bloom [30] and Anderson and Sosniak [31]. Although the reports featured display similar activity profiles, recognition for the need to consider learning taxonomies and the importance of domains of learning (cognitive, affective and psychomotor) when designing continuation training programmes is considered essential. In addition, organisations are not required to have a formal mechanism of assessing efficacy, instead many take comfort in national, European and international holistic safety reports as a means of gauging their performance as part of the collective. Assuming the purpose of learning objectives is to assist with the delivery and measurement of the effectiveness of learning actions, developing an overview of a harmonised taxonomy is helpful in this regard.

In Table 1 above, potential causation factors for each of the 15 selected incidents were matched with the twelve elements of the 'Dirty Dozen' human factor taxonomy. In order to prevent an over-simplification of each event's contributing factors, the authors were careful not to be seduced into seeking a singular root cause. Therefore, it was decided to include both primary and secondary human factor elements so that causation could be considered in a holistic manner. The following paragraphs (a–h) and Figure 3 give a breakdown of the issues emerging from the assessment of the mandatory occurrence reports (MOR's) as seen through the lens of association with a taxonomy.

a. Lack of knowledge features as a primary element in 13 (87%) of 15 occurrences. This can be closely related to the competence required to perform the task as it relates to aircraft maintenance and continuing airworthiness management activities which are defined as comprising of 'knowledge, skills and attitude/ability' [4]. As a secondary potential contributing element, it relates to only 1 (7%) of the 15 occurrences.
b. Lack of awareness is highlighted as a primary potential causation factor in 9 (60%) of the 15 reviewed occurrences. This element can be closely related to competence, communication and teamwork. As a secondary contributing factor, lack of awareness was noted during the review in 5 (33%) of 15 reviewed occurrences.
c. Lack of resources were recorded in 3 (20%) of 15 events. Adequate resources are required in order for an operator to adequately staff an organisation so that an aircraft can be maintained to the correct standard and when required. EU 1321/2014 [4] mandates that a manpower plan is maintained in support of ensuring adequate levels of staff are consistently available. As a secondary issue, lack of resources appeared as an issue in 5 (33%) of 15 cases. Ultimately, accountable managers are the key to ensuring sufficient resources are made available so that the organisational elements continue to remain compliant and effective in this respect.
d. Norms accounted for 3 (20%) of 15 reports examined. Norms are often viewed as behaviours that are developed and accepted within a group. However, when the resulting behaviour requires a deviation from approved procedural function, the consequences are often unknown. Although such actions may offer short-term productivity gains, they may also introduce active and latent safety hazards. In the case of secondary causation, norms are associated with 8 (53%) of 15 assessed occurrences.
e. Lack of communication was found to be evident in 3 (20%) of 15 occurrences in the study. Communication in aircraft maintenance and management activities is a vital element in the release of a safe product. Poor communication can amplify many other elements of the human factors leading to a deterioration in human performance, Chatzi [32], Chatzi et al. [33]. 2 (13%) of the 15 reviewed communication-related occurrences were recorded as contributing to secondary event causation.
f. Complacency was revealed as a primary factor in the causation of 1 (7%) of 15 events studied. However, as a secondary contributing factor it accounted for 5 (33%) of 15 reports. Stress levels associated with a task can diminish performance if one becomes complacent. Its presence can contribute in concert with other elements capable of setting the scene for an unwelcome event.
g. Stress as a primary factor appeared in 1 (7%) of the 15 reviewed events. However, it was associated with 2 (13%) of 15 reports as a secondary issue. Stress can be both a by-product and an enabler of other Dirty Dozen elements. Fatigue for example can be closely coupled to stress and displayed similar pattern in the study with 7% and 13% respectively of prevalence in the reports reviewed.
h. Lack of assertiveness was evident as a primary and as a secondary causation factor in both cases and occurring at rate of 1 (7%) of 15 events under review. Distraction and lack of teamwork appeared in similar proportions in the review results.

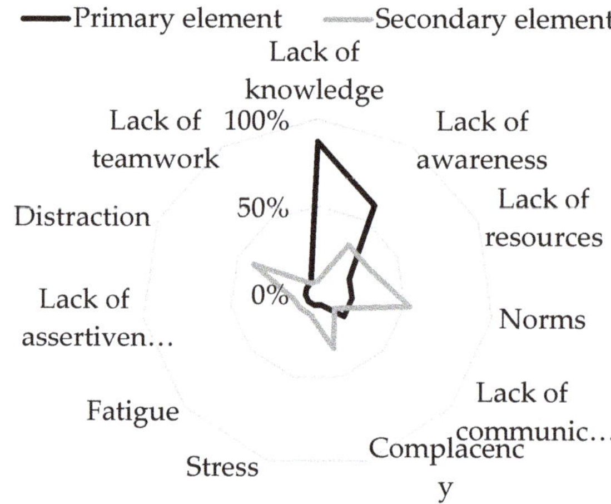

Figure 3. Representation of primary and secondary causation in the reviewed occurrence reports.

5. Discussion

Recalling the causal factors attributed to the featured occurrence reports in the paragraphs above, it is easy to appreciate their relationships with the 'Dirty Dozen' example of human factor elements. For example, lack of resources can be a major constraint when it comes to providing adequate levels of appropriately qualified competent staff. Pressures exerted upon staff in a dynamic industry sector to absorb additional workload can of course have a potentially detrimental effect on safe operations. Competent and available supervision of maintenance and inspection staff is a core requirement of a quality mission in aircraft maintenance and continuing airworthiness management operations. In many regions the maintenance requirements (e.g., EU regulation 1321/2014 [4]) stipulate a process whereby all staff must meet the qualification criteria and be deemed competent before unaccompanied work can take place. For the purpose of the discussion, key elements of the incident cycle components are examined through pertinent elements identified during the analysis. The iterative approach suggested during the management of the incident information is supported by the context outlined below. Understanding the relevance of each of the sections is intended to support more effective learning outcomes. The following paragraphs discuss the incident cycle from the perspective of developing a sound learning product.

5.1. Acquiring, Processing and Storing Incident Data

According to Garvin [34], a clear definition of learning has proven to be elusive over the years. Garvin suggests *'a learning organization is an organization skilled at creating, acquiring and transferring knowledge and at modifying its behaviour to reflect new knowledge and insights'*. Figure 1 illustrates the evolution of an incident as it is managed through its cycle. The incident/occurrence will need to be detected if it is to possess any potential for learning. Acquiring information in support of learning is one of the key actions. Such learning material originates from compliance audits, amended regulatory requirements, best practice, and incidents and occurrence reports. Within the greater area of aircraft maintenance and continuing airworthiness management, details of incidents and occurrences tend to be reported soon after an event. Reporting requirements are normally timebound (i.e., 72 h). Most organisations endeavour to notify the necessary stakeholders as soon as possible, often by telephone in the first instance. As many airline staff are employed on a shift work basis, the window of 72 h is useful in support of administering the reporting function. It is not unusual to have numerous points of contact for reporting within organisations. However, reporting generally follows a consistent route regardless of who the initial point of contact is. Some organisations appear to empower and encourage the submission of reports by any individual. Other organisations appear to endorse reporting through a 'chain of command'. Regardless of the chosen initial reporting route input, all reports are progressed to a 'gate-keeper' within an organisation. The people responsible initially for examining the validity and completeness of submitted reports often hold a key position in either the quality assurance, technical services or maintenance departments. Generally, there is a strong awareness of the need to report incidents and occurrences classified as mandatory. There may be numerous motivational reasons to report, such as ethical, safety, compliance with regulatory requirements and best practice for example. Those submitting reports embrace mandatory reporting as an obligation underpinned by the cultural norms of aircraft maintenance and continuing airworthiness management. When an issue is discovered, it is progressed through the reporting system regardless of its status. Many organisations welcome all reports including non-mandatory events that are highlighted through voluntary reporting streams. They evidently see value in including them in their analysis of events and the subsequent learning opportunities the reports may offer.

5.2. Single, Double and Triple-Loop Learning

From an organisational point of view, single-loop learning can be experienced when an error is detected and corrected but little else changes, Argyris and Schön [19] (p. 18). In aircraft line maintenance environments where a 'find and fix' ethos prevails, single-loop learning is often evident. It is not unusual for technical issues to befall an aircraft's departure time. Such pressure points often associated with fulfilling contractual obligations may have a negative impact on the potential for learning from a related event. In such cases, if issue arises the matter may be resolved without any further recorded action. Because of the terse nature of the experience for an individual concerned, the opportunity for further learning may not extended beyond the single loop. Argyris and Schön [19] (p. 21) and Lukic et al. [35] proffer double-loop learning as learning that takes place and results in organisational norms and theory in use being altered. Presently, aircraft certifying, and support staff are obliged to continuously preserve an adequate understanding of the aircraft being maintained and managed along with associated regulations and procedures. A desired outcome of double-loop learning is often witnessed for example through the adjustment of environmental, behavioural and procedural norms. Instances of double-loop learning can be evident following unsuccessful attempts through single-loop learning. In-service continuation training is an effective enabler that is capable of supporting double-loop learning. Organisations are also required by EU 1321/2014 [4] to establish and maintain a continuation training programme for staff. A primary pillar of continuation training syllabi is the use of incidents and occurrences as lesson content for influencing organisational norms and behaviour in support of preventing recurrence of incidents and occurrences. Deutero-learning (triple-loop) relates to when members of an organisation reflect upon previous learning and sets about to improve how the organisation can refine and improve the process of learning from events, Argyris and Schön [19] (p. 29), Bateson [36]. This could also be stated as learning how to learn by seeking to improve single and double loop learning. Although deutero-learning may be considered as a natural extension of other levels of learning, the concept does not feature as a requirement in aircraft maintenance and continuing airworthiness management regulatory codes.

5.3. Learning Product

Aircraft maintenance and management regulatory codes require reporting of 'any identified condition of the aircraft or component that has resulted or may result in an unsafe condition that hazards seriously the flight safety' [4]. Generally, a learning product can originate from numerous information sources within the aircraft maintenance and continuing airworthiness management arena. Specifically, GM1 145.A.30(e) [4] requires the use of accident and incident reports in support of the mandatory human factors training content. The intent of this material is to ensure information is imparted upon the organisations' staff in support of preventing the subject event reoccurrence. Such continuation training is mandated by European requirements for all aircraft maintenance and continuing airworthiness management organisations. Continuation training is also a product as well as a medium for imparting learning from incidents. Inputs to continuation training syllabi often feature learning from incidents and experience augmented by safety notices, toolbox talks and are recognised as a means of presenting the learning product to operational staff. Drupsteen and Guldenmund [16] cite, 'Lampel et al. [37] where they use the term "learning about events"'. This is further explained as 'information about events is shared and diffused to help create new ideas', in this case in the support of safe operations.

5.4. Effectiveness of Learning

The evaluation of any initiative's success is much more straight forward when clear objective indicators (learning outcomes) are employed. In the case of learning in an aircraft maintenance and management environment, organisations can generally employ indicators such as inspection non-compliance, audit findings and rates of incident reoccurrence in support of gauging the effectiveness of learning. Probing salient aspects such as timely investigation of incidents, assessing the learning content and feedback are a starting point for assessing effectiveness. Cooke [20] concludes the absence of or poor information can compromise the effectiveness of feedback. He also suggests that if the feedback cycle is ailing, the climate may deteriorate and have a negative impact upon organisational safety. From a commercial viewpoint, it is perhaps understandable that aircraft tend to only generate revenue when flying. However, airline operators need to maintain a balance between safe operations and productivity. It is essential that incident causal factors are fully identified and adequate time and resources are available to support this important aspect of learning. Cooke [20] endorses a suggestion that increased reporting of incidents enhances continuous improvement in high reliability industries. However, establishing adequate causation is also an attribute capable of supporting effective learning from an event in dynamic environments.

The importance also of just culture as an enabler for incident reporting and subsequent effective learning cannot be ignored. Under-reporting of events resulting from a single-loop learning experiences amongst operational maintenance staff and production pressures can also impact negatively upon efforts to propagate a learning environment. McDonald [38] suggests from their analysis, '*that there is a strong professional sub-culture, which is relatively independent of the organization. One implication of this finding is that this professional sub-culture mediates the effect of the organizational safety system on normal operational practice*'. Von Thaden and Gibbons [39], conclude safety culture '*refers to the extent to which individuals and groups will commit to personal responsibility for safety; act to preserve, enhance and communicate safety information; strive to actively learn, adapt and modify (both individual and organizational) behaviour based on lessons learned from mistakes ... ...* '. A just culture is defined in the affecting regulation EU 376/2014 [6] as, 'a *culture in which front line operators or other persons are not punished for actions, omissions, or decisions taken by them, that are commensurate with their experience and training, but in which gross negligence, wilful violations and destructive acts are not tolerated*'. Accordingly, a just culture is a fair culture. The effectiveness of the learning system can also be compromised by its efficiency as well as its inadequacies. The volume of information that staff must process and assimilate is continually increasing. Guardians of learning outcomes should be mindful that staff risk becoming information weary as a result of the ever-increasing demands on their cognitive abilities.

5.5. *Types of Knowledge*

This relates to; conceptual, dispositional, procedural and locative knowledge forms [40]. One of the key objectives of learning from incidents is to identify the type of knowledge needed to prevent an issue recurring. When a reportable issue for example is discovered, the submitted report will identify 'what' happened. Subsequent follow-up will set out to determine 'why' the issue occurred. The guiding principles of 'how' to perform the task or operation are often contained in procedures or data particular to the task. The information contained in procedures will enable a person to utilise other forms of knowledge. Prevailing culture within an organisation will have an impact on learning from incidents. If a strong commercial culture exists, this may have an impact on for example the depth and breadth of learning from incidents within the company. Induction and initial training for new staff is an important element for demonstrating where organisational sources of information can be accessed. Accident data repositories contain many well documented examples of human factor related precursors to incidents. Many of which may have originated in poor access to approved data and culminated in serious and possibly preventable incidents. Acknowledging and addressing the limitations related to the types of knowledge when developing continuation training programmes would have a positive impact on participants. The enabling industry requirements do not specify any discernible differences in how the types of knowledge are differentiated. A review of the human factors syllabus requirements did not highlight a need to appreciate or account for these human centred limitations when designing and delivering training lessons.

6. Conclusions

It has been highlighted during this research that the opportunity to learn from incidents is not being fully embraced in the aircraft maintenance and continuing airworthiness management segment of the industry. While the idea of eliminating all incidents is a fallacy, reducing their numbers and potential for harm is a reality. Air travel is on the increase and it is envisaged that current sectors flown will have doubled within the next two decades. If current levels of safety were to remain stagnant with a doubling in activity, twice the current fatality rate would surely not be acceptable. Many people relate safety to freedom from risk and danger [41]. Unfortunately, risk and danger are often ubiquitous in the presence of aircraft maintenance and continuing airworthiness management activities. Managing sources of risk and danger is a tall order for some organisations. Document 9859 [42] recognises that *'aviation systems cannot be completely free of hazards and associated risks'*. However, the guidance does acknowledge that as long as the appropriate measures are in place to control these risks, a satisfactory balance between *'production and protection'* can be achieved. Perrow [43] (p. 356) acknowledges that *'we load our complex systems with safety devices in the form of buffers, redundancies, circuit breakers, alarms, bells, and whistles'* because no system is perfect.

Detecting and identifying hazards highlighted through incident reporting systems is recommended by ICAO standards and recommended practices as an effective means of achieving practicable levels of safe operations. Therefore, objective data mined from a reporting system offers the potential to enlighten aviation stakeholders and to illuminate weakness that may be present. Such information can assist with a better understanding of events and augment mitigating measures against the potential effects of these hazards. When incidents occur, this can be an indication of a failure in an organisation's process and/or practice. Because of continuous challenges faced by organisations in the aviation industry, there is still potential to learn from resulting incidents and pre-cursors. The learning is based on the potential new knowledge available from the associated collection, analysis and interventions for these events. Effective learning can be considered as a successful translation of safety information into knowledge that actively improves the operating environment and helps prevent recurrence of unwelcome events.

The paper features a brief exercise to demonstrate how safety information can be translated into lessons capable of augmenting knowledge within an aircraft maintenance and management organisation. To support this, fifteen occurrences drawn from an ECCAIRS incident database portal were analysed. The result of the analysis along with potential causation factors are presented. Additionally, a simple mechanism in support of the delivery of associated safety lessons was developed and is presented in Table 1 above. Integrating the known causal factors with the 'Dirty Dozen' taxonomy which is already associated with this aviation segment provides a useful template for continuation training in the segment. The emerging incident/occurrence themes related to the featured events are briefly discussed and presented within the document. The publication also introduces a framework that assembles and explains the main elements of an incident within its lifecycle. The purpose of this is to illustrate tacit aspects of an incident that have the potential to augment learning within the process. In order to leverage the maximum benefit from details of an incident, learning processes must recognise the existence of these event components. There can therefore be a formal approach to gauging the effectiveness of learning and a means of identifying underperforming elements of the learning process.

This publication could assist subject organisations with a review of their management of incident information when developing continuation training material and learning outcomes.

Author Contributions: Conceptualization, J.C. and K.I.K.; methodology, J.C.; formal analysis, J.C.; investigation, J.C.; validation, J.C and K.I.K.; data curation, J.C.; writing—original draft preparation, J.C. and K.I.K.; writing—review and editing, J.C. and K.I.K. All authors have read and agreed to the published version of the manuscript.

Funding: This research received no external funding.

Institutional Review Board Statement: Not applicable.

Informed Consent Statement: Informed consent was obtained from all subjects involved in the study.

Data Availability Statement: The data presented in this study are available on request from the corresponding author.

Conflicts of Interest: The authors declare no conflict of interest.

References

1. Silva, S.A.; Carvalho, H.; Oliveira, M.J.; Fialho, T.; Guedes Soares, C.; Jacinto, C. Organizational practices for learning with work accidents throughout their information cycle. *Saf. Sci.* **2017**, *99*, 102–114. [CrossRef]
2. Akselsson, R.; Jacobsson, A.; Bötjesson, M.; Ek, Å.; Enander, A. Efficient and effective learning for safety from incidents. *Work* **2012**, *41*, 3216–3222. [CrossRef] [PubMed]
3. Document 32018R1139. Regulation (EU) 2018/1139 of the European Parliment and the Council of 4 July 2018 on common rules in trhe field of civil aviation and establishing a European Union Aviation Safety Agency, and amending Regulations (EC) No 2111/2005, (EC) No 1008/2008, (EU) No 996/2010, EU376/2014 and Directives 2014/30/EU of the European Parliment and the Council, and repealing Regulation (EC) No 216/2008 of the European Parliment and of the Council and Council Regulation (EEC) No 3922/91. In *Official Journal of the European Union*; European Commission: Brussels, Belgium, 2018; pp. 1–122.
4. Document 32014R1321. Commission Regulation (EU) No 1321/2014 of 26 November 2014 on the continuing airworthiness of aircraft and aeronautical products, parts and appliances, and on the approval of organisations and personnel involved in these tasks. In *Official Journal of the European Union*; European Commission: Brussels, Belgium, 2014; pp. 1–194.
5. Harvey, C.; Stanton, N.A. Safety in System-of-Systems: Ten key challenges. *Saf. Sci.* **2014**, *70*, 358–366. [CrossRef]
6. Document 32014R0376. Regulation (EU) No 376/2014 of the European Parliament and of the Council of 3 April 2014 on the reporting, analysis and follow-up of occurrences in civil aviation, amending Regulation (EU) No 996/2010 of the European Parliament and of the Council and repealing Directive 2003/42/EC of the European

Parliament and of the Council and Commission Regulations (EC) No 1321/2007 and (EC) No 1330/2007. In *Official Journal of the European Union*; European Commission: Brussels, Belgium, 2014; pp. 18–43.
7. Transport Canada. *Human Performance Factors for Elementary Work and Servicing*; Canada, T., Ed.; Transport Canada: Ottawa, ON, Canada, 2003.
8. FAA. *120-92B-Safety Management Systems for Aviation Service Providers*; Federal Aviation Administration: Washington, WA, USA, 2015.
9. ICAO. ICAO Annex 13 to the Convention on International Civil Aviation: Aircraft Accident and Incident Investigation. In *Issue 10 Amendment 14*; ICAO: Montreal, QC, Canada, 2010.
10. ICAO. *Convention on International Civil Aviation*; Doc 7300; ICAO: Montreal, QC, Canada, 1944.
11. ICAO. *ICAO Accident/Incident Reporting Manual*, 3rd ed.; ICAO: Montreal, QC, Canada, 2014.
12. Boeing. *Current Market Outlook 2015–2034*; Boeing: Chicago, IL, USA, 2015.
13. Leveson, N. A new accident model for engineering safer systems. *Saf. Sci.* **2004**, *42*, 237–270. [CrossRef]
14. Deming, W. *The New Econonics for Industry, Government, Education*, 2nd ed.; MIT Press: Cambridge, MA, USA, 2000.
15. Bond, J. A Janus Approach to Safety. *Process. Saf. Environ. Prot.* **2002**, *80*, 9–15. [CrossRef]
16. Drupsteen, L.; Guldenmund, F.W. What is Learning? A Review of the Safety Literature to Define Learning from Incidents, Accidents and Disasters. *J. Contingencies Crisis Manag.* **2014**, *22*, 81–96. [CrossRef]
17. Hovden, J.; Størseth, F.; Tinmannsvik, R.K. Multilevel learning from accidents-Case studies in transport. *Saf. Sci.* **2011**, *49*, 98–105. [CrossRef]
18. Jacobsson, A.; Ek, Å.; Akselsson, R. Method for evaluating learning from incidents using the idea of "level of learning". *J. Loss Prev. Process Ind.* **2011**, *24*, 333–343. [CrossRef]
19. Argyris, C.; Schön, D.A. *Organizational Learning II: Theory, Method, and Practice*; Addison-Wesley Publishing Company: Boston, MA, USA, 1996.
20. Cooke, D.L. A system dynamics analysis of the Westray mine disaster. *Syst. Dyn. Rev.* **2003**, *19*, 139–166. [CrossRef]
21. ED Decision 2020/002/R. Amending the Acceptable Means of Compliance and Guidance Material to Annex I (Part-M), Annex II (Part-145), Annex III (Part-66), Annex IV (Part-147) and Annex Va (Part-T) to as well as to the articles of Commission Regulation (EU) No 1321/2014, and issuing Acceptable Means of Compliance and Guidance Material to Annex Vb (Part-ML), Annex Vc (Part-CAMO) and Annex Vd (Part-CAO) to that Regulation. In *Official Journal of the European Union*; European Commission: Brussels, Belgium, 2020.
22. ICAO. Annex 19 to the Convention on International Civil Aviation Safety Management. In *Safety Management*; ICAO: Montreal, QC, Canada, 2013.
23. Document 52011DC0670. Communication from the Commission to the Council and the European Parliament Setting up an Aviation Safety Management System for Europe. In *Setting up an Aviation Safety Management System for Europe*; European Commission: Brussels, Belgium, 2011.
24. Cooke, D.L.; Rohleder, T.R. Learning from incidents: From normal accidents to high reliability. *Syst. Dyn. Rev.* **2006**, *22*, 213–239. [CrossRef]
25. Drupsteen, L.; Groeneweg, J.; Zwetsloot, G.I. Critical steps in learning from incidents: Using learning potential in the process from reporting an incident to accident prevention. *Int. J. Occup. Saf. Ergon.* **2013**, *19*, 63–77. [CrossRef] [PubMed]
26. Liberati, A.; Altman, D.G.; Tetzlaff, J.; Mulrow, C.; Gøtzsche, P.C.; Ioannidis, J.P.A.; Clarke, M.; Devereaux, P.J.; Kleijnen, J.; Moher, D. The PRISMA statement for reporting systematic reviews and meta-analyses of studies that evaluate healthcare interventions: Explanation and elaboration. *BMJ* **2009**, *339*, b2700. [CrossRef] [PubMed]
27. Paradies, M.; Busch, D. Root cause analysis at Savannah River plant (nuclear power station). In Proceedings of the Conference Record for 1988 IEEE Fourth Conference on Human Factors and Power Plants, Monterey, CA, USA, 5–9 June 1988; pp. 479–483.
28. Card, A.J. The problem with '5 whys'. *BMJ Qual. Saf.* **2017**, *26*, 671–677. [CrossRef] [PubMed]
29. Ohno, T. *Toyota Production System: Beyond Large-Scale Production*; CRC Press: Boca Raton, FL, USA, 1988.

30. Bloom, B.S. *Taxonomy of Educational Objectives*; Cognitive Domain; Edwards Bros: Ann Arbor, MI, USA, 1956; Volume 20, p. 24.
31. Anderson, L.W.; Sosniak, L.A. *Bloom's Taxonomy*; Chicago Press: Chicago, IL, USA, 1994.
32. Chatzi, A.V. The Diagnosis of Communication and Trust in Aviation Maintenance (DiCTAM) Model. *Aerospace* **2019**, *6*, 120. [CrossRef]
33. Chatzi, A.V.; Martin, W.; Bates, P.; Murray, P. The unexplored link between communication and trust in aviation maintenance practice. *Aerospace* **2019**, *6*, 66. [CrossRef]
34. Garvin, D.A. *Building a Learning Organization*; Harvard Business Review July-August: Brighton, MA, USA, 1993; Volume 71.
35. Lukic, D.; Littlejohn, A.; Margaryan, A. A framework for learning from incidents in the workplace. *Saf. Sci.* **2012**, *50*, 950–957. [CrossRef]
36. Bateson, G. *The Logical Categories of Learning and Communication. Steps to an Ecolology of Mind*; Ballantine Books: New York, NY, USA, 1972; pp. 279–308.
37. Lampel, J.; Shamsie, J.; Shapira, Z. Experiencing the improbable: Rare events and organizational learning. *Organ. Sci.* **2009**, *20*, 835–845. [CrossRef]
38. McDonald, N.; Corrigan, S.; Daly, C.; Cromie, S. Safety management systems and safety culture in aircraft maintenance organisations. *Saf. Sci.* **2000**, *34*, 151–176. [CrossRef]
39. Von Thaden, T.L.; Gibbons, A.M. *The Safety Culture Indicator Scale Measurement System (SCISMS)*; National Technical Information Service Final Report; Office of Aviation Research and Development: Washington, DC, USA, 2008; pp. 1–57.
40. Thorndike, E.L. Fundamental theorems in judging men. *J. Appl. Psychol.* **1918**, *2*, 67. [CrossRef]
41. Reason, J.T. *Managing the Risk of Organisational Accidents*; Ashgate Publishing: Farnham, UK, 1997.
42. ICAO. *Safety Management Manual*; Doc 9859; ICAO: Montreal, QC, Canada, 2013.
43. Perrow, C. Normal Accidents Living with High-Risk Technologies. In *With a New Afterword and a Postscript on the Y2K Problem*; Perrow, C., Ed.; Princeton University Press: Princeton, NJ, USA, 1999.

MDPI
St. Alban-Anlage 66
4052 Basel
Switzerland
Tel. +41 61 683 77 34
Fax +41 61 302 89 18
www.mdpi.com

Aerospace Editorial Office
E-mail: aerospace@mdpi.com
www.mdpi.com/journal/aerospace

www.ingramcontent.com/pod-product-compliance
Lightning Source LLC
LaVergne TN
LVHW070056170726
843515LV00007B/1545

* 9 7 8 3 0 3 6 5 0 7 4 4 6 *